REFRACTORIES FOR THE STEEL INDUSTRY

Proceedings of a conference held in Luxembourg, 7–8 September 1989, and organised by the Commission of the European Communities, Directorate-General Science, Research and Development, in cooperation with the European Steel Industry and the European Federation of Manufacturers of Refractory Products (PRE)

REFRACTORIES FOR THE STEEL INDUSTRY

Proceedings of a conference held in Luxembourg, 7–8 September 1989, and organised by the Commission of the European Communities, Directorate-General Science, Research and Development, in cooperation with the European Steel Industry and the European Federation of Manufacturers of Refractory Products (PRE)

REFRACTORIES FOR THE STEEL INDUSTRY

Edited by

R. AMAVIS

*Commission of the European Communities,
Directorate-General Science, Research and Development,
Brussels, Belgium*

ELSEVIER APPLIED SCIENCE
LONDON and NEW YORK

ELSEVIER SCIENCE PUBLISHERS LTD
Crown House, Linton Road, Barking, Essex IG11 8JU, England

Sole Distributor in the USA and Canada
ELSEVIER SCIENCE PUBLISHING CO., INC.
655 Avenue of the Americas, New York, NY 10010, USA

WITH 58 TABLES AND 157 ILLUSTRATIONS

© 1990 ECSC, EEC, EAEC, BRUSSELS AND LUXEMBOURG

British Library Cataloguing in Publication Data

Refractories for the steel industry.
1. Steel. Production. Refractories
I. Amavis, R. II. Commission of the European Communities
669.142

ISBN 1-85166-491-2

Library of Congress CIP data applied for

Publication arrangements by Commission of the European Communities, Directorate-General Telecommunications, Information Industries and Innovation, Scientific and Technical Communication Unit, Luxembourg

EUR 12541

Neither the Commission of the European Communities nor any person acting on behalf of the Commission is responsible for the use which might be made of the following information.

No responsibility is assumed by the Publisher for any injury and/or damage to persons or property as a matter of products liability, negligence or otherwise, or from any use or operation of any methods, products, instructions or ideas contained in the material herein.

Special regulations for readers in the USA

This publication has been registered with the Copyright Clearance Center Inc. (CCC), Salem, Massachusetts. Information can be obtained from the CCC about conditions under which photocopies of parts of this publication may be made in the USA. All other copyright questions, including photocopying outside the USA, should be referred to the publisher.

All rights reserved. No part of this publication may be reproduced, stored in a retrieval system, or transmitted in any form or by any means, electronic, mechanical, photocopying, recording, or otherwise, without the prior written permission of the publisher.

Printed in Northern Ireland by The Universities Press (Belfast) Ltd.

PREFACE

Two of the main objectives of the ECSC steel research programme are the achievement of improved cost-effectiveness and enhanced product quality in iron and steelmaking. In this context, refractories play a significant role since they have a direct bearing on production costs and on steel cleanliness. As a consequence, investigations into the utilisation and performance of refractories in steelmaking form an important part of this ECSC research effort and the accumulated results of the work supported in recent years have made a valuable contribution to progress in steelworks refractories technology. It is against this background that the Commission of the European Communities has organised this conference in collaboration with the Community's steel industry and the European Federation of Manufacturers of Refractory Products(PRE).

The aim of the conference has been to present the results of this ECSC research to a wider audience while the participation of refractory producers has provided a timely opportunity to hear also about some of their recent R&D activities.

The scope and quality of the technical papers contained in these proceedings reflect not only the level of interest in steelworks refractories but also the vital contribution R&D is making to progress in this field. Furthermore, the constructive dialogue and fruitful exchange of views that has taken place between the users and the suppliers of refractories has been valuable in highlighting future requirements and challenges for these materials.

Finally, may I, on behalf of the organisers, express thanks to all the authors, session chairmen and participants for ensuring the success of the conference and the opportunity it has provided to review and to disseminate the results of our collaborative research effort.

A. GARCIA ARROYO
Director

Organizing Committee

R. AMAVIS
Commission of the European Communities
Directorate-General Science, Research and Development
Rue de la Loi 200
B-1049 Brussels

P. ARTELT
Commission of the European Communities
Directorate-General Science, Research and Development
Rue de la Loi 200
B-1049 Brussels

C. GUENARD
Institut de recherches de sidérurgie française
Voie Romaine
BP 64
F-57210 Maizières-lès-Metz

M. KOLTERMANN
Hoesch Stahl AG
Abt. Feuerfeste Baustoffe
Postfach 10 50 42
D-4600 Dortmund

M. LEFEBVRE-JON
Fédération européenne des fabricants de produits réfractaires
44, rue Copernic
F-75116 Paris

E. MARINO
Centro sviluppo materiali SpA
Via di Castel Romano, 100-102
I-00129 Roma

G.C. PADGETT
British Ceramic Research Ltd
Queens Road Penkhull
Stoke-on-Trent ST4 7LQ
United Kingdom

J. PIRET
Centre de recherches métallurgiques (CRM)
Rue Ernest Solvay 11
B-4000 Liège

Publication arrangements and registration
D. NICOLAY
Commission of the European Communities
Directorate-General Telecommunications, Information Industries & Innovation
JMO B4/087
L-2920 Luxembourg

Local organization
A. POOS
Commission of the European Communities
Directorate-General Personnel and Administration
JMO B1/39
L-2920 Luxembourg

CONTENTS

Preface v

OPENING SESSION

WELCOMING ADDRESS
    A. GARCIA-ARROYO, Director for Technical Research
    Directorate-General, Science, Research and Development
    Commission of the European Communities 3

INTRODUCTORY TALK: REVIEW OF RECENT RESEARCH CARRIED OUT
UNDER THE ECSC RESEARCH PROGRAMME ON REFRACTORIES
    J. PIRET, Chairman of the Executive Committee on
    ECSC Research into Refractories, CRM, Liège, Belgium 7

SESSION I

STRESS CALCULATIONS FOR REFRACTORY MATERIALS WITH HELP
OF THE FINITE ELEMENT METHOD
    J.A.M. BUTTER and J. DE BOER, Refractories R&D,
    Hoogovens Groep BV, IJmuiden, The Netherlands 11

WEAR MECHANISMS OF CONVERTER AND STEEL LADLE LININGS:
REPAIR METHODS
    C. GUENARD and P. TASSOT, IRSID - French Iron and
    Steel Industry's Research Institute,
    Maizières-Lès-Metz Cedex, France 22

THE USE OF UNFIRED REFRACTORY BRICKS IN TORPEDO LADLES
AND CASTING LADLES
    M. KOLTERMANN, Hoesch Stahl AG, Dortmund,
    Federal Republic of Germany 31

BORON NITRIDE-ENRICHED SUBMERGED NOZZLES FOR
CONTINUOUS CASTING
    J. PIRET, CRM, Liège, Belgium 47

USE OF CALCIUM OXIDE AS REFRACTORY MATERIAL IN STEELMAKING
PROCESSES
    E. MARINO, Centro Sviluppo Materiali SpA,
    Rome, Italy 59

ESTABLISHMENT OF OPTIMUM HEATING RATES FOR REFRACTORY
STRUCTURES IN THE IRON AND STEEL INDUSTRY
    D. WOLTERS, Thyssen Stahl AG, Duisburg,
    Federal Republic of Germany 69

INSTALLATION OF INSULATING PROTECTION LININGS FOR
CONTINUOUS CASTING TUNDISHES BY FLAME GUNNING
    J. PIRET, CRM, Liège, Belgium 81

LIST OF REFRACTORY RESEARCH SUPPORTED ON THE
ECSC STEEL RESEARCH PROGRAMME   92

SESSION II

THE EUROPEAN FEDERATION OF MANUFACTURERS OF REFRACTORY
PRODUCTS (PRE): ITS TECHNICAL ACTIVITY AND ITS ROLE IN
INTERNATIONAL STANDARDIZATION
    M. LEFEBVRE, Technical Secretary, European
    Federation of Manufacturers of Refractory Products,
    Paris, France   97

UNSHAPED REFRACTORIES
AN EXAMINATION OF ASSESSMENT AND QUALITY CONTROL METHODS
    B. CLAVAUD, Director for Research and Development,
    Lafarge Réfractaires Monolithiques
    M. JACQUEMIER, Head of Applied Research,
    Lafarge Réfractaires Monolithiques
    H. LE DOUSSAL, Head of the Refractory Service and
    Technical Ceramics, Société Française de Céramique, France   109

DATABASE ON REFRACTORY MATERIALS FOR IRON AND STEELMAKING
    E. CRIADO and A. PASTOR, Instituto de Cerámica y Vidrio,
    CSIC, Arganda del Rey, Madrid, Spain
    R. SANCHO, Instituto de Información y Documentación
    en Ciencia y Technologia, CSIC, Madrid, Spain   118

SESSION III

REQUIREMENTS FOR REFRACTORY MATERIALS IN MODERN STEEL PRODUCTION
    G. KLAGES, Refractory Technology Centre,
    Thyssen Stahl AG, Duisburg, Federal Republic of Germany
    H. SPERL, Steelworks Committee and Refractories
    Technology, Association of German Metallurgists,
    Düsseldorf, Federal Republic of Germany   133

FUTURE STEELMAKING REQUIREMENTS AND IMPLICATIONS ON REFRACTORIES
    R. BAKER, British Steel Technical, Swinden Laboratories,
    Moorgate, Rotherham, UK   142

EVOLUTION IN THE USE OF REFRACTORIES IN THE FRENCH IRON
AND STEEL INDUSTRY
    M. BEUROTTE, Sollac Dunkerque, France   152

SESSION IV

ADVANCES IN RAW MATERIALS AND MANUFACTURING TECHNOLOGY
IN THE PRODUCTION OF REFRACTORY LININGS FOR PRIMARY
AND SECONDARY STEELMAKING OPERATIONS
    P. WILLIAMS, D. TAYLOR and J.S. SOADY, Steetley
    Refractories Limited, Worksop, Nottinghamshire, UK   163

HOW DOLOMITE REFRACTORIES CONTRIBUTE TO CLEAN STEELMAKING
    R.D. SCHMIDT-WHITLEY, Director of Development,
    Didier SIPC, Paris, France   175

USE OF NONOXIDE CERAMIC MATERIALS IN METALLURGY
    T. BENECKE, Elektroschmelzwerk Kempten GmbH,
    Munich, Federal Republic of Germany 187

EVOLUTION OF THE TECHNOLOGY AND USE OF REFRACTORIES IN
ITALY AND IN OTHER EUROPEAN COUNTRIES
    P.L. GHIROTTI, Nuova Sanac SpA, Genova, Italy 195

REFRACTORIES TO MEET FUTURE STEELMAKING REQUIREMENTS
    C.W. HARDY, British Steel - Technical, Teesside
    Laboratories, Grangetown, Teesside, UK
    P.G. WHITELEY, GR-Stein Refractories Ltd., Central
    Research Laboratories, Worksop, Nottinghamshire, UK 211

ADVANCED MATERIALS FROM MICROCRYSTALLINE MAGNESITE
FOR THE MODERN STEELMAKING INDUSTRY
    Z.E. FOROGLOU, FIMISCO, Athens, Greece 225

THE USE OF METAL POWDERS IN CARBON-CONTAINING REFRACTORIES
    T. RYMON-LIPINSKI, Forschungsinstitut der
    Feuerfest-Industrie, (Refractory Industry Research
    Institute), Bonn, Federal Republic of Germany 236

A NEW GENERATION OF PRE-CAST REFRACTORY PRODUCTS
FOR THE STEEL INDUSTRY
    J. SCHOENNAHL and C. NATUREL,
    Savoie Réfractaires, Venissieux Cedex, France 246

SUMMARY OF THE CONTRIBUTIONS FROM REFRACTORY PRODUCERS
AND OF THE SESSION ON STANDARDIZATION
    G.C. PADGETT, British Ceramic Research Ltd.,
    Stoke on Trent, UK 255

CONCLUDING ADDRESS: THE POINT OF VIEW OF THE STEEL
INDUSTRY AND OF STEEL RESEARCH
    J. PIRET, Chairman of the ECSC Executive Committee
    'Refractories' 257

CLOSING ADDRESS
    R. AMAVIS, Directorate Technical Research,
    Directorate-General Science, Research and Development,
    Commission of the European Communities 260

ZUSAMMENFASSUNGEN 261

RÉSUMÉS 277

LIST OF PARTICIPANTS 295

INDEX OF CONTRIBUTORS 309

Opening Session

WELCOMING ADDRESS

A. GARCIA-ARROYO
Director for Technical Research
Directorate-General, Science, Research and Development
Commission of the European Communities

Ladies and Gentlemen,

On behalf of the Commission of the European Communities, I welcome you to Luxembourg for this International Conference on 'Refractories for the Steel Industry', which is being held by the Commission of the European Communities as part of its ECSC steel research programme.

The choice of Luxembourg as a venue for this Conference is quite a symbolic one, since the ECSC, the European Coal and Steel Community, was set up here 37 years ago in 1952.

For the European steelmakers in particular, it is hence an opportunity to make a 'pilgrimage' to the place of origin of European cooperation.

The appropriateness of the subject chosen for this Conference has been confirmed by the considerable interest it has aroused in the iron and steel industry - some 300 participants from 21 countries are attending this meeting. From the outset, the organisers have been concerned to limit participation, first of all to ensure that the scientific and technical content of these proceedings will be of the highest quality and, secondly, because the accommodation facilities available at this time of year are limited.

As you can see, the participants present today fill the conference room to capacity.

All the Member States of the European Community are, of course, well represented.

It gives me special pleasure, however, to draw your attention to the participation of experts from Canada, the United States of America, Mexico, Africa (Zimbabwe) and many other countries outside the European Community: Sweden, Finland, Switzerland, Yugoslavia, Austria and Norway. To all of them, I should like to say that we are especially happy and proud to be their hosts.

This Conference was organized in close cooperation with the European Federation of Manufacturers of Refractory Products. My thanks are due in particular, to that body's Secretary-General, Mrs Lefebvre-Jon who has been associated from the start with the Organizing Committee and who participated in the selection of objectives and priority subjects for the discussions.

It should be pointed out that, in implementing the ECSC steel research programme, we attach special importance to consultation between producers and consumers. This is in keeping with the medium-term 'guidelines' for our steel research policy and serves the twofold purpose of responding more effectively to current market demand and of determining future requirements. Where this Conference is concerned, such consultation is a dialogue between suppliers and producers (refractory materials suppliers and steel producers): hence the unique nature of this event and the justification for active participation by representatives of the refractory products industry.

Under the Treaty of Rome, refractory products, such as raw materials for the production of cast iron and steel, are not to be included in the action of the European Coal and Steel Community from the standpoint of their manufacture, but from that of their use in the activities of the iron and steel industry properly speaking. It is in this special context of the use of refractories that research work has been and continues to be supported financially by the ECSC.

The dual basis of this event, moreover, provides an additional assurance that the objectives of this international Conference as set out below will be attained:

- presentation and analysis of the results of the ECSC research conducted in this field over the past ten years,
- as a function, on the one hand, of all the projects relating to refractories and, on the other hand, of future requirements in the iron and steel industry, characterization of future technological developments regarding the use of refractories in the steel industry and research and development projects which would have to be undertaken.

These objectives recur among the subjects to be dealt with in the various sessions during the two days of this Conference.

The intention is:

1) To describe the research already carried out on the use of refractories with the financial support of the ECSC. A general introductory talk will be given by Mr Piret, Chairman of the Executive Committee, entrusted by the ECSC with the task of monitoring the progress of research work. This talk will be followed by papers dealing with more specific research.
2) To describe the work carried out in the field of refractories standardization, which is of fundamental importance as regards the characteristics of these materials from the standpoint of their manufacture and from that of their use in modern steel-making processes, particularly in the development of techniques for the continuous casting of high-quality steels.
3) To attempt to determine the future requirements for refractories in the steel-making industries. As yet it is impossible to envisage producing cast iron and steel without using refractories.
4) To provide some information on the development of new refractory materials with two objectives in mind: one, to obtain steel of still higher grade and, two, to improve the cost-effectiveness of using these materials, mainly by achieving greater purity of the product and greater specificity.

In the 1970s, the consumption of refractories in steel making amounted to more than 20 kg per tonne of steel, but we have now achieved, in certain cases, a consumption level below 10 kg per tonne.

This reduction results chiefly from the use of higher quality materials, that is to say, materials of greater mechanical strength and with greater resistance to heat, and from an easier and more effective application technique. Furthermore, the development of new steel-making methods (for example, continuous casting, which is becoming increasingly widespread) will require still more sophisticated refractories.

The programme for the two days of this Conference is certainly an ambitious one, but I believe that the quality of the speakers and the presence of so many high level experts in the field of refractories production - as in that of steel making - are two guarantees of quality in the scientific exchanges that will take place during the discussions on the papers and will also be continued outside the official sessions. In this connection, the reception this evening, to which you are all invited, will doubtless provide an additional occasion for establishing relationships and contacts between yourselves.

The dialogue between suppliers and producers, which of course already exists, will be intensified to the benefit of both partners and with a view to the development of new techniques for achieving ever greater scientific progress.

It is incumbent on the steel industry to make careful preparations for its future in the coming decades, and the role of the refractories industry is particularly important in this context. This Conference must provide an occasion to take stock of the present situation and, above all, to determine what the future will hold in order to establish a pattern for research and development work in this field, particularly for ECSC steel research.

In conclusion, I should like to thank all those who have worked so hard to organize this meeting, the members of the Organizing Committee, the Chairmen of the sessions and the speakers, and our Commission colleagues in Directorate-Generals IX and XIII. I trust that this Conference will be attended by every success and that your stay in this lovely city of Luxembourg will be both pleasant and fruitful.

INTRODUCTORY TALK: REVIEW OF RECENT RESEARCH CARRIED OUT
UNDER THE ECSC RESEARCH PROGRAMME ON REFRACTORIES

J. PIRET
Chairman of the Executive Committee on ECSC Research into Refractories
Centre de Recherches Métallurgiques (CRM), Abbaye du Val-Benoît
B-4000 Liège, Belgium

General talks on refractory materials often begin with a graph showing production trends over the previous few years. Fortunately or unfortunately (depending on your point of view), this downward curve is more or less offset by a curve in the opposite direction showing the costs of refractories. I would, however, like to draw attention to the correlation with trends in steel production, as this too sometimes fluctuates. Whilst it is true that all professions go through good and bad patches, there can be few examples of two trades dealing with such different materials, each with its own restrictions, whose destinies are so closely interlinked.

Even with water cooling, which may sometimes be used as an alternative, the saying, "No refractories, no steel", will hold true for some time to come. The converse of this is, of course, "No steel production, hardly any production of refractories".

This was why the Research Committee on Refractories was set up more than 15 years ago under the ECSC Research agreements of the Commission of the European Communities. It was realized that refractory materials are just as important for the iron and steel industry as raw materials, and that good or bad results from refractory materials do not depend solely on the manufacturers.

We should like to take this opportunity to thank the Commission's Directorate-General for Research for the continued support it has given to our work, as well as the Directorates for Information and Technology Transfer and Personnel for organizing today's Conference, which is the first to bring together so many representatives of both sectors under the European banner.

Since it was set up, the Committee on Refractories has monitored the progress of some 15 research contracts.

We shall not go through all these reports now, as the most recent and striking results will be presented in a few moments. We are simply going to point out the main areas of interest of our work. You will note that these are often common to several research programmes and so are worth discussing on a group basis within the Committee. As a result of the recent reorganization of the meetings of the 'Refractories', 'Steelworks' and 'Technology' Committees, we can broaden the scope of our discussions further still. You will also note that, as is usually the case, research reflects the concerns of those who use refractory materials rather than those who produce them.

The first family of refractories consists of materials with a very long lifespan, and which are regarded as investments: this includes most of the material used in blast furnaces, cowpers, mixers and torpedo ladles.

Numerous parties are involved in the making of these: producers of refractories, those in charge of engineering, fitters and users. Mistakes prove very costly and can result in reduced output over a period of years, production losses due to repairs, or premature shutdown. The highly detailed study carried out by THYSSEN into the heating of steel furnaces shows that firing time for some furnaces can be reduced, but highlights the precautions necessary to prevent stresses which may damage the metal casings and insulating layers.

The second family of refractories accounts for the main bulk of consumption and has been the subject of most of the work done by our Committee. These are refractories which wear out.

The main objectives of research are as follows:

1) The study of **wear mechanisms**, which vary greatly according to individual applications. Attention has once again focussed primarily on **thermomechanical phenomena**: by studying thermal stress mathematically, HOOGOVENS has been able to estimate the stress limits of materials on the basis of their properties as measured in the laboratory, the practical upshot of which has been an improvement in the life of coke oven burners and the porous components of converters. IRSID has also highlighted the need to reduce thermal stress in the bottom sections of furnaces fitted with combined blowing tuyeres, and in steel ladles. Solutions have been put forward for both of these: a practical study of wet or flame gunning of protective linings, evaluation of the lower cooling limit for basic ladles between melts, and experiments with ladle linings which are less sensitive to thermal shock.

As regards wear by **chemical corrosion**, CRM has developed a special test involving both the slag and metal phases. This has led to a substantial improvement in the life of continuous casting nozzles.

2) **Reducing the cost of refractories** has obviously featured prominently in much of the work done; particular mention should be made of the study carried out by HOESCH, which has shown that improved wear resistance may sometimes be obtained by using unfired refractories; these may require up to ten times less energy to produce than their fired equivalents.

3) **Improving the quality of steel** is the priority objective of some research: work carried out by the CSM (Centro Sviluppo Materiali) has shown that calcium oxide can be used as a refractory material in steel ladles or calibrated nozzles. This leads to increased desulphurization and allows continuous casting of aluminium-killed steels, which is difficult or impossible to achieve using current methods; CRM, working towards the same objective, has carried out research into protective linings (applied by flame gunning) in order to produce hydrogen-free casts with reduced superheating.

4) Finally, it will be noted that much of our research is concerned with **improving the performance and reliability of equipment in the iron and steel industry**. This question is obviously of fundamental importance at a time when much fewer production lines are being used, and we will return to it at the end of the Conference, after we have heard our colleagues from the iron and steel industry and the various refractory firms.

Session I

Chairmen: P. ARTELT (*Commission of the European Communities*)
Mrs M. LEFEBVRE (*PRE, Paris, France*)

STRESS CALCULATIONS FOR REFRACTORY MATERIALS WITH HELP OF THE FINITE ELEMENT METHOD.

J.A.M. Butter, J. de Boer
Refractories R & D
Hoogovens Groep BV
P.O. Box 10.000, 1970 CA IJmuiden, The Netherlands

Summary

Thermal spalling is an important wear phenomenon in connection with ceramic refractory materials. In some cases the consequences of failure due to internal stresses are far-reaching; even a shut down of a complete installation belongs to the possibilities.
A research has been carried out in order to quantify the phenomenon. The objective was to develop a flexible model in which temperature dependent material properties can be applied. The output should be evaluated in such a way that an easy interpretation is possible.
The model is based on standard Finite Element Method software. However the software had to be optimised for the specific properties of porous material.
The final model has been tested with help of simulative laboratory measurements and with measurements in the plant.
The model has been applied to several critical lining positions of furnaces in the integrated steel plant. As an example the results of an ECSC-project on the use of a mathematical model for the ceramic burner of a coke oven is given. The second example shows recent results of a research on the wear of stirring elements used in basic oxygen furnaces. The latter concerned a cooperative ECSC-project of British Steel and Hoogovens.

Introduction

An important disadvantage of refractory material is its brittle behaviour. Local high stresses can not be released by plastic deformation. Fatal cracking or a substantial decrease of strength is the result.
Under influence of transient heat flow the refractories of furnaces can crack due to internal stresses within the separate bricks. In general the phenomenon is named: spalling (1,2,3).
The major reason why this phenomenon of spalling shows up in a pronounced way in integrated steel plants is mainly caused by the big temperature changes during the production processes. Expansion variations between 0.3 % and 1.5 % are not unusual. The material maximum strain does contrast with these expansion variations. Usually the strain potential under tensile conditions is only 0.1 % till 0.3 %.
The inability of the maximum strain to cope with the big expansion variations is the source of spalling problems.
Since the last twenty years interests and possibilities in the steel industry have increased the need to develop materials with better spalling resistance. This need is mainly found at sections of the integrated steel plant which are involved in cokemaking, ironmaking and steelmaking. At vital spots in these production units refractory materials with a good spalling behaviour are necessary.
Based on the above mentioned findings it was decided to describe the spalling process as a whole. In order to translate the findings to

industrial situations it is necessary to incorporate process conditions and brick geometries together with the material properties.
In such a case simulation by calculation or by laboratory reproduction of the process offers a solution for the description of spalling.
A model based on a standard Finite Element software was developed. It has been optimised for the specific properties of porous ceramic material (4).

Model description

The mathematical model is based on a Finite Element Method (FEM) program. The program is applicable to stresses in steel constructions and heat flow situations with moderate temperatures. As conditions and materials are rather different a number of additional programs are necessary for calculations on refractory materials. The main reasons are:

* Ceramic materials do have more complicated elastic behaviour as compared to metals.
* Change of the material properties as a function of temperature must be incorporated.

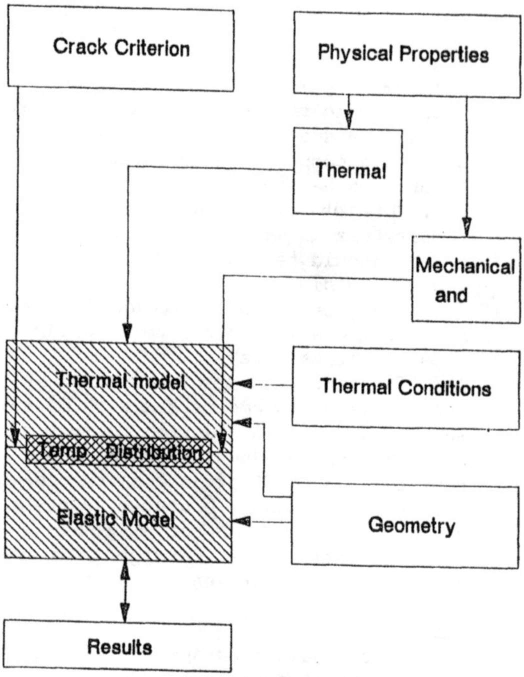

The model roughly exists out of 4 parts which are passed through sequentially (Figure 1).
These parts are:
a **block generator** which

Fig. 1 Input output survey of model

generates the geometry of the bricks; a **thermal program** which calculates the three-dimensional temperature distribution as a result of a transient heat transfer; a **stress program** which delivers a three-dimensional stress distribution and a **evaluation program** which converts bulk data output in such a way that access to the results can be improved.
Material properties are necessary input for the transient heat transfer calculation and the stress calculation.
The heat transfer program needs values for:

- Thermal conductivity (5,6)
- Specific heat
- Bulk density

The stress program needs values for:

- Strength
- Young's modulus
- Coefficient of linear expansion

- Poisson ratio

These properties change markedly with temperature. The model must contain the possibility to handle the material properties as temperature dependent properties. Many of the above properties can be determined by methods which are either standardised by PRE or ISO.

Strength and Young's modulus can be described with help of the stress-strain diagram of refractories as determined in our laboratory for ceramic materials. Dynamic methods can not be used for the determination of the Youngs modulus as can be seen from Figure 2. It appears that an important discrepancy exists between static and dynamic methods. Since spalling is a static phenomenon values from dynamic origin must be abandoned (7). As a consequence values from 3-point bend tests are used. Figure 3 is a schematic of the applied method.

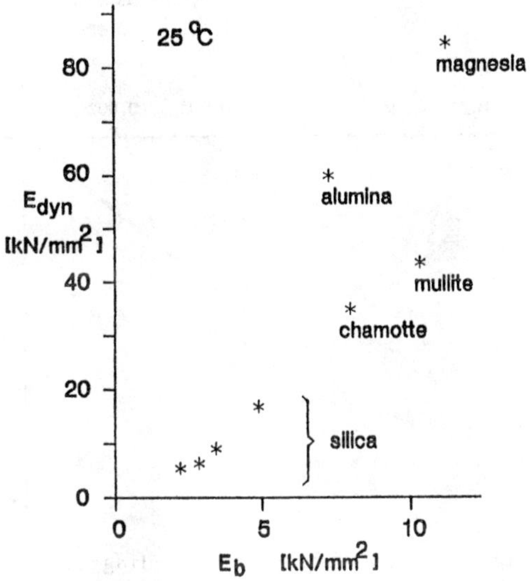

Fig. 2 Dynamical modulus as a function of static modulus.

Figure 4 is an example of a stress-strain diagram. It has been derived from several tests, including the 3 point bend test. The difference between the tensile mode and the compressive mode is striking and has two aspects.

- The values are substantially higher at the compressive section (negative coordinates) of the diagram than at the tensile section (positive coordinates).
- A discontinuity is found at the origin. In other words: The Young's modulus at the tensile section appears to be of a lower level than the Young's modulus at the compressive section.

The above findings had direct consequences for the model calculations. Precautions have been taken in order to correct the calculations for the nonlinear

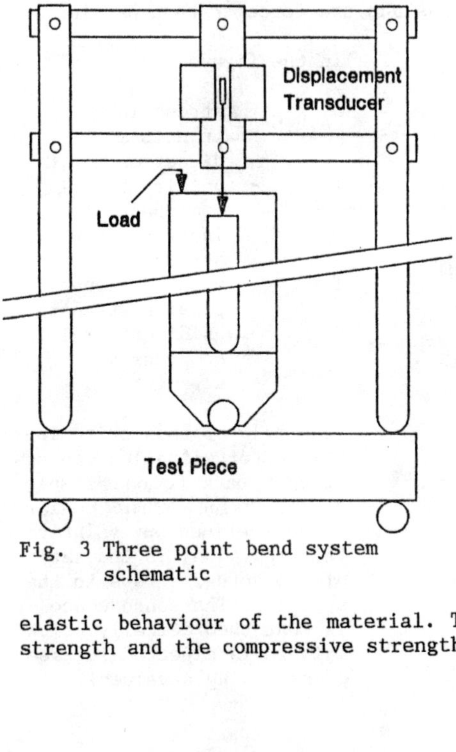

Fig. 3 Three point bend system schematic

elastic behaviour of the material. The big difference between the tensile strength and the compressive strength had to be implemented with help of

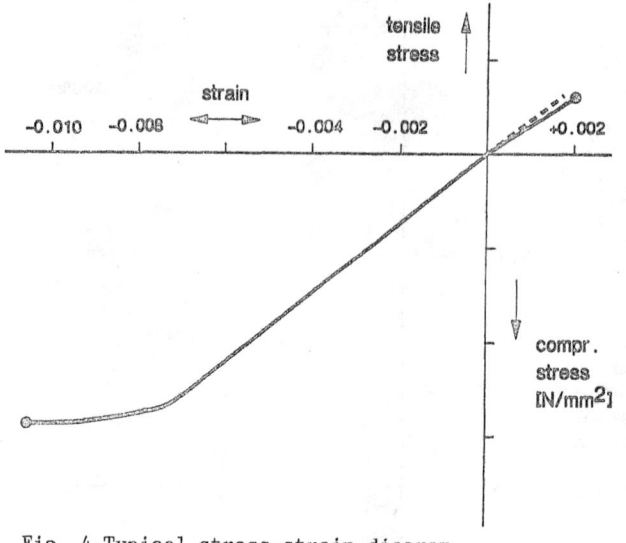

Fig. 4 Typical stress-strain diagram

the correct choice of the crack criterion. Because of the compact shape of bricks many combinations of stress vectors can be expected. That is why a three-dimensional crack criterion is absolutely necessary in order to achieve a consistent agreement with the actual stress situation. The applied yield criterion is based on the maximum distortion energy theory (8,9). Because of the evident difference of tensile strength a variant has been chosen which is known as the Drucker-Prager criterion. Figure 5 visualises this criterion. It shows some cuts of a spatial cone. The cone exists in a three-dimensional coordinate system where the axes are principal stress axes. Any combination of the 3 principal stresses which is positioned outside the cone leads to material failure. As long as the stress description of the situation is at the inside of the cone the material will withstand the force at that particular point.

In the figure:

$\sigma_{A,B,C}$ - stress in principal direction A,B,C
I - unidirectional component of isostatic tensile strength
T - unidirectional tensile strength
D - unidirectional compressive strength

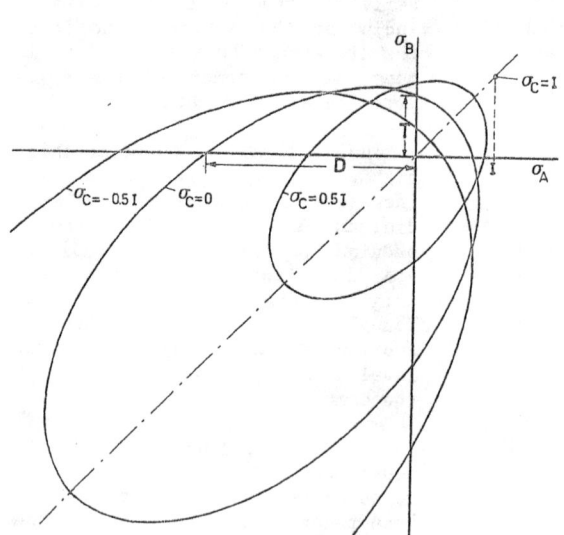

Fig. 5 Drucker-Prager criterion

With help of this criterion the vector situation can be brought back to one simple scalar value. This scalar value (defined as relative failure value) is the ratio of the actual stress to the strength. The consequence is that theoretically brick failure is expected in case of exceeding a value 1.

Simulation test

The model has been tested with help of a laboratory spalling furnace (15,16,4). The furnace as shown in figure 6 can submit 8 bricks to a unidirectional transient heat load. There is an option to heat the bricks to a higher steady state temperature with help of a second furnace. By means of displacement transducers the moment of cracking can be recorded.
In order to get an idea of the position of crack initiation the transducers are placed along the axis of the brick as well as along the edge of the brick. Separately a model calculation

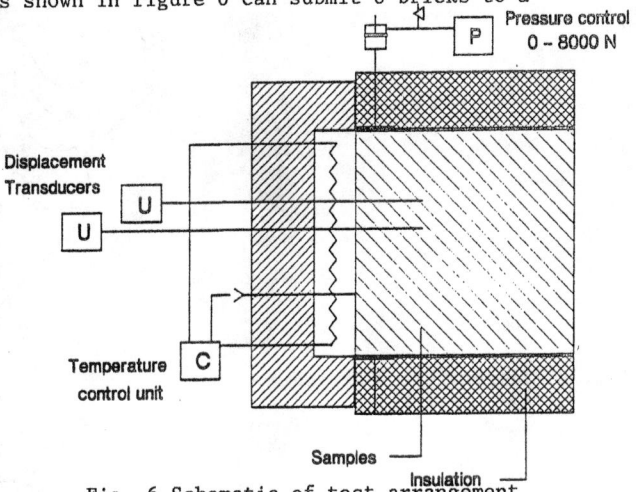

Fig. 6 Schematic of test arrangement.

has been executed on these bricks under the same conditions.
From the results it is concluded that not the theoretical level 1 of the relative failure is the critical value for failure but the level 0.7. This level shows to be consistent.

Application of the model on the ceramic burner inlet of a coke oven

Because of poor combustion performance the interior of coke plant II at IJmuiden was inspected visually by means of an endoscope (Figure 7). These inspections showed that the first ceramic burner inlet (silica) of many coke gas channels was damaged. With help of the model, calculations of the existing situation have been executed (17). For this purpose the silica brick with maximum dimensions at the inlet of the burner was chosen. In this case again the calculations were supported by temperature measurement in the brickwork. However the results of these measurements were ignored initially. The process conditions (heat transfer coefficients) were calculated on basis of process events by means of gas flow parameters.

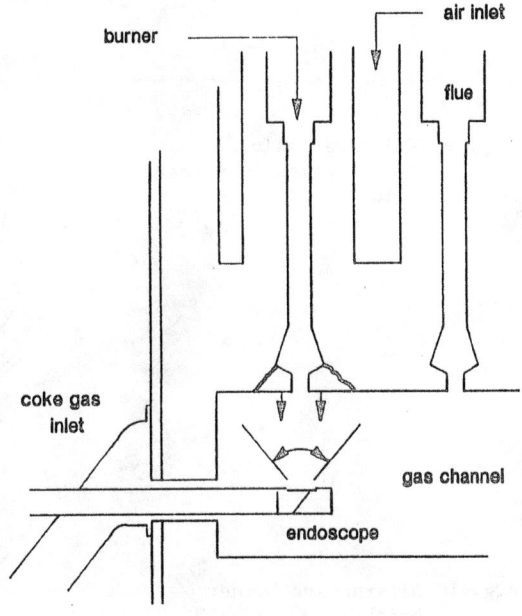

Fig. 7 Ceramic burner inlet in coke oven.

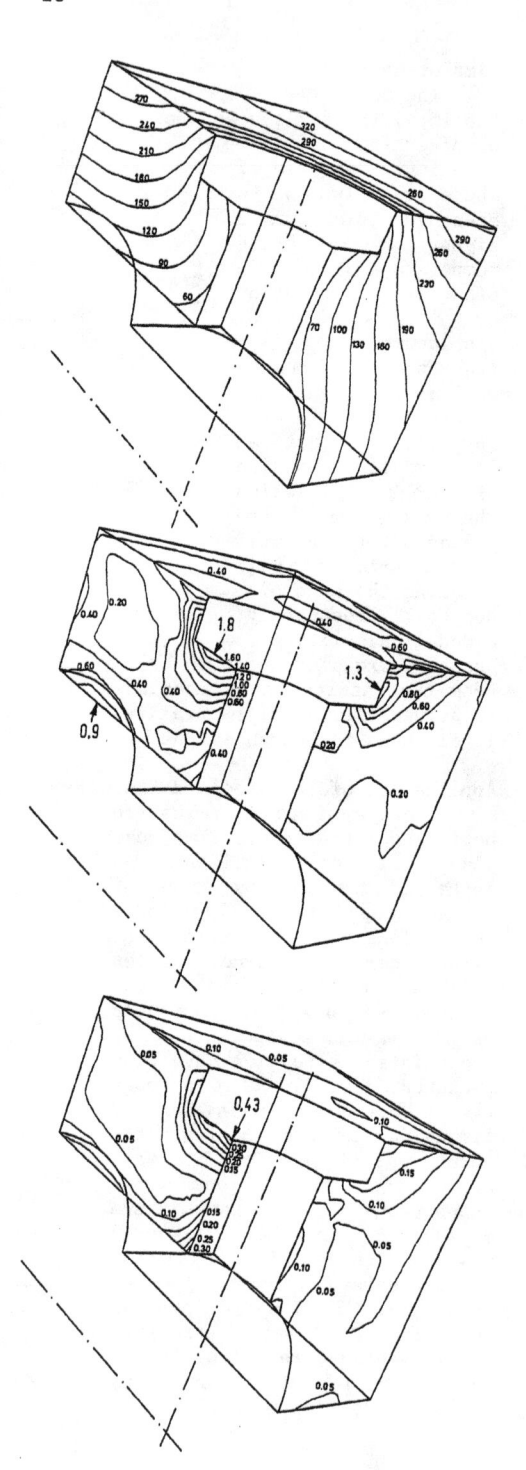

Fig. 8 Temperature distribution of burner brick.

Fig. 9 Relative failure distribution in burner brick.

Fig. 10 Alternative burner brick.

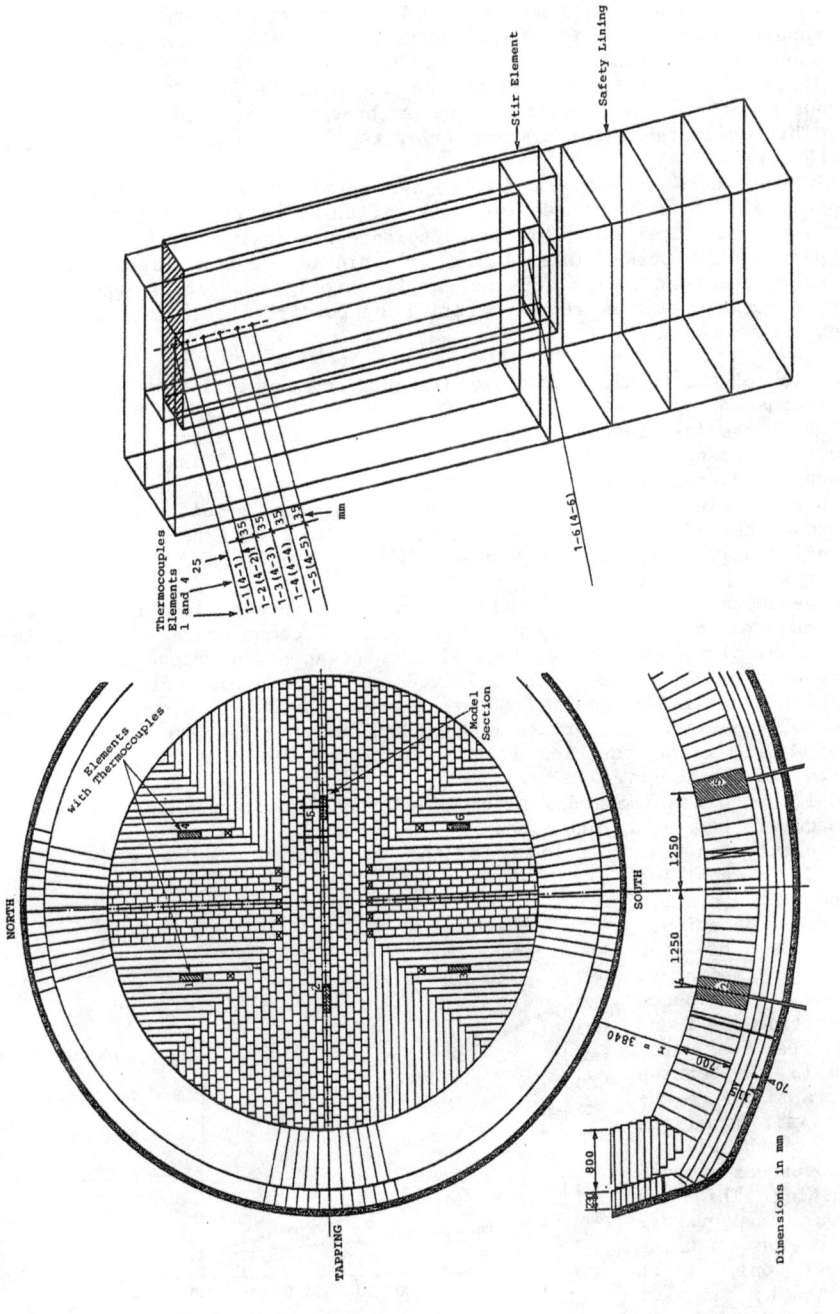

Fig. 11 Position of the stirring elements at Hoogovens BOS 2

Fig. 12 Detail of situation

The process conditions could be described in such an accurate way that the model results were in good agreement with the temperature measurements without necessity of tuning.

From Figure 8 it can be seen that the isotherm of 200 °C, which is a dangerous temperature for silica, wanders through the centre region of the brick. This unfavorable temperature level results into a very high stress level (Figure 9).

From these results it appears that under normal operation circumstances some parts of the brick exceed the critical relative failure level of 0.7 up to a maximum level 1.8. This leads to severe cracking especially at the upper part of the burner inlet brick. By coincidence this is exactly the place where the restriction for the gas flow to the burner is positioned. After replacement of the silica material by cordirite (see Figure 10) no damages were found.

Application of the model to stirring elements for combined blowing in basic oxygen furnaces

During the last 2 years BSC and Hoogovens extended their collaboration in the development of the Bath Agitation Process (BAP). This was done by an ECSC-supported research on the wear behaviour of bottom stirring elements. Through these elements inert gas is injected into the liquid steel in order to increase the mixing of the bath. This leads to a bath composition which is significantly closer to the thermodynamical equilibrium. Some important metallurgical advantages are the result.

The research was focused on the thermal stresses in the element bricks and bricks adjacent to these elements. With help of temperature measurements, model calculations and a mineralogical investigation on recovered material important information has been collected on the wear mechanism of bottom stirring elements. The applied materials were varied from magnesia-resin bricks to magnesia-carbon bricks with carbon contents between 10% and 20%.

Figure 11 shows the position of the stirring elements in the 300 ton converter of the Hoogovens BOS 2.

Figure 12 shows the the model section and the thermocouple positions. The thermocouples had to approach the hot face because model calculations led to the conclusion that only 150 mm of the total lining thickness is involved in spalling because of dynamic temperature behaviour. During one of the numerous temperature measurements evidence for spalling was found as can be seen from Figure 13. In between the "3rd stop O_2" and "SIT" (wait state at the end of the blow) a slag cover was removed by melting. This led to a temperature distribution which could only be explained by the fact that approximately 50 mm of material was removed from the hot

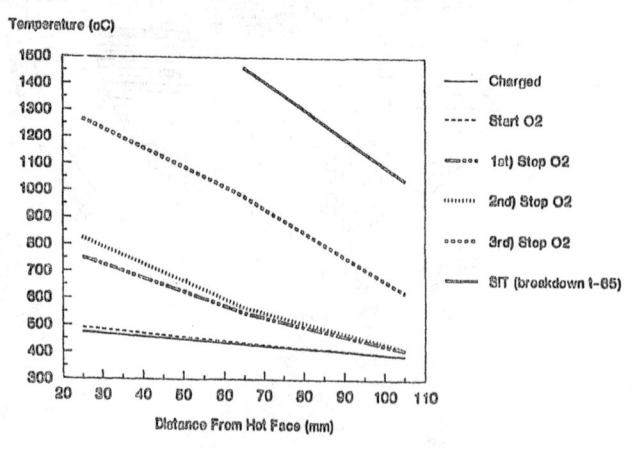

Fig. 13 Temperatures during spalling.

face. A similar situation has been calculated with help of the model. The circumstances were less severe because of a less thick slag cover at the start brought the hot face temperature to 800 °C instead of 400 °C. In Figure 14 the temperatures at various depths are plotted.

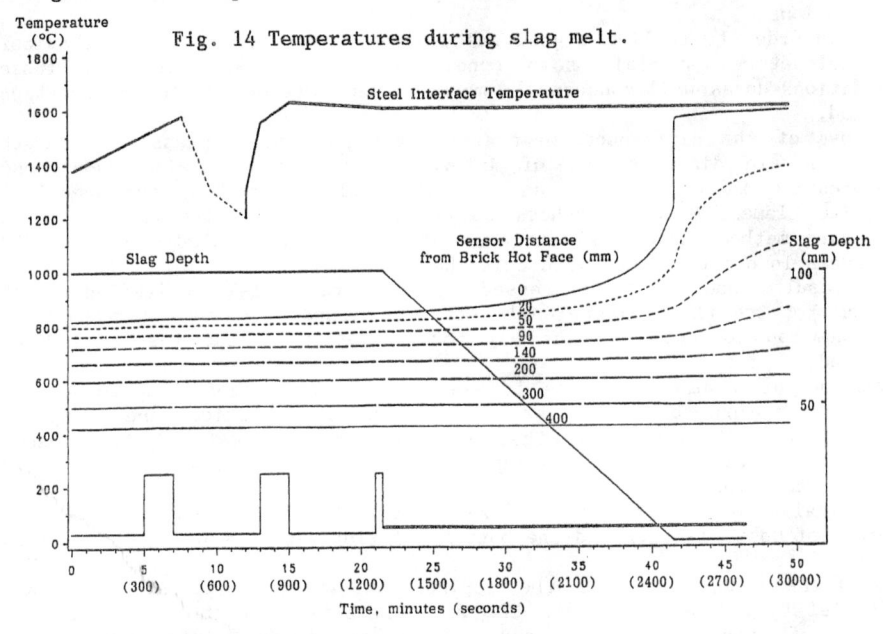

Fig. 14 Temperatures during slag melt.

The period up to 22 minutes is part of the standard converter cycle. From time 22 minutes the original slag cover of 100 mm melts away with a melting rate of 5 mm/min. Excessive temperature rise is seen during the last minutes of the slag melting. From the moment that the slag cover has disappeared the stresses in the brick and the position of the highest stress level are rapidly changing (Figure 15). The relative failure value exceeds the critical value during a period of 8 minutes. This should result in crack formation in the hot face during the first 2 minutes and loss of 35 mm of material in the following stage.

From this research important conclusions were drawn in connection with materials, operations and design. During the research stirring elements at

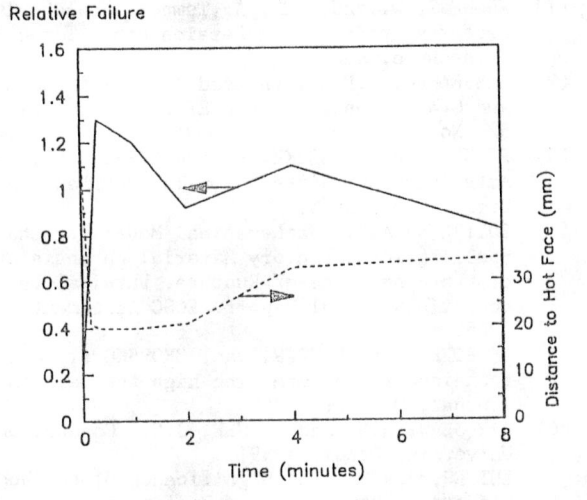

Fig. 15 Relative failure levels after the slag melt.

Hoogovens were changed from canned elements to a slit type element. The lifetime of the elements improved such that the availability of the stir system is the same as the lining life of the converter (55 to 60 days).

Conclusions

In order to realise a good calculation method for spalling behaviour of refractory materials some conditions had to be fulfilled. These conditions do as well concern the material test methods as the calculation method.

Because of the difference in tensile strength and compressive strength stress-strain diagrams are of interest under both tensile mode and compressive mode. Up to now analytical calculation methods and numerical (Finite Element Method) methods do not have possibilities as such as a standard method. Dedicated programs had to be implemented into the FEM programs in order to cope with this phenomenon.

The results had to be processed by a correct yield criterion which transforms the three dimensional vector to a scalar value. By processing the numerous output data an easy interpretation of the results can be reached.

By means of simulative tests it was possible to establish a connection between rupture behaviour as experienced in practice and computer predictions. It was found that in case of well defined homogeneous refractory materials practical rupture appeared at a stress level of about 70% of the theoretical values.

The final model outfit has been used next to others to study the ceramic burner of coke plant II and the bottom stirring elements of the converters of BOS 2.

The first example led to the replacement of the original refractory material by another material. The second example led to changes in the design of stirring elements and operational changes. Extension of the lifetime of the stir elements is the result.

References

(1) KRöNERT, W. and BöHM, A.,Temperaturwechselverhalten tonerdereicher feuerfester Steine im Bereich hoher Temperaturen. Westf. Techn. Hochschule, Aachen.
(2) HASSELMAN, D.P.H., Unified Theory of Thermal Shock Fracture Initiation and Crack Propagation in Brittle Ceramics. JACS, November 1969, Vol. 52, No. 11.
(3) FIEDLER, U., JESCHKE, P. and KIENOW, S., Zulässige Aufheizgeschwindigkeit von Torpedopfannen- und Mischersteinen. Tonind. Ztg., 1976, Nr. 5.
(4) BUTTER, J.A.M., Mathematical Model for the Determination of Thermal Spalling in Refractory Material on Basis of the Practical Relationship of the appearance of Rupture, Physical Properties and Physical Conditions. Final Report. ECSC Agreement No: 7220-EB/603. 5 April 1985.
(5) DE BOER, Jur, BUTTER, Jan, GROSSKOPF, Bernd, JESCHKE, Peter, Hot wire technique for determining high thermal conductivities. Refractories Journal, October 1980.
(6) CARLSLAW, H.S. and YEAGER, J.C., Conduction of Heat in Solids. Oxford University Press, 1959.
(7) BUTTER, J.A.M., The Significance of the Modulus of Elasticity for Refractory Materials and Engineering. SIPRE-meeting, Noordwijkerhout, may 1983.
(8) TIMOSHENKO, S.P., History of strength of materials. McGraw-Hill, New

York, Toronto and London, 1953.
(9) TIMOSHENKO, S. and GOODIER, J.N., Theory of elasticity. McGraw-Hill, 1951.
(10) BROEK, David, Elementary Engineering Fracture Mechanics. Noordthof Int. Publ.,Leyden, 1974.
(11) DAVIDGE, R.W., Mechanical Behaviour of Ceramics. Cambridge University Press, 1979.
(12) UCHIYAMA Shoichi. Views on Fracture Behaviour of Refractories. Transl. Taikabutsu, 10-1980.
(13) HOOGLAND, R.G., MARSHALL, C.W. and DUCKWORTH, W.H. Reduction of Errors in Ceramic Bend Tests. JACS, June 1976, Vol. 59, Nr. 5-6.
(14) BELL, D.A., PALIN, F.T., PADGETT, C.C. Effect of rapid heating on blast furnace refractories. Steel times, February 1984.
(15) DE BOER, J., Essai de résistance au choc thermique dans un four spécial. L'Industrie Céramique, Nr. 764, 9/82.
(16) BROERSEN, P., THIJSSEN, N.J.W., The Repair of Coke Chamber Walls of a Coke Battery in Operation. Final Report. ECSC Agreement 7220-EB/601. 30 October 1981.
(17) BUTTER, J.A.M., Calculation of Stresses in Coke Oven under Process Circumstances with a Mathematical Model. ECSC-Round Table Coke. Luxemburg, 27-28 October 1983.
(18) BUTTER, J.A.M. and RENGERSEN, J., Temperature Distribution and Thermal Stresses in Refractory Material. Inst. Metals, Sutton Coldfield, October 1987.
(19) NORMANTON, A.S., BUTTER, J.A.M., Thermal Stressing in Stirring Elements used in Basic Oxygen Furnaces. Final Technical Report. ECSC Agreement Nos 7210.CB/807, 7210.CB/601. June 1989.
(20) DENIER, G., GROSJEAN, J.C. and ZANETTA, H, Heat transfer in tuyeres for oxygen bottom blowing converters. Iron and Steelmaking, 7 (3), 1980.
(21) MORIMOTO T., MATSUO A., MIYAGAWA S., OGASAHARA, K., TACHIBANA, R. and KUWAYAMA, M., Progress in refractory techniques for combined blowing system (LD-KGC) with wide flow rate range. 2nd International Conference on Refractories, Tokyo, Nov. 10-13 1987.
(22) SHARMA, S.K., Criteria for Predicting Penetration of Molten Steel in Tuyeres During Inert Gas Injection. ISS Transactions, volume nine 1988-13.

WEAR MECHANISMS OF CONVERTER
AND STEEL LADLE LININGS ; REPAIR METHODS

C. GUENARD and P. TASSOT

IRSID - French Iron and Steel Industry's Research Institute
B.P. 320 - 57214 - MAIZIERES-LES-METZ CEDEX - FRANCE

Summary

IRSID has undertaken, with the financial support of ECSC, some studies intending to :
- understand the wear mechanisms, and especially thermomechanical phenomenas, of bottom blowing converter and steel ladle linings,
- improve knowledge of gunning systems for lining-repair : semi- wet and flame-gunning.

These studies allowed to reduce operating costs by decreasing the number of bottoms required in each campaign of converters and by improving the stability of ladle linings.

INTRODUCTION

1 - Bottom-blowing processes as OBM or LWS, are handicapped when compared with other blowing processes because of the shorter life time of the bottom linings which leads to an increase of consumption and losses of productivity.

IRSID has consequently undertaken a study intending to limit the number of bottom replacements during each campaign by means of :
- improved understanding of the mechanism of bottom wear in the form of a study carried out in collaboration with UNIMETAL Rehon that has allowed a decrease of the number of bottoms required in each campaign,
- improved knowledge of gunning systems of repair :
 the traditional semi-wet technique, firstly, and flame-gunning, secondly, a technique that has required the development of burners and repair materials.

2 - Secondary metallurgy treatments needed for steel cleanness and purity entail an increasing severity of refractory material utilization conditions. An intense calling into question of the concept of steel ladle linings has led to the use of basic products (MgO, dolomite, magnesite-chrome) or products having a high alumina content.

Thermomechanical stress constitutes an important agression during ignition and during different ladle temperature cycles.

Accordingly a study aimed at a better understanding of the thermomechanical behaviour of steel ladle refractory linings by means of the following approach :
- reproduction of temperature cycles in a test furnace and continuous measurement of material deterioration in a laboratory study,
- industrial situation analysis at SOLLAC Dunkerque, Fos and Florange.

1. CONVERTER LININGS
1.1. Study of bottom-wearing mechanism

This study started in collaboration with the OBM steelshop of UNIMETAL Rehon where the life of converter bottoms at the time being was about 250 heats, requiring three or even four bottom replacements during each campaign.

The aim was at this time to increase the life of the first bottoms up to 400 heats per campaign and also to reduce the operating costs by attempting particularly to improve the tuyere technology.

Some 20 parameters that influence the life of bottoms were listed and it was soon apparent that the main cause of wear was the very considerable temperature variations of the tuyere-holding bricks.

A number of measures were taken in order to reduce thermal shocks :
- controlling the propane temperature,
- improving the tuyere technology, jointly with a reduction of the annular space and of propane flowrate,
- reducing the thermal cycles imposed on the refractories by heating the tuyere protection gases (nitrogen, compressed air),
- protecting the bottoms against thermal shocks by systematically coating after each heat,
- reducing the risks of tuyere clogging by reducing the flowrate of nitrogen used fot bath-stirring,
- controlling the equal distribution of propane between the nozzles,
- controlling the quality of the propane used.

At the end of three years of efforts, the initial target was reached : 250 heats with the average life of the first bottoms reaching 400 heats whereas, during the same time, the average life of the linings was being reduced in spite of the practice of maintenance by gunning and selection of refractory materials of best quality ; this was due to changes in steelmill operation with the introduction of continuous casting, installation of vacuum ladle processes and of outlet nozzles.

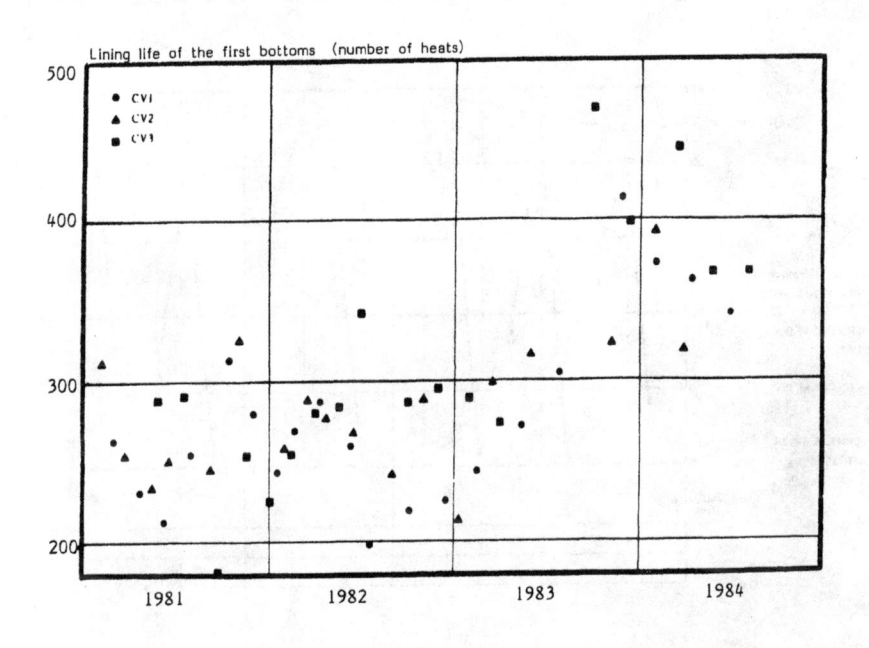

Figure 1 : Evolution of the lining life of the first bottoms.

1.2. REPAIR BY SEMI-WET GUNNING

The characteristics of a series of gunning products has been determined in laboratory (table 1).

PRODUCTS	WEIGHT ANALYSIS %						
	MgO	CaO	SiO$_2$	Al$_2$O$_3$	Fe$_T$	Na$_2$O	P$_2$O$_5$
A	82.05	2.13	7.67	0.52	2.90	1.88	traces
B	80.24	3.30	5.41	2.07	2.11	1.64	2.50
C	85.80	2.70	5.40	0.63	2.80	0.35	traces
D	80.39	4.17	3.39	0.82	4.50	0.69	1.45
E	78.78	4.40	5.97	1.14	3.77	0.67	traces
F	81.24	6.80	2.37	0.24	2.05	1.75	3.78
G	77.81	10.81	2.60	0.88	0.72	1.39	3.45
H	67.08	16.65	3.22	1.54	1.10	1.09	2.82

Table 1 : Chemical analysis of the gunning products studied

Some of these products have been tested under industrial conditions in order to correlate their physical and mechanical characteristics with their operating life.

Test panels were produced by gunning on the linings of the LBE converters of steel shop N°1 at SOLLAC Dunkerque and Fos. Changes in the thickness of these panels were determined by means of laser-interferometer measurements.

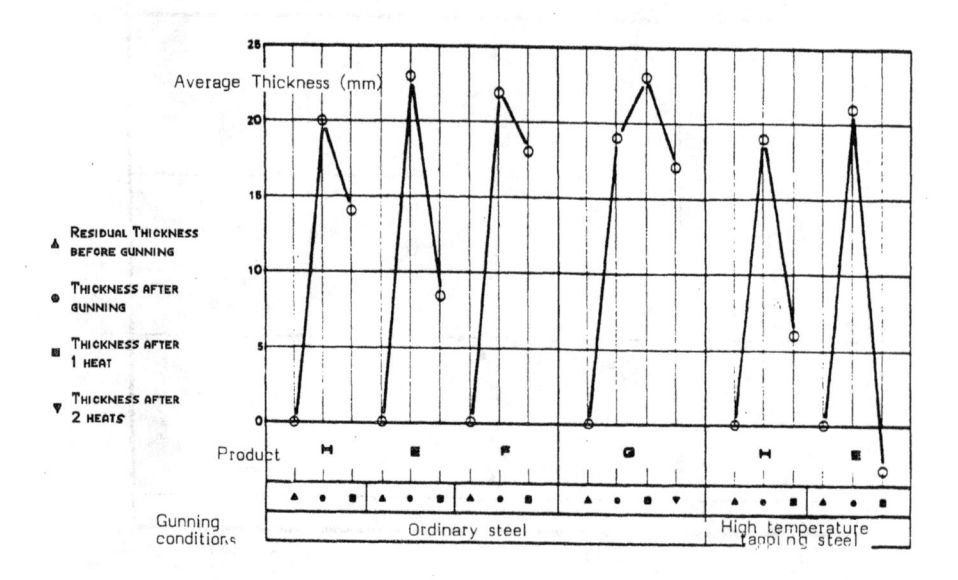

Figure 2 : Example of thickness evolution of gunned panels.

These tests led to the selection of gunning materials that provided the longest lifes of bottoms when repaired.

1.3. REPAIR BY FLAME-GUNNING

Various types of burners were tested and it was finally a multijet burner which was chosen because it allows good thermal exchange between flame and particles.

Two generations of burners then followed, the first with a power of 500 kW allowing the gunning of 3 kg of powder per minute and the second with a power of 2500 kW with a powder flow of 15 kg per minute (table 2).

CHARACTERISTIC		BURNER N°1	BURNER N°2
Power	(kW)	500	2500
Powder flow rate	(kg/min)	2 to 4	10 to 20
Gas flow rate (PCI = 7.5 th/m³)	(m³/min)	1	5
Oxygen flow rate	(m³/min)	2	10
Distance lance-wall	(m)	0.7 to 0.9	0.4 to 0.6
Cooling water flow rate	(m³/min)	-	0.3
Powder feeding		\multicolumn{2}{c}{Full pneumatic distribution and transport}

Table 2 : Characteristics of multijet burners

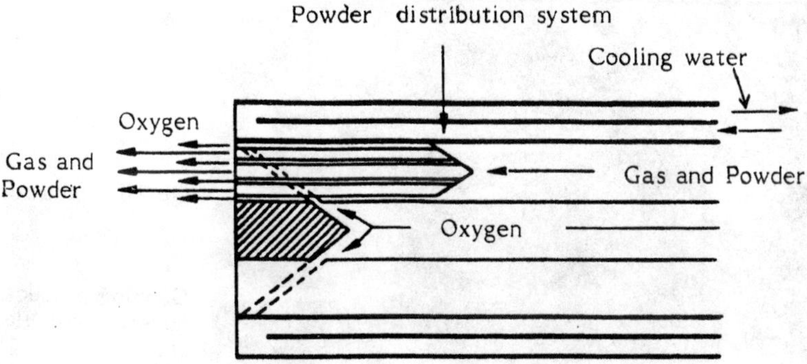

Figure 3 : Multijet burner structure

These burners were used to develop a range of products containing the main granulate, magnesia, a binding agent : phosphate or calcium phosphate or silicate and an additif intended to increase the refractoriness of the material after gunning.

The optimal characteristics of gunning were determined, particularly, the relation between density, liquid phase content of the product and specific power (kg of powder per kW of power).

Liquid phase content (%)

Specific power (kW/kg)

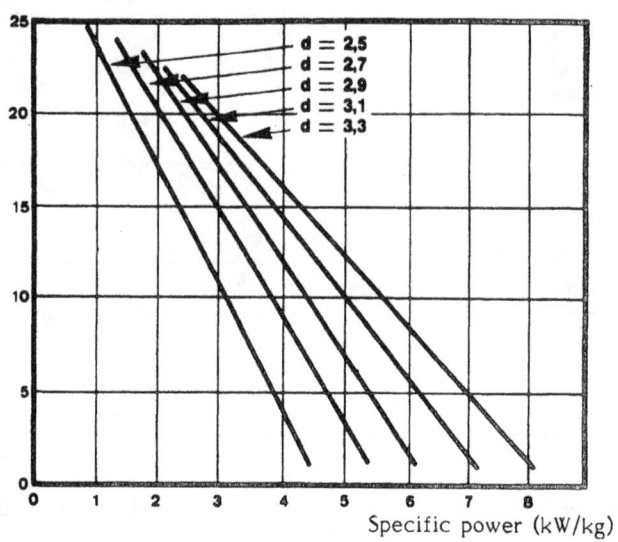

Figure 4 : Density in relation to specific power and liquid phase content

Gunning series were conducted with lances and these products on converters of various sizes ; the most recent gunning was carried out on the bottom of the IRSID pilot converter equipped with LWS tuyeres (figure 5).

Tuyere

Metallic Mushroom

Gunning product covered with slags

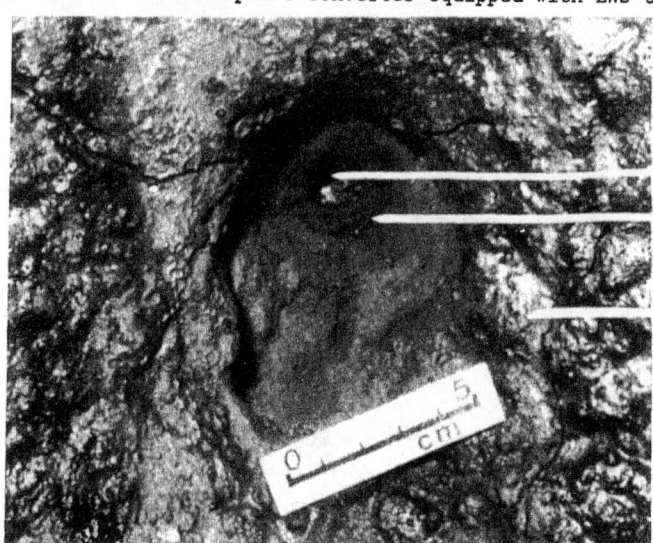

Figure 5 - Gunning on a L.W.S. bottom

This gunning experiments provided confirmation :
- that repair by flame-gunning of bottoms fitted with tuyeres was possible without damaging the latter,
- that the lives of the layers gunned was very much longer than that of the semi-wet gunning materials.

The result of the study has improved knowledge of the mechanisms of bottom wear and also the development of a flame-gunning process suited for repair of bottoms of bottom-blowing converters.

2. STEEL LADLE LININGS
2.1. Laboratory study : thermal cycle simulation trials

Before thermal stress tests, materials commonly employed in steel ladles were characterized.

A new-design furnace (figure 6) was constructed which allows the reproduction of steel ladle temperature cycles and a continuous measurement of deterioration by tracking the material's self-resonance frequency (figure 7). Characterization of the internal state of the products is possible by this measurement which is sensitive to the number and size of cracks.

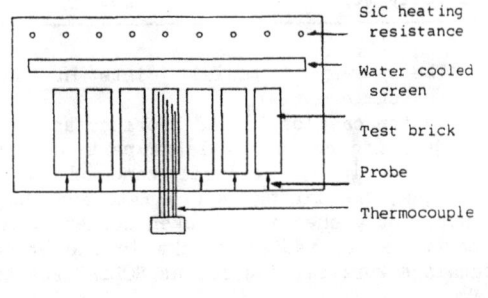

Figure 6
Thermal shock furnace

Figure 7
Thermal shock furnace
Operation diagram

The study was mainly focused on magnesite-chrome type materials used in ladle-furnace at SOLLAC Florange.

Determination of the critical temperature difference which can be supported by this type of material (Tc ≈ 760°C) was made using cycles of increasing temperatures.

Using fractographic analysis of the tested samples, a verification was made showing that rebonded materials retain the best resistance to cracking at low temperature. On the other hand, directly bonded materials derived from sintering seem to progressively acquire a stronger crack propagation resistance with temperature rise. It is this last type which must be privileged from a strictly thermomechanical point of view for industrial applications. It was possible to verify these results by performing postmortem analyses of samples taken from SOLLAC Florange.

Theoretical considerations of steel ladle thermomechanical stress

SOLLAC steelplant treatment ladles are lined in spiral according to the indications given in Table 3.

	SOLLAC	Dunkerque	Florange	Fos
W E A R	Slag line	MgO	MgO-Cr	MgO-Cr
	Wall	Dolomite	Alumina-chrome	HTA
Jointing of joints between wear bricks		no	yes	yes (wall)
Type of lining work		spiral	spiral	spiral
Number of wall layers		4	3	2
Mean diameter (m)		3.8	4.0	4.4
Capacity (t)		230	240	310

The number of radial joints in the three plants is sufficient to limit circumferential.

In the case of SOLLAC Florange and Fos, the lining is made of a point flat jointing cement. A large part of the stress is absorbed and induces few cracks in the central zone of the lining bricks. In the case of Dunkerque, the lining is dry laid and the gaps due to the joints are thus reduced. At preheating, broken corners can be found on the hot face. After a dozen casts, radial cracks in the bricks can be found. Two industrial trackings were carried out at SOLLAC Fos and at SOLLAC Dunkerque.

2.2. Study of steel ladle buildup fracturing at SOLLAC Fos

Rotation of the ladles which is disturbed by cracking and detachment of the buildup formed led to the study of the process of development of this phenomenon.

Postmortem analysis has shown that this buildup is made of a layer of high refractoriness composed of spinels and calcium aluminates whose bonding agent (gehlenite) fuses at high temperature. This layer forms, with the impregnated part of the high alumina content wear bricks, a fragile layer which can crack. The metal could then penetrate and form, cast after cast, a buildup of twelve centimeters thickness (figure 8).

Taking into account its refractoriness, this buildup could detach at contact with the steel. It is therefore necessary to act upon slag composition at bath surface.

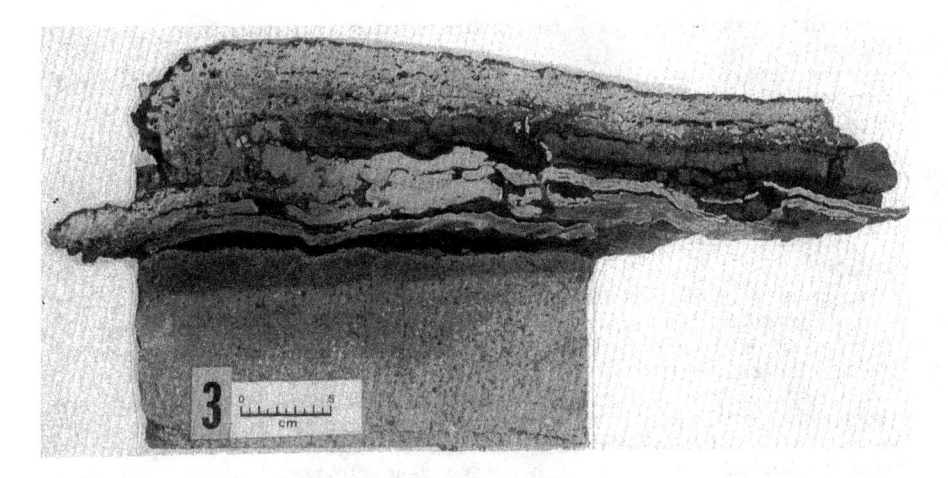

Figure 8 : Buildup on steel ladle wall at SOLLAC Fos

By ajusting the quantity of lime added in the ladle, the composition of calcium aluminates can be adjusted permit liquefying this layer and washing the wall.

The ideal situation consists of keeping a coating containing in the order of 70 % solid phase and 30 % liquid phase. This protects the wall from chemical corrosion and limits thermal shocks at ladle filling.

This optimal situation is difficult to maintain in practice however, giving rise to an alternance of corrosion and buildup periods.

2.3. Improvement of steel ladle refractory lining resistance behaviour at SOLLAC Dunkerque

In order to improve the performances of steel treatment, ladle dolomitic linings, tests were made with carbon enrichment of the material (using graphite and carbon black). These additions were done to improve the thermal shocks resistance and the dolomite corrosion resistance, while at the same time preserving a good price/quality ratio.

The average life time of the ladles has been increased by at least 20 % due to the use of dolomite-graphite (9 % total carbon) and dolomite-carbon black (6 % total carbon) (figure 9).

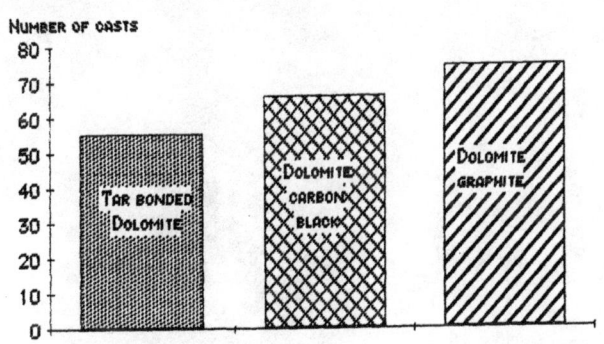

Figure 9 : Ladle performance at SOLLAC Dunkerque

Postmortem analysis has shown that the addition of carbon permitted the following :
- lower the number of cracks parallel to the hot face leading to product peeling,
- reduce the number of cracks perpendicular to the hot face.

Insofar as the mechanical strength of the materials is sufficient, and the higher carbon content of the bricks does not disturb low carbon content steel production planning in a steelplant, carbon enrichment is a good technical-economical solution to improve dolomitic lining potential.

CONCLUSIONS

The results of these studies have allowed :

1) to improve knowledge of the mechanisms of bottom wear and also the development of a flame-gunning process suited for repair of bottoms of bottom-blowing converters.

The work performed in collaboration with UNIMETAL Rehon highlighted the considerable effect of temperature variations on the bottoms life, which can be minimized by taking action on tuyere technology, on the characteristics of the tuyere protecting gasesflowrate and temperature ; and by protecting the bottom by coating them with the refining slag.

A pilot converter was used to show that is was possible to repair the bottoms of bottom-blowing converters ; the tuyeres were not damaged and the life of the protection coats is very much longer than that of traditional gunning materials.

2) a better knowledge of steel ladle refractory stress :
- a laboratory study permitted setting up and exploitation of a method of hot characterization of material deterioration during cyclic thermal shocks,
- two indutrial site studies permitted improvement of the stability of linings having high alumina and dolomite contents in the SOLLAC Fos and in the SOLLAC Dunkerque steelplants.

THE USE OF UNFIRED REFRACTORY BRICKS
IN TORPEDO LADLES AND CASTING LADLES

MANFRED KOLTERMANN

Hoesch Stahl AG, Dortmund

Summary

The objective of this research project was to explore the scope for using unfired alumina silicate bricks in torpedo and steel casting ladles. The properties of these bricks were to be compared with those of fired bricks. Phosphate and pitch-bonded andalucite bricks with added carbon were tested in the laboratory at temperatures of 1700°C.

The results of service tests on these bricks in 200-t torpedo ladles showed strength characteristics similar to those of fired bricks. In the bottoms of ladles with basic linings phosphate-bonded andalucite bricks showed the same strength characteristics as fired andalucite bricks. Further tests with unfired andalucite - zirconium silicate bricks proved unsuccessful.

The cost savings possible with the use of unfired bricks will be determined largely by the price levels for energy and bonding agents.

Energy consumption for the production of refractory bricks in GJ/t is somewhere between 5 and 80 - determined by the temperature and duration of firing, which can be between 1300 and 1900°C. Fusion-cast bricks and special materials such as directly bonded SiC and picrochromite $MgO.Cr_2O_3$ require much greater amounts of energy. Virtually all the raw materials must be prefired before the bricks can be manufactured - Fig. 1.

The only raw materials that do not require prefiring are the following:

SiO_2 (quartz)
$Al_2O_3.SiO_2$ (andalucite)
$ZrSiO_4$ (zirconium silicate)
$(Mg, Fe)_2SiO_4$ (olivine)

Energy requirements are considerably lower at <5 JG/t for the manufacture of unfired, chemically bonded bricks. Consequently, the use of bricks for which neither the raw material nor the brick itself needs firing and for which thermal treatment between 100 and 500°C produces sufficient strength, offers considerable advantages.

This research project centred on the following:

- the scope for using unfired refractory bricks in the iron and steel industry with particular reference to torpedo and steel casting ladles;

- the technical and chemical/mineralogical properties of unfired bricks;

- a comparison of laboratory and service test results for fired and unfired refractory bricks.

Energy problems and refractory materials have been discussed by K.K. Kappmeyer and D.H. Hubble (1). Data and model calculations relating to energy savings have been discussed in other works (2), (3). These subjects were discussed in depth at the 24th International Refractory Symposium held in Aachen in 1981, at which 17 related papers were presented (4).

Experience with the use of unfired refractory bricks in torpedo ladles has been reported in two papers (5), (6). The laboratory tests and service tests described in (5) formed the basis of this research project.

The use of olivine has already been described in some detail (7), (8).

1. THE TECHNOLOGICAL AND CHEMICAL/MINERALOGICAL PROPERTIES OF UNFIRED BRICKS

1.1 Tests on unfired andalucite bricks

Three types of bricks were tested:

- andalucite brick - chemically bonded (phosphate-bonded) and tempered
- andalucite brick - pitch-bonded and tempered
- andalucite brick - with added carbon - phosphate-bonded and tempered.

The results of tests on these types of bricks are shown in Table 1. The behaviour of the carbon-enriched bricks in oxidizing atmospheres is of particular interest. The carbon content is intended to increase the bricks' slag resistance. If the carbon burns off, for example when it is heated up in the torpedo ladle or casting ladle, the technological properties of the bricks undergo substantial change.

Fig. 2 shows the behaviour for a holding time of two hours at a range of temperatures. It can be seen quite clearly that at temperatures above 800°C the carbon burns out and that at 1200°C there is no evidence of carbon inside the opened test piece (in the form of a cube with 50 mm long edges). Both types of brick show similar reactions. Even the use of various types of graphite, carbon and pitch has little effect on this process.

Fig. 3 shows the effect of this process on the bulk density and open pores while Fig. 4 gives a comparison of the effect on hot compression strength in reducing and oxidizing atmospheres.

Of particular relevance is the length change under load (compression creep characteristics) illustrated in Figs. 5 and 6. Both graphs show that stabilization occurs even after a relatively short time. The similarity of the graphs for oxidizing and N_2 atmospheres is surprising.

1.2 Tests on refractory bricks made from andalucite zirconium silicate mixes

A synopsis of zirconium and zirconium-enriched materials was produced by P. Wecht (9). Detailed reports on the use of zirconium silicate in casting ladles were published by R.J. O'Brien; E. Tauber (10) and M. Pauline; R.A. James (11). Both papers consider in detail fired bricks made from zirconium silicate and zirconium silicate with added pyrophylite. Particularly in Japan, Australia, Korea and Taiwan such bricks are used extensively in casting ladles, a principal reason being the availability in the Far East of large deposits of pyrophylite.

Table 2 contains the data on the bricks tested. It can be seen that compared with fired bricks the chemically bonded ones show very low hot compression strength at 1500°C and evidence of compression creep is clear even at 1400°C — Fig. 7.

In addition to the type of bonding agent used, the proportion of $ZrSiO_4$ is itself a very important factor. Dissociation in $ZrSiO_4$ commences at temperatures as low as 1400°C and increases with temperature and holding time (12). ZrO_2 is also destabilized by phosphate bonders at around 1000°C (13). Mixes containing andalucite and zirconium silicate show strong reactions after firing for two hours at 1650°C. The test pieces became deformed and porous. The additional conversion of andalucite to mullite and SiO_2 should also be borne in mind:

$$3 (Al_2O_3 \cdot SiO_2) \longrightarrow 3 Al_2O_3 \cdot 2 SiO_2 + SiO_2.$$

Fig. 8 shows the pyrometric cone softening point for various mixes containing and andalucite and zirconium silicate. There is a clear drop in the softening point at and above 20% $ZrSiO_4$.

The reactions between $ZrSiO_4$, the andalucite and the phosphate bonding agent in unfired bricks are the cause of the reduced hot compression strength. Wide-ranging tests on the reactions between andalucite and zirconium silicate led O. Koegel (14) to the conclusion that the properties of andalucite could not be improved by the addition of $ZrSiO_4$. Any mix of the two raw materials produces values for the technical application which are poorer for the mix than for the raw materials.

2. RESULTS OF SERVICE TESTS WITH FIRED AND UNFIRED ANDALUCITE BRICKS IN TORPEDO LADLES AND IN THE BOTTOMS OF CASTING LADLES WITH BASIC LININGS

The following types of refractory bricks were selected for the service tests:

No 1	fired andalucite brick	9 torpedo ladles
No 2	fired andalucite brick	36 torpedo ladles
No 3	unfired andalucite brick with phosphate bonding	4 torpedo ladles
No 4	unfired andalucite brick with added carbon and phosphate bonding	5 torpedo ladles
No 5	unfired andalucite brick with pitch bonding	3 torpedo ladles

No 6   unfired andalucite corundum brick with
       phosphate bonding   6 torpedo ladles

The test data for the bricks are shown in Table 1.

2.1 Results of service tests on 200-t torpedo ladles

Detailed information on the test conditions can be found in (5). The operating conditions remained unchanged throughout the tests. The hot metal was held at temperatures between 1480 and 1520 °C and the $CaO:SiO_2$ ratio in slag samples from the torpedo ladle averaged 0.6. No in-ladle metallurgical treatment took place. Over the testing period from 1.1.1983 to 30.6.1985 all six types of brick were tried. 45 ladles were lined with two types of fired brick and 18 ladles with four types of unfired brick.

The safety lining used was high-grade CO-resistant fireclay brick.

Fig. 9 shows the number of journeys for each ladle during the testing period. The two types of fired refractories are linked by broken lines. Fig. 10 shows the average throughput in terms of hot metal per ladle and type of brick. Fig. 11 gives a comparison of costs.

Types 3, 4 and 5 of the unfired bricks proved successful. Type 6 with added corundum proved less successful, probably because of the type of corundum used, the granulometry of the raw materials or the reactions occurring between the andalucite, corundum, and phosphate bonding agent at temperatures between 1400 and 1500°C.

In conclusion it can be said that unfired refractory bricks have more or less the same strength characteristics as fired bricks.

2.2 The results of service tests in the bottoms of ladles with basic linings.

This series of tests centred on brick types 2 (fired) and 3 (unfired, phosphate bonded, no added carbon).

The tests were carried out in 180-t ladles equipped with 3 gas flushers.

The wear lining in the wall of the ladle was pitch-bonded, tempered dolomite, behind which there were 30 mm dolomite slabs (pitch bonded, tempered), 40 mm fireclay (ladle brick quality) and 32 mm light refractory brick. Around the slag line pitch-bonded tempered magnasite bricks were used. Some tests were carried out using synthetic resin bonded magnasite-carbon bricks.

Three flushing units and a perforated brick were located in the bottom, which was composed as follows:

andalucite (fired and unfired) 210 mm (brick dimensions B 2, 210 x 187 x 155 mm). Magnecite slabs of 32 mm. Tests with forsterite slabs were carried out and led to cost savings. Fireclay (ladle brick quality): 32 mm.
Light-weight refractory brick: 32 mm.

Some 165 andalucite bricks weighing in all approximately 2.6 t were used to line the bottom of each ladle.

The results are shown in Fig. 12. 83 bottoms lined with fired bricks showed an average life of 17.8 charges. 56 bottoms composed of unfired brick showed an average life of 17.4 charges, in other words the unfired brick had a life span similar to that of the fired bricks.

2.3 Requirements with regard to phosphate-bonded andalucite bricks for torpedo ladles and steel casting ladles

Laboratory and service tests showed that the two types of brick should offer the following test values:

— torpedo ladles
cold compression strength: >30 N/mm^2 <100 N/mm^2
hot compression strength at 1500°C: >4 N/mm^2 and <10 N/mm^2.
after-expansion at 1500°C — 12 hours: $<2\%$.

— steel casting ladles
Cold compression strength: >30 N/mm^2 and <100 N/mm^2
Hot compression strength at 1500°C: >4 N/mm^2 and <10 N/mm^2.
After-expansion at 1500°C — 12 hours: $<2\%$.

Compression creep: at a temperature of 1600°C, a load of 0.2 N/mm^2 and a hold time of 24 hours, compression creep should be $<0.15\%$ per hour. Compression creep graphs for 1500°C and 1600°C are shown in Fig. 13 as examples.

After-expansion for various types of brick at 1700°C is shown in Table 3 in percentages.

3. TRENDS

Trends in the development of torpedo ladle linings in the Federal Republic of Germany are shown in Fig. 14. The methods of lining and the types of bricks used following completion of this research project have been discussed in two publications (15) and (16). The use of synthetic resin-bonded andalucite, corundum and bauxite bricks with added carbon and added SiC in heavy-duty areas of torpedo ladles is on the increase. Fig. 15 shows two types of lining for different sorts of application.

In addition to fired, phosphate-bonded, alumina silicate bricks in casting ladles, synthetic resin-bonded bricks with added carbon have also proved satisfactory.

A synopsis of the types of application and properties of chemically bonded bricks, including the various types of magnasite bricks, has been published (17).

Cost savings with the use of unfired bricks will be determined largely by price trends for energy and bonding agents.

The price movements for the period 1974 to 1988 shown in Fig. 16 indicate how good energy prices have been over the past few years. Energy will, without doubt, become more expensive over the next 10 to 15 years and it should be borne in mind, for example, that to produce a fired andalucite brick about 20 GJ/t energy are required while only 2 GJ/t is needed to manufacture a chemically bonded brick.

An additional factor is that the high level of capital investment and maintenance costs required for high-temperature furnaces will fall or disappear as chemically bonded bricks only require tempering at temperatures of between 150 and 500°C.

BIBLIOGRAPHY

(1) KAPPMEYER, K.K., D.H. HUBBLE: Trends and challenges in the future of steel plant refractories. Ironmaking and Steelmaking (1976) No. 3, p. 113/28

(2) KOLTERMANN, M.: Feuerfeste Baustoffe — Produktions- und Verbrauchszahlen, Energiefragen, Entwicklungsrichtungen. Radex-Rundschau, Heft 4 (1979), p. 1 120/27

(3) JESCHKE, P., G. KONIG, A. MAJDIC: Rohstoffsituation in der Feuerfest-Industrie. Stahl und Eisen 102 (1982), p. 435/40.

(4) Feuerfeste Baustoffe fur Industrieöfen und ihr energiesparender Einsatz. 24. Internationales Feuerfest-Kolloquium, 1981, Aachen. Published papers.

(5) KOLTERMANN, M.: Ungebrannte feuerfeste Baustoffe in Torpedopfannen. Stahl und Eisen 101 (1981), p. 33/36

(6) HOFGEN, H., M. SEEGER, H. COORDES: Erste Betriebserfahrungen mit einem neuartigen Feuerfest-Werkstoff beim Einsatz in Roheisen-Transportgefässen. Stahl und Eisen 102 (1982), p. 213/16

(7) KOLTERMANN, M., S. SELTVEIT, R. SCHEEL: Olivin als feuerfester Baustoff. Stahl und Eisen 105 (1985), p. 683-688

(8) KOLTERMANN, M.: Olivine as a Refractory Material in the Steel Industry. Taikabutsu Overseas Vol. 8, No 1, 1988 p. 3-8

(9) WECHT, P;: Feuerfeste Zirkon- und zirkonhaltige Materialien mit Rohdichte 3.9 g/cm^3. Glas-Email-Keramo-Technik 23 (1972), p. 363/367

(10) O'BRIEN, R.J., TAUBER, E.: The use of zircon in steel ladle linings. Zweiter Internationaler Kongress "Industrial Minerals", Munich 1976. Published papers.

(11) PAULINE, M., JAMES, R.J.: Zircon-Pyrophyllite- An economical steel ladle refractory. Steelmaking Proceedings, Volume 62, Detroit 1979, p. 246/50

(12) HARDERS-KIENOW: Feuerfestkunde. Springer-Verlag Berlin, 1960

(13) WILSON, H.H.: Destabilization of Zirconia by Phosphoric Acid. Bull. Amer. Ceram. Soc. 57 (1978), p. 455/458

(14) KOEGEL, O.: Mineralogisch-technologische Untersuchungen an Andalusit-Zirkon-Werkstoffen. Dissertation submitted to the Fachhochschule für Keramik, Hohr-Grenzhausen, 1987

(15) KOLTERMANN, M.: Torpedo ladle refractories in West-Germany. Taikabutsu Overseas, Vol. 5, No 2, p. 35-40 (1985)

(16) Refractories in the manufacture and transport of pig iron. 31. Internationales Feuerfest-Kolloquium, Aachen 1988

(17) KOLTERMANN, M., P. BARTHA: Recent Trends in Unfired Energy Saving Refractories for the Steel Industry. Proceedings of International Symposium on Refractories, Hangzhou, China, November 1988, p. 333-354

Chemical composition (%)	1	2	3	4	5	6
SiO_2	34.8	35.9	33.1	37.6	38.4	29.8
Al_2O_3	61.8	61.7	61.6	57.3	59.0	66.1
TiO_2	0.7	<0.3	0.3	0.5	0.2	0.3
Fe_2O_3	0.9	0.7	0.6	1.1	0.7	1.1
CaO	0.1	0.2	0.1	0.2	0.1	0.2
MgO	0.4	0.2	0.2	0.4	0.4	0.2
Na_2O	0.1	0.1	0.1	0.1	0.1	0.1
K_2O	0.4	0.4	0.1	0.2	0.3	0.2
P_2O_5	0.1	<0.1	1.8	1.9	--	1.4
C	--	--	--	9.7	8.0	--
Ignition loss	0.1	0.3	2.0	10.5	8.4	1.6
Bulk density g/cm^3	2.55	2.59	2.67	2.46	2.74	2.76
Open pores %	16.7	11.6	15.3	16.6	5.0	14.9
Cold compression strength N/mm^2	79	110	42-45	21	45-55	40-50
Hot compression strength, oxidizing, 1500°C, N/mm^2	4-6	14-16	10-11	3-5	4-5	11-12

Table 1: Properties of fired (1, 2) and unfired andalucite bricks (3-6)

	Type of refractory brick			
	1	2	3	4
SiO_2	36.0	36.5	27.8	34.6
Al_2O_3	54.0	51.5	17.0	36.6
TiO_2	0.25	0.2	0.2	0.8
Fe_2O_3	1.30	1.3	0.1	0.5
MnO	<0.1	<0.1	n.a.	<0.1
CaO	0.2	<0.2	0.15	<0.2
MgO	0.11	<0.1	<0.1	<0.1
Na_2O	0.16	0.2	0.4	<0.1
K_2O	0.17	0.2	0.2	<0.2
P_2O_5	0.06	0.04	1.7	2.1
ZrO_2	6.8	9.4	54.2	25.4
1. Bulk density g/cm^3	2.63	2.69	3.55	3.06
2. Open pores %	14.5	11.9	14.0	11.7
3. Cold compression strength N/mm^2	100–120	120–135	50–60	78–80
4. Compression creep 1500–1000–1500°C Load 0.2 N/mm^2 in %/h	0.011	0.009	n.a.	n.a
5. After-contraction/after-expansion % 12 h x 1500°C	± 0	± 0	− 0.2	+ 0.5
6. Thermal shock resistance 1350°C ← → CU plate	>20	4/20	n.a.	>20
7. Hot compression strength at 800°C	—	—	—	46
1200 n/mm^2,	—	—	—	14
1300	—	—	9.1	—
1400	—	—	3.9	6
1500	8–9	3–5	1.1	1

Table 2: Test data for refractory bricks containing ZrO_2

		%
Corundum	98% Al$_2$O$_3$	± 0
Bauxite	85% Al$_2$O$_3$	− 0.7
Andalusite	60% Al$_2$O$_3$	+ 0.3
Forsterite	56% MgO	± 0
Magnesite-chrome	62% MgO	+ 0.2
Andalusite, Phosphate-bonded, tempered	60% Al$_2$O$_3$	+ 1.3

Table 3: After-expansion / after-contraction for various types of brick at 1700°C – holding time = 4h

Silica brick
High-density fireclay brick (43% Al$_2$O$_3$)
Corundum sliding gate (92% Al$_2$O$_3$)
Chemically-bonded MC brick (70% MgO)
Fired MC brick (70% MgO)
Directly bonded MC brick
Hot-fired MC fused particle brick
Silica glass immersed nozzle
Corrundum – graphite immersed nozzle
Self-bonded SiC brick
Plastic fireclay mass
Fireclay refractory concrete
Pi-alumina plastic mass
Injected dolomite mass
Corrundum SiC spout material

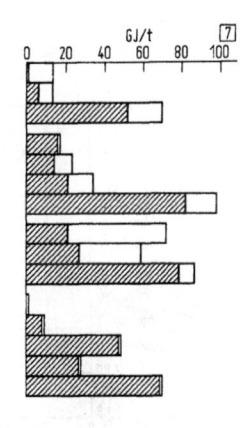

◨ Energy consumption for production of raw material
◻ Energy consumption for production of finished products

Fig. 1: Examples showing specific energy consumption for the production of refractory products and the raw materials used
(Stahl und Eisen 9 (1982), p. 439)

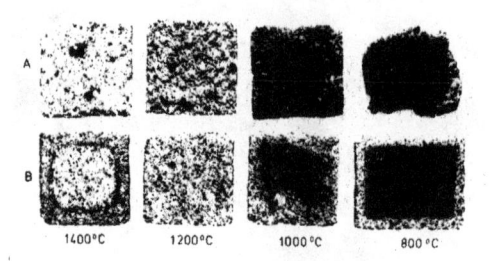

Fig. 2: Carbon-enriched andalucite bricks in an oxidizing atmosphere – holding time: 2h

A — Pitch-bonded
B — Phosphate-bonded with additional graphite

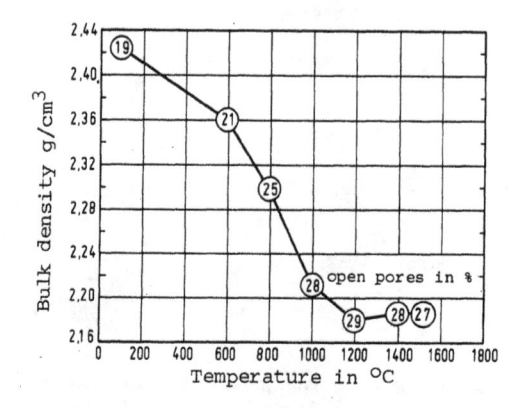

Fig. 3: Unfired carbon-enriched andalucite bricks in an oxidizing atmosphere Changes in bulk density as a function of temperature at a holding time of 4h

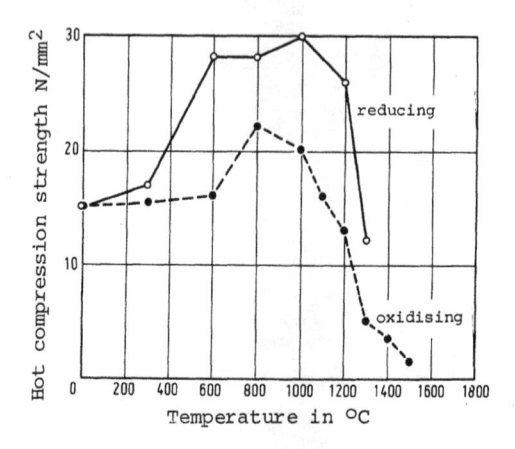

Fig. 4: Hot compression strength of unfired pitch bonded andalucite bricks

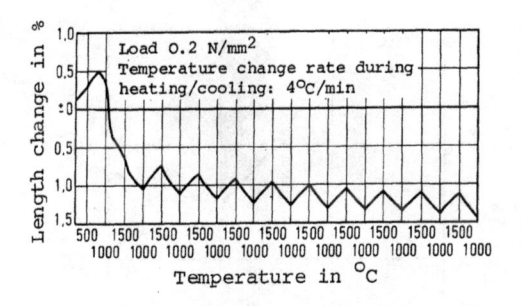

Fig. 5: Compression creep of pitch-bonded andalucite brick. Prior to the test C was burned out at 1200°C.

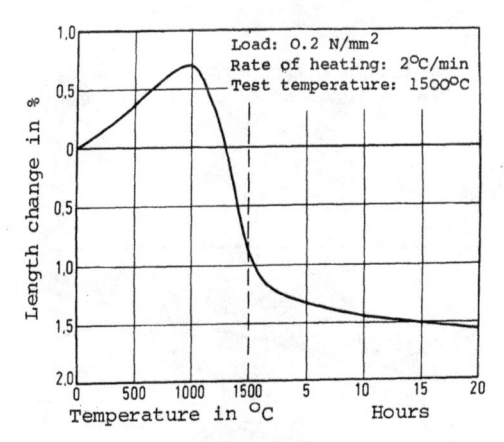

Fig. 6: Compression creep for an unfired carbon-enriched andalucite brick in a nitrogen atmosphere

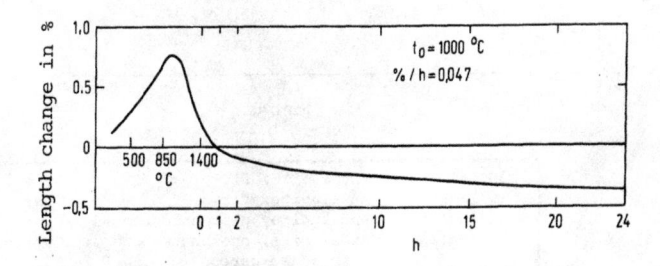

Fig. 7: Compression creep for andalucite-zirconium silicate brick (phosphate-bonded); temperature 1400°C; load 0.1 N/mm^2; holding time 24h

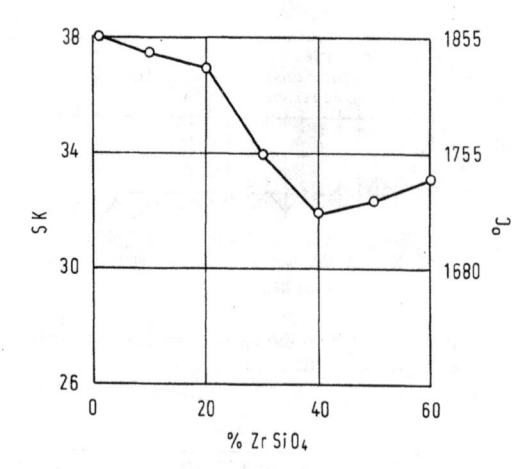

Fig. 8: Behaviour of andalucite — zirconium silicate test specimens. Cone softening point as a function of $ZrSiO_4$ content

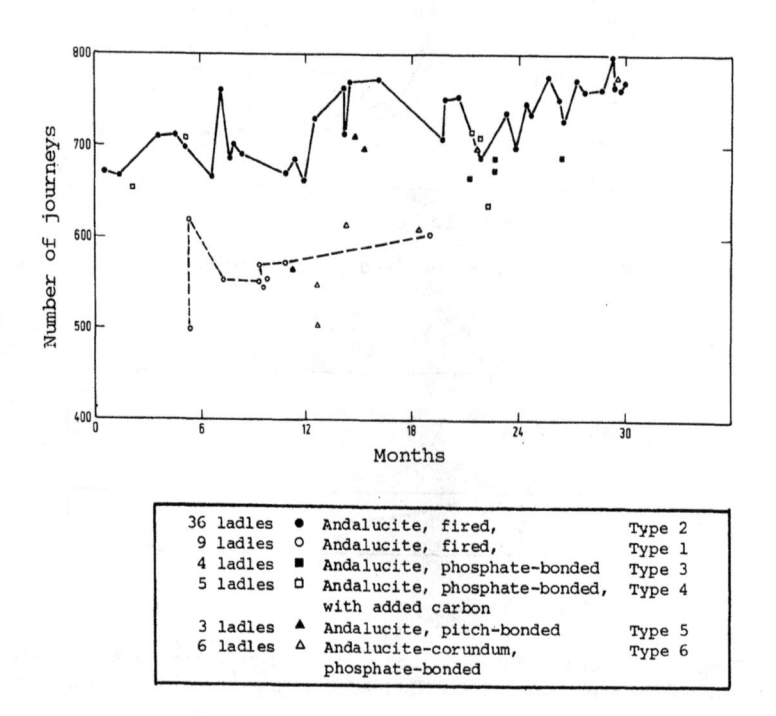

Fig. 9: Service life of torpedo ladles from 1.1.83 to 30.6.1985

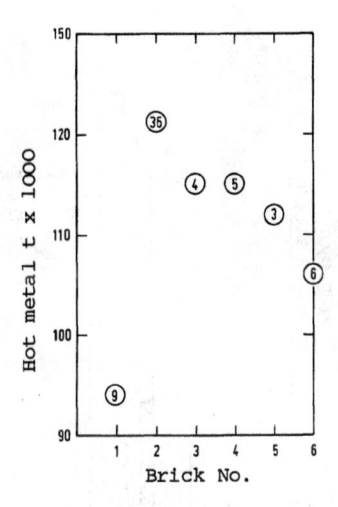

Fig. 10: Torpedo ladles — average throughput per ladle in a comparison between fired (1 and 2) and unfired bricks (3-6)
① = Number of ladles used

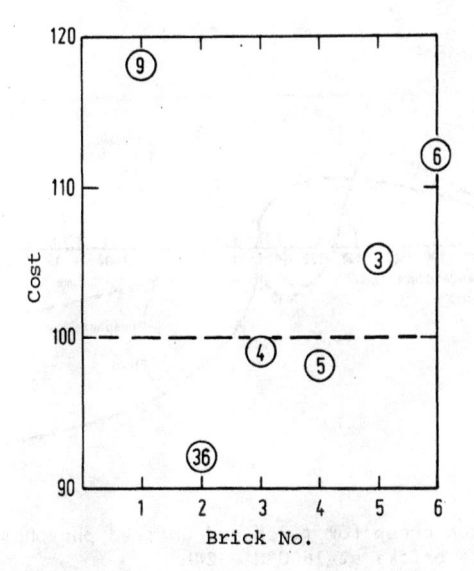

Fig. 11: Torpedo ladles — comparison of costs for fired and unfired bricks (1 and 2 = fired, 3-6 = unfired types)
① = Number of ladles used
Cost reference level = 100

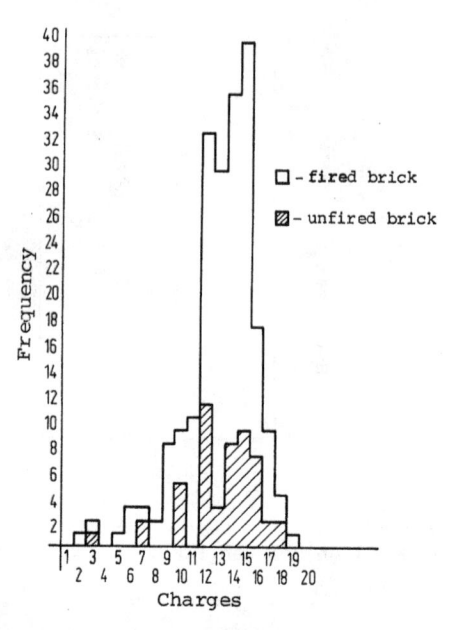

Fig. 12: Service life of ladle bottoms with fired and unfired andalucite bricks (180-t ladle; testing period 1.3-30.11.83)

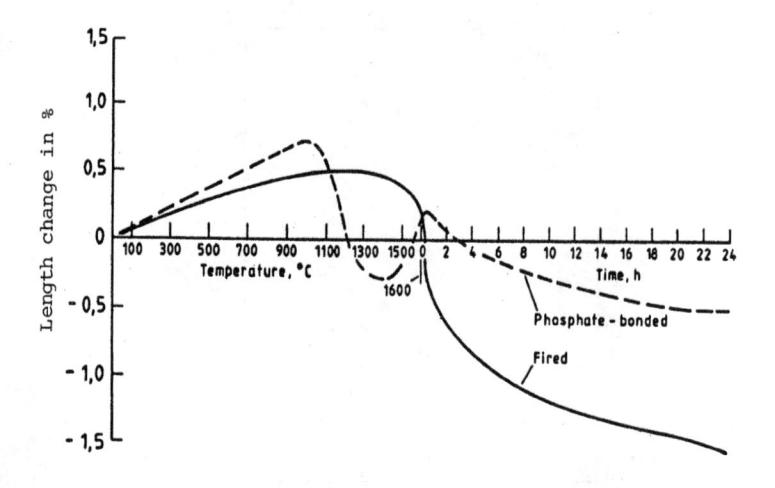

Fig. 13: Compression creep for fired and unfired phosphate bonded andalucite bricks at 1600°C - 24h Load 0.2 N.mm^2

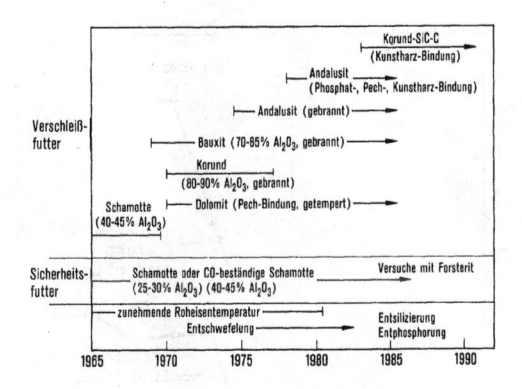

KEY TO DIAGRAM

Korund	: corundum
Kunstharz-Bindung	: synthetic resin-bonded
Phosphat-, Pech-, -Bindung	: phosphate/pitch-bonded
gebrannt	: fired
Verschleissfutter	: wear lining
Schamotte	: fireclay
getempert	: tempered
Sicherheitsfutter	: safety lining
CO-beständig	: CO-resistant
Versuche mit Forsterit	: tests with forsterite
zunehmende Roheisentemperatur	: increasing hot metal temperature
Entschwefelung	: desulphurization
Entsilizierung	: desiliconization
Entphosphorung	: dephosphorization

Fig. 14: Linings for torpedo ladles in the Federal Republic of Germany

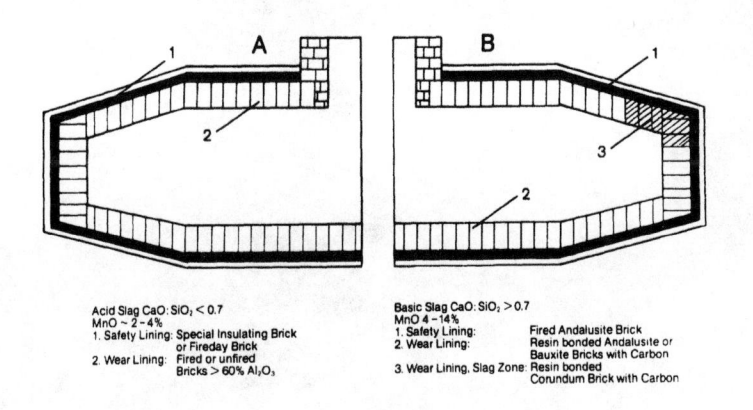

A
Acid Slag CaO: SiO$_2$ < 0.7
MnO ~ 2 - 4%
1. Safety Lining: Special Insulating Brick or Fireday Brick
2. Wear Lining: Fired or unfired Bricks > 60% Al$_2$O$_3$

B
Basic Slag CaO: SiO$_2$ > 0.7
MnO 4 - 14%
1. Safety Lining: Fired Andalusite Brick
2. Wear Lining: Resin bonded Andalusite or Bauxite Bricks with Carbon
3. Wear Lining, Slag Zone: Resin bonded Corundum Brick with Carbon

Fig. 15: Possible linings for torpedo ladles

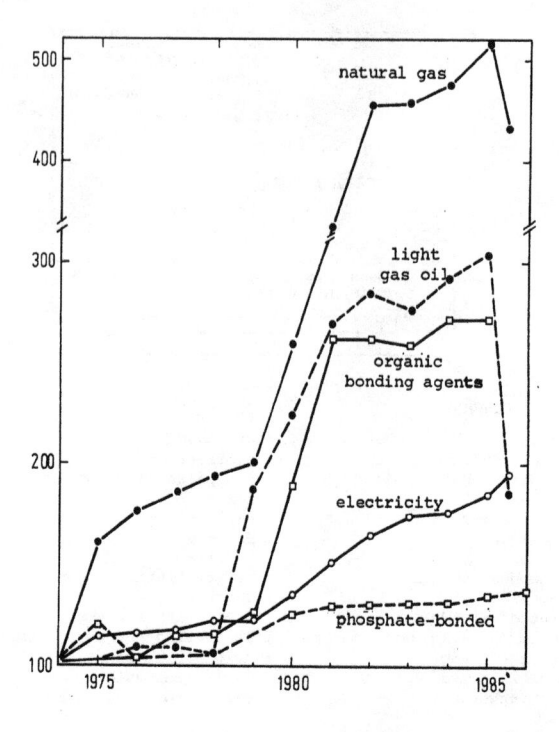

Fig. 16: Prices of energy and bonding agents (1974 = 100)

BORON NITRIDE-ENRICHED SUBMERGED NOZZLES FOR CONTINUOUS CASTING

Jacques PIRET
C.R.M.
Abbaye du Val-Benoit
B-4000 LIEGE (BELGIUM)

Summary

We started by laboratory tests on the pattern of wear for alumina graphite-type nozzles in the slag lubrification zone. Our accelerated corrosion test showed that the addition of boron nitride to the bulk mix greatly reduced nozzle wear. Service tests confirmed the accuracy of the laboratory test results. These nozzles are now in current use. Apart from their resistance to slag corrosion, their performance in service has shown that they are desirable for other reasons, in particular the quality of the cast product.

I. INTRODUCTION

The submerged nozzle is a critical component in the continuous caster: the number of casts possible per tundish and, in more general terms, the smooth operation of the caster are linked to its behaviour and reliability.

Some years ago submerged nozzles were used exclusively for casting slabs and blooms. Now their use has been extended to include billets to meet increasing demand for quality.

These components must satisfy a series of requirements (Fig. 1):

— attack by slag formed by melted powder from the ingot mould;

— internal corrosion caused by certain types of steel, for example steel with high manganese content;

— thermal shock at start of operation.

In addition, the two factors described below limit the strength required of nozzles:

- firstly, clogging by deposits of alumina or mixtures of steel and alumina calls for a minimum internal cross-section;

- secondly, thin cross section, such as in the case of billets, or thickness in the case of slabs, impose limits on the external dimensions of the nozzles to prevent the risk of bridges forming by solidifying steel between the nozzle and the wall of the ingot mould.

In what follows we will consider mainly the problems connected with corrosion in the molten slag zone at the metal-slag lubrification interface and we will show how a study of the pattern of nozzle attack allowed us, with financial support from the European Community, to develop a much more corrosion-resistant nozzle .

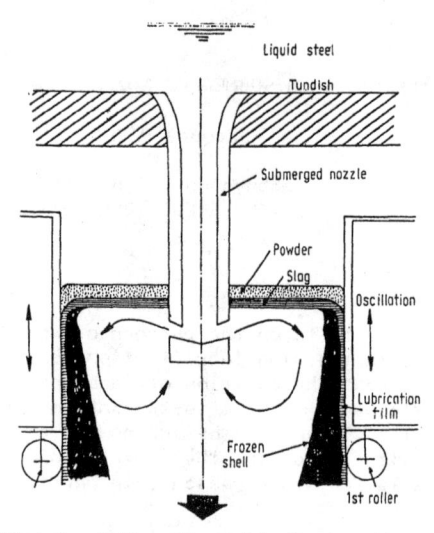

Fig. 1. Sketch of the CC mould showing the working
conditions of a submerged nozzle

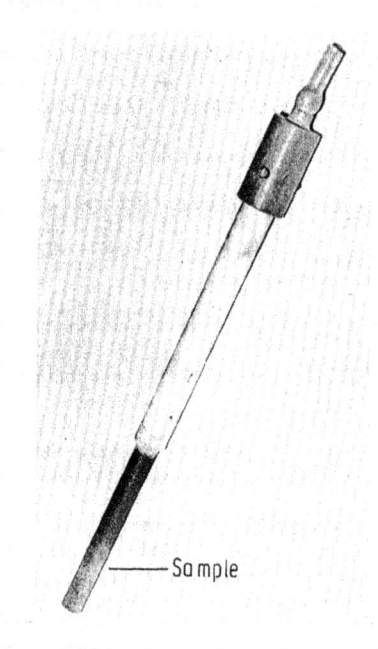

Fig. 2a. Tammann furnace and pneumatic
sample hanger

Fig. 2b. Close view of a sample
bored out a black submerged nozzle

Fig. 2. Test developed by CRM to simulate the working conditions
of submerged nozzles in the steel/slag interface

2. SERVICE CONDITIONS IN THE SLAG ZONE AND SELECTION OF CORROSION TEST

Fig. 1 shows the working conditions in the slag zone:

- the nozzle is in contact with a covering powder and a slag arising when it melts; since the purpose of the slag is basically to act as a lubricant, there is little scope for changing the way it attacks the nozzle;
- the oscillating movement of the ingot mould and the upward direction of the flow of liquid steel generate some agitation in the bath and contact by the nozzle alternately with the lubricating slag and the hot metal.

3. SELECTION OF CORROSION TEST FOR OPTIMUM SIMULATION OF WORKING CONDITIONS

Given that the working conditions are relatively complex to simulate CRM developed an original test which is a by-product of the "hanger test". The test apparatus is shown in Fig. 2a and comprises essentially:

- a Tammann furnace for heating the aggressive agents placed in a Al_2O_3 crucible;
- a pneumatic device to move the sample in the following way:
  * rotation about its own axis to accelerate corrosion,
  * vertical oscillation to simulate the movements in the bath inside the ingot mould.

The tests were carried out at 1535°C normally for a duration of 15 minutes, during which time the sample, of which Fig. 2b shows details, is submerged in the corrosive environment (after pre-heating).

3. LABORATORY TEST OF THE PATTERN OF CORROSION OF SUBMERGED NOZZLES FOR CONTINUOUS CASTERS

Our tests covered corrosion in the slag zone for black alumina-graphite nozzles and white silica-glass nozzles. However, since the former are becoming increasingly important, they were studied at greater length so that a way could be found to extend their service life and we will therefore restrict our comments to that type of nozzle.

3.1 Aspects of quality with regard to corrosion in the slag zone

Our earlier corrosion tests took the form of melting a sample of covering powder by itself in the Tammann furnace to convert it to aggressive slag. The results obtained were quite surprising: no attacks. This led us to conclude that the presence of metal under the liquid slag had an effect on its corrosion resistance whereas, frequently, chemical corrosion was considered to be caused solely by the slag.

We then revised our tests and included two further cases:

— attack uniquely by the metal bath;
— attack by metal covered with molten slag.

In the case of the black nozzles tested for corrosion solely by the metal, no evidence of wear was found. It is only when the metal and the slag are combined that highly corrosive attack occurs (Fig. 3). We studied this phenomenon extensively.

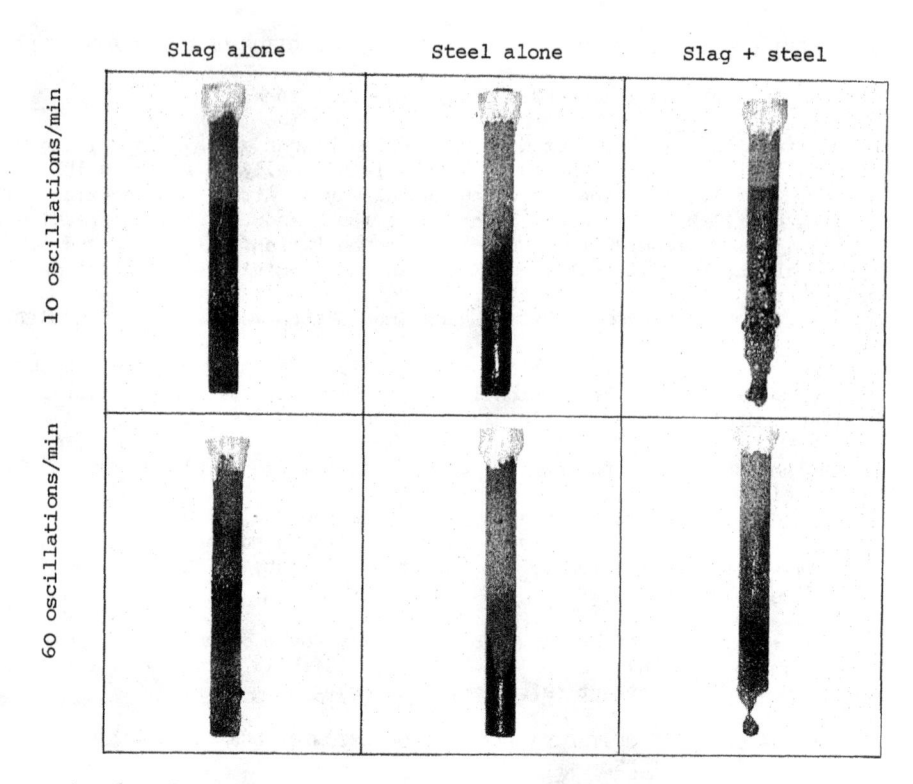

	Slag alone	Steel alone	Slag + steel
10 oscillations/min			
60 oscillations/min			

Fig. 3. Table showing the behaviour of black submerged nozzles in various laboratory test conditions

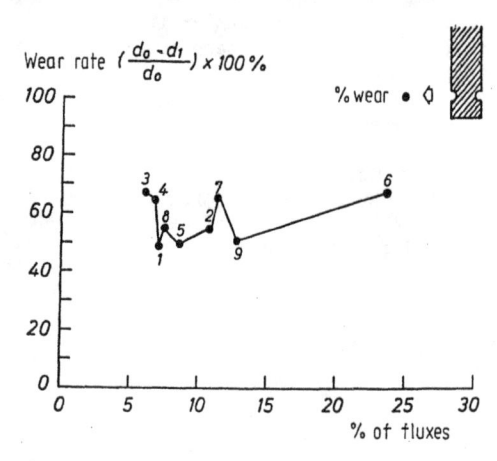

Fig. 4. Influence of the total content of fluxes of the powder on the wear rate (% reduction of the sample diameter) The numbers refer to the powder identification

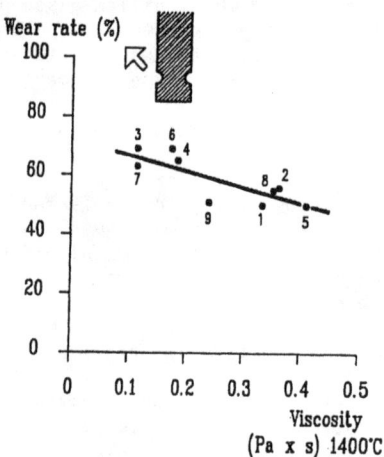

Fig. 5. Relation between the wear rate of Al_2O_3-C submerged nozzles and the viscosity (at 1400°C) of the molten powder (% wear rate = % diameter reduction of the sample) (Numbers: powder identification)

3.2 Quantitative aspects of corrosion in black nozzles

3.2.1 Speed of oscillation

It quickly became clear that the rate of corrosion was largely determined by phenomena linked to the kinetics of reaction: Fig. 3 shows that a switch from a frequency of 10 to 60 oscillations per minute doubles the rate of corrosion.

3.2.2 Flux content of the slag

Erosion in a grade of isostatically pressed nozzle (see Annex) was studied systematically using slag produced from various molten powders that are commercially available. Erosion was expressed as a proportion, measuring the diameter of the sample before and after testing:

$$\frac{d_0 - d_1}{d_0} \times 100 \text{ (in \%)}$$

Since most covering powders contain varying quantities of flux, the nature of which differs from case to case (NA_2O, K_2O, F^-, Li_2O, B_2O_3) we sought to ascertain whether certain fluxes were more destructive than others. We established that there was no clear-cut link between the flux content and corrosion (Fig. 4). Similarly, it was impossible to gather more evidence against one type of flux than against another. In simple terms we suggest that the use of K_2O may help to reduce corrosion.

3.2.3 Viscosity of slag

By way of contrast, a link was definitely established between the rate of erosion and the viscosity of the slag, as shown by Fig. 5 for example, which shows viscosity measured at 1400°C. However, the link was also established when the viscosity was measured at 1300 or 1500°C. It can be seen that since the samples oscillated vertically (or, to be more exact, the bath oscillated) superficial cooling of the wall of the nozzle takes place by radiation or contact with the unmelted powder and the real temperature of corrosion is not actually known. Some measurements carried out in situ suggest that a temperature of 1400°C on the external surface of the nozzle is fairly typical.

3.2.4 Understanding the pattern of wear in black nozzles

The increased rate of wear caused by the combined presence of steel and slag (in the case of the heterogenuous material of which the alumina graphite nozzles are composed) suggest that corrosion follows a two-stage pattern:

— the metal dissolves the graphite around the grains of alumina and thereby greatly increases the porosity of the material (porosity of material when new: 16%; graphite content: approximately 30% by weight);

— this increased porosity allows much faster penetration by the slag into the refractory component, which is heated from within by the flow of steel.

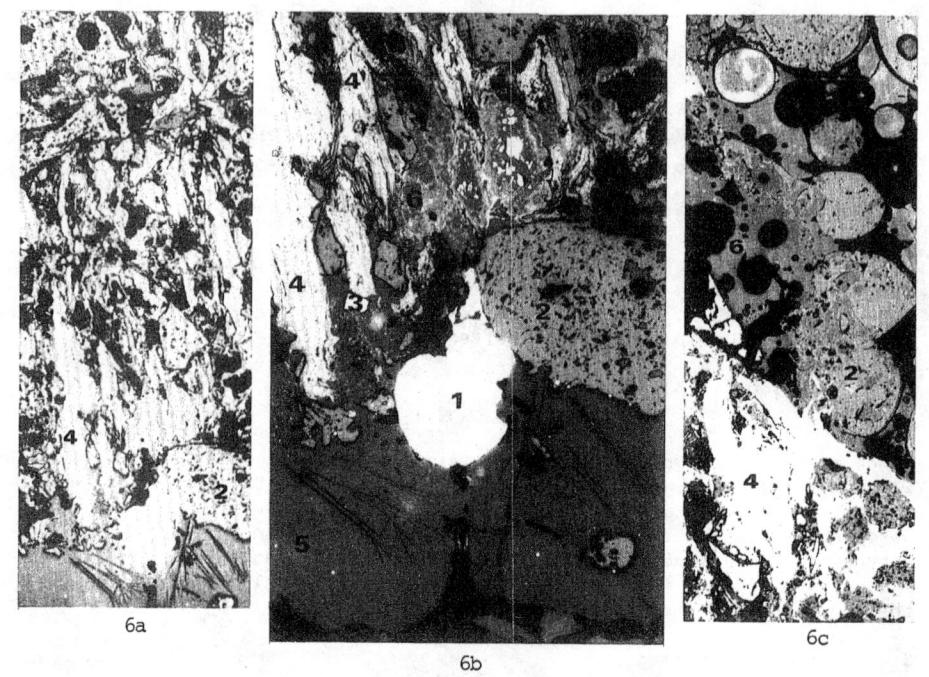

Fig. 6. Microphotographs showing the corrosion of black submerged nozzles

6a (65x) and 6b (135x): corrosion by slag + steel
6c (135x): corrosion by slag alone

1 : Steel
2 : Alumina grain
3 : SiC grain
4 : Graphite flake
5 : Slag
6 : Infiltrated slag

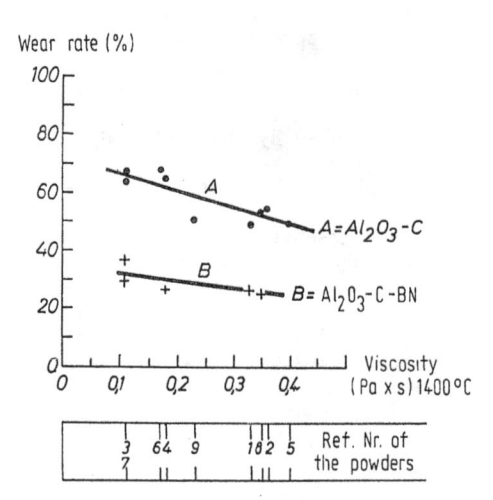

Fig. 7. Influence of the addition of BN on the wear rate of Al_2O_3-C submerged nozzles

This pattern of corrosion is observed by a look at the micro-photographic sections (Fig. 6). The first two (6a and 6b) show the slag-nozzle interface in the presence of the melt. The penetration by the slag into the space between the grains of Al_2O_3 normally occupied by the graphite flakes is evident. The grains of Al_2O_3 located at the interface show little sign of attack, which demonstrates rapid corrosion of the inter-granular matrix. Small drops of metal are also visible. On the other hand, where attack is attributable to the slag alone (Fig. 6c), the surface area from which the graphite has disappeared contains all its grains of Al_2O_3, which are now very rounded, proving that they have had the time to be attacked to an extent short of the destruction of the inter-granular matrix (slow attack).

3.2.5 Reducing the rate of wear in black nozzles

Since the dissolution of the graphite by the steel is the prime mover in the process of corrosion in black nozzles we sought to tackle the problem from this angle. This approach led us in the direction of an insoluble "graphite", in other words boron nitride. This material is structurally identical to graphite but is insoluble in metal.

This in turn prompted the adoption of solutions in which a part of the graphite was replaced by boron nitride (1).

It was thus possible to retain the advantages of graphite (lubrification during the manufacture of the nozzles, thermal conductivity) and also to protect the boron nitride from oxidation by air and so prevent its conversion to B_2O_3.

As regards the reduction of corrosion around the slag zone, tests in the laboratory showed that the rate of wear (Fig. 1) could be halved.

4. INDUSTRIAL APPLICATION

Some submerged nozzles incorporating enrichment, at least around the molten slag corrosion zone, have been manufactured jointly with a producer of refractory materials in Germany.

Some results of comparative tests with non-enriched black nozzles can be seen below. The results obtained in service proved better than those from laboratory tests since wear has been reduced to a third or a quarter of its former level (Fig. 8).

Experience gained has also shown that it was possible to enrich only a part of the nozzle with nitride, for example the slag/steel interface area, since no problems - of cohesion or whatever - were encountered in the transition zone in the case of the non-enriched base material (Al_2O_3-C). Consequently this good compatibility of the basic mix and the enriched mix meant that it was no longer necessary to design a means of strengthening the corrosion zone in the form of a nozzle sleeve, which would have led to excessive thickness.
Moreover, during a number of tests it was noted that clogging cause by trapped alumina particles was less evident.

We feel it would be of relevance to report here some of the results recently obtained by other research teams, which show the practical importance of boron nitride-enriched nozzles (2), (3):

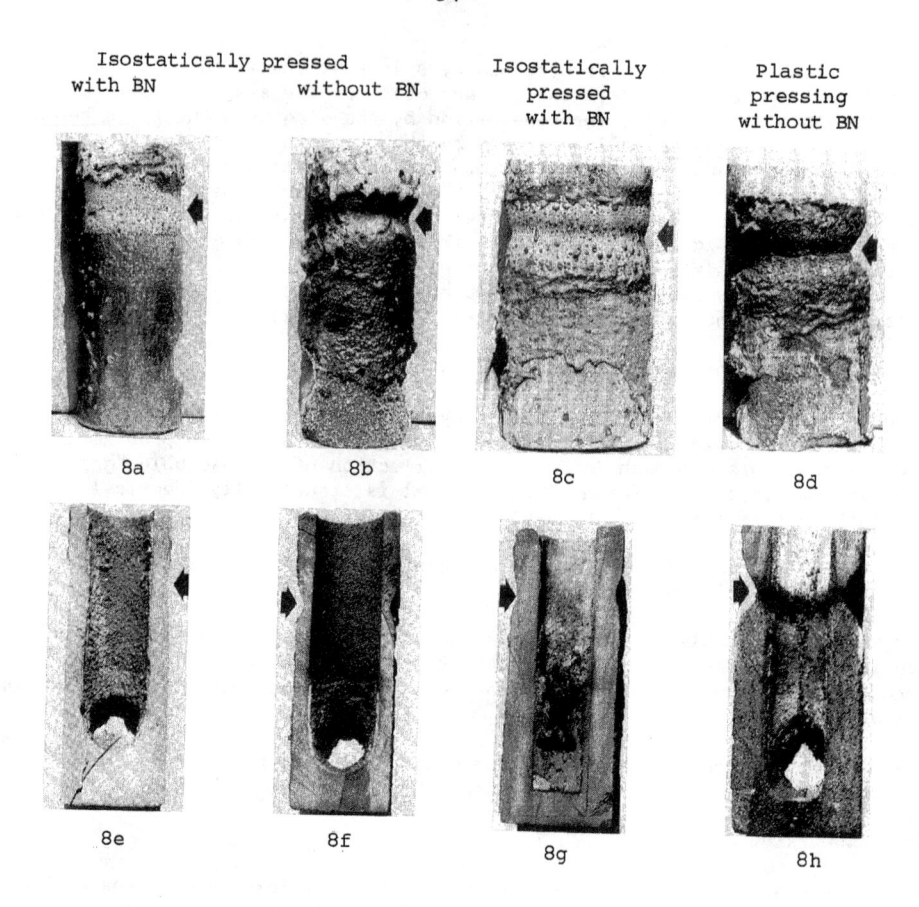

Fig. 8. Comparative trials on continuous casting machines of BN enriched and standard Al_2O_3 submerged nozzles

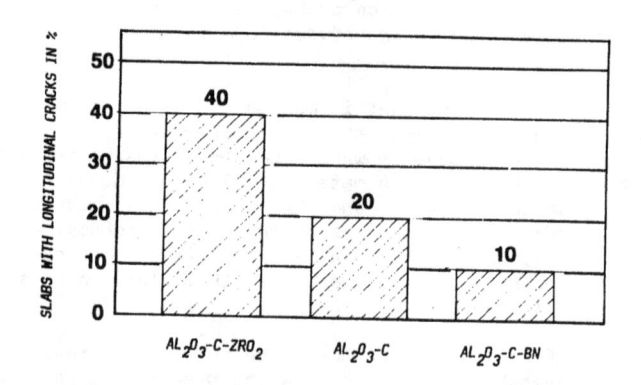

Fig. 9. Influence of refractory material in the bath level zone on the occurrence of longitudinal cracks (from ref.2)

- Thyssen (Fig. 9) has shown that all other things being equal the increased thermal conductivity in the nozzle due to the addition of boron nitride led to a halving of the frequency of the formation of longitudinal cracks in the steel compared with Al_2O_3-C nozzles (and quartered it compared with the Al_2O_3-C-ZrO_2 nozzles). It was also apparent that the replacement of the bulky oval nozzles made from traditional materials by round, boron nitride-enriched nozzles of a smaller size reduced the rate of occurrence of longitudinal cracks in slabs by a factor of between 2 and 20 depending on the composition of the steel;

- Didier and Dillinger Hüttenwerke (Table 1) established by an exhaustive comparison that the Al_2O_3-C-BN nozzles were those which most satisfied all the following criteria: thermal shock resistance; resistance to corrosion by steel only or by the combined action of slag and steel and satisfactory performance as regards alumina clogging.

5. Conclusions

The enrichment of alumina-graphite nozzles with boron nitride offers steelmakers a more powerful tool for continuous casting.

If the wall thickness is kept the same as in conventional nozzles service life increases by 3 or 4 times, which greatly reduces production costs (cost savings on refractories for tundishes, preheating and reduced metal loss).

If the thickness of the nozzle wall is reduced it is possible to increase the internal section (for example as a way of preventing alumina clogging) and reduce the external dimensions. This latter option is particularly interesting since:

- it can help to improve metal quality;
- it offers scope for casting smaller and thinner sections and clearly has a future as part of new continuous casting processes.

****

This work was carried out jointly with Mr B. Mairy, former engineer at C.R.M.

****

We wish to thank Messrs H. Lax from Thyssen Stahl A.G. and F. Schruff from Didier-Werke A.G. for their kind permission to reproduce data from their work.

6. Bibliography

(1) Patent C.R.M. No 842.477

(2) E. Hoeffken, H. Lax, G. Pietzko
Development of Improved Immersion NOzzles for Continuous Slab Casting
Preprints No 2 of the 4th Intern. Conference Continuous Casting,
pp. 4610479 (ICC Brussels 1988). Ed. Stahleisen

(3) W. Parbel, F. Schruff, B. Bergmann, N. Weiler,
High-Quality Refractory Materials. The key to Modern Continuous Casting Technology.
Ibid (ICC 88) Preprint No 2, pp 483-494.

TABLE 1 - COMPARISON OF PERFORMANCES OF VARIOUS SUBMERGED NOZZLES (REF.3)

MATERIAL	PRICE INDEX	THERMAL SHOCK RESISTANCE	CORROSION RES. AGAINST STEEL	CORROSION RES. AGAINST STEEL/ MOULD POWDER	BEHAVIOUR AGAINST CLOGGING
Al_2O_3 graphite	100	1	2	4	4
Al_2O_3 C - BN	250	2	2	2	2
BN sintered	4 000	2	3	-	1
Al_2O_3 densely sintered	400	4	3	3	-
MgO graphite	100	3	1	4	4
MgO C - BN	250	3	-	4	-
MgO sintered	45	5	1	3	-
ZrO_2 graphite	200	3	3	2	4
ZrO_2 C - BN	350	3	3	2	3
ZrO_2 sintered	400	4	2	1.5	3
ZrO_2 dense	2 000	5	1	1	4
fused silica	100	1	2 - 5	3	2

1 very well   3 sufficient   5 very insufficient
2 well   4 insufficient   — not yet tested

price index = 100 for alumina - graphite

ANNEX

WORKING CONDITIONS FOR THE HANGER TEST APPLIED BY
CRM ON SUBMERGED NOZZLES FOR CONTINUOUS CASTING

1. Analysis of steel (in 10^{-3}%)

   C   : 70
   MnO : 510
   Si : 245
   Al : 245
   P   : 21
   S   : 14

2. Powders: see Table A.1.

3. Black nozzle: isostatically pressed nozzle

   Al_2O_3 : 71 – 73%) excluding graphite

   SiO_2 : 32 – 33%) excluding graphite

   Graphite   : 29 – 31%
   Density   : 2.6
   Porosity   : 16%

4. Main test parameters

   Samples   : 15 mm diameter, 150 mm high

   Vertical movement   : amplitude: 10 mm
                       : frequency: up to 60 cycles/min

   Movement of rotation: amplitude :90°
                         frequency: 20 cycles/min

   Temperature   : 1535°C

   Duration   : from 15 to 30 min. depending on the type of product.

TABLE A.1. Characteristics of commercial powders used to test attacks on nozzles

CHARACTERISTICS POWDERS	1	2	3	4	5	6	7	8	9
Chemical analysis (% w/w)									
CaO	32.8	29.1	34.2	33.4	29.2	28.7	31.5	30.6	30.7
MgO	0.6	0.5	-	-	-	0.9	0.5	0.4	0.4
SiO_2	28.4	30.9	29.1	30.9	29.8	26.1	24.5	32.1	32.6
Al_2O_3	11.8	9.5	5.6	6.1	7.7	15.0	6.1	7.3	6.0
Na_2	2.8	4.7	4.8	2.3	6.7	13.6	5.5	3.8	3.6
K_2O	0.2	2.5	0.3	0.3	0.7	0.3	0.6	0.1	0.5
F	4.4	5.4	1.2	4.4	1.3	6.7	5.5	3.6	9.0
LiO_2	-	-	-	-	-	2.8	-	-	-
B_2O_3	-	-	-	-	-	0.4	-	-	-
Ctot	9.8	8.7	5.6	6.9	9.4	4.6	11.5	9.2	9.1
CO_2	13.0	10.8	8.5	7.1	4.4	0.2	8.9	11.8	11.3
Viscosity at (°C) (Pa x s)									
1 250	1.25	1.11	0.33	0.66	1.41	0.34	0.70	1.20	0.80
1 300	0.80	0.78	0.21	0.42	0.88	0.24	0.31	0.76	0.55
1 350	0.52	0.53	0.15	0.26	0.60	0.17	0.22	0.54	0.40
1 400	0.33	0.36	0.11	0.18	0.40	0.10	0.16	0.40	0.30
1 450	0.20	0.26	0.08	0.12	0.27	0.05	0.13	0.29	0.23
1 500	0.13	0.18	0.60	0.11	0.19	0.01	0.10	0.23	0.16

USE OF CALCIUM OXIDE AS REFRACTORY MATERIAL IN STEEL MAKING PROCESSES

E. Marino

Centro Sviluppo Materiali SpA
C.P. 10747 - 00100 ROMA-EUR (ITALY)

Summary

Owing to its stability and refractoriness, calcium oxide could provide an excellent refractory materials for steelmakers.

The preparation and application of calcium oxide refractories have been hindered, however, by the tendency of CaO to hydrate.

During two research projects conducted with financial assistance from the ECSC, refractory components have been developed using stabilized calcium oxide having characteristics suitable for manufacture, storage and utilization without any relevant hydration difficulties.

The first project concerned the production and works testing of ladle bricks made with stabilized calcium oxide. These tests also provided confirmation of the positive influence of this new type of lining in the ladle-treatment of steel.

The other project included development of and trials with lime metering nozzles for continuous casting. The work performed showed that with these nozzles it is possible to continuously cast Al-killed steel without encountering the clogging difficulties caused by alumina deposits. Numerous trials performed on continuous-billet casters have demonstrated the efficiency of these nozzles which permit the complete casting of steel heats in a manner that is quite impossible when normal nozzles are employed.

1. INTRODUCTION

A list of characteristics of an ideal refractory material for use in steel-making processes would have to include the following:
- High refractoriness
- Thermodynamic stability
- Resistance to attack by slags and steels
- Stability in reducing atmospheres and under vacuum
- Ready availability
- Low cost of raw material
- Ease of storage and application.

Of the many possible refractory materials, one of those that best meets this list of requirements is certainly calcium oxide. However, the biggest problem to be overcome when using this materials is how to eliminate or at least attenuate the tendency of products consisting mainly of CaO to hydrate. This can be done by high-temperature treatment of calcium compounds (oxide, hydrate or carbonate) mixed with additives capable of producing a calcium-oxide sinter characterized by improved hydration resistance.

During the course of two separate research projects performed with financial aid from the European Community, the CSM has developed and tested two different types of refractory compounds made with calcium oxide.

The first project concerned the preparation of and trials with bricks to be used for lining ladles employed in steel-refining treatments, while the other involved the development of and trials with lime metering nozzles for continuous casting.

In both cases stabilized calcium oxide was required as the raw material for the production of components (brickes and nozzles) having sufficient hydration-resistance to permit easy storage and application.

2. PREPARATION OF STABILIZED CALCIUM OXIDE

The research was preceded by the development of a method for preparation of the stabilized raw material. It is easy to find on the market hydrated lime of the desired purity ($Al_2O_3+SiO_2<2\%$) which gives rise to a stable oxide when mixed with suitable additives and calcined at high temperature.

Of the possible additives, it has been ascertained that calcium chloride possesses very good sintering properties and does not introduce elements that may adversely affect the refractoriness of the final product.

In order to sinter the additive-treated hydrated lime it is granulated in a disc pelletizer. Fig. 1 illustrates the main stages in the preparation of the stabilized sintered material.

The material is sintered at 1600°C, since it has been ascertained that the higher the treatment temperature the lower the susceptibility of the product to hydration.

The final grain-size distribution of the material depends, of course, on the kind of product it is wished to make (bricks or nozzles).

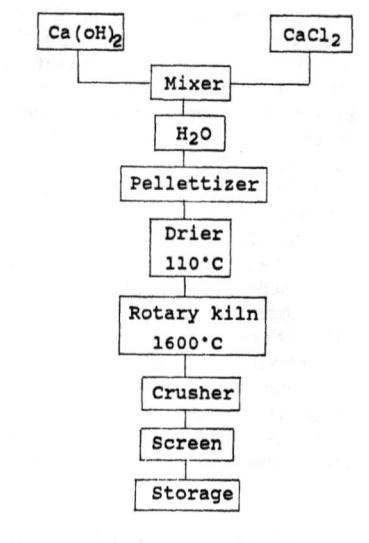

Fig. 1 - Flowsheet for preparation of stabilized calcium oxide

3. USE OF CALCIUM OXIDE FOR LINING LADLES FOR STEEL REFINING TREATMENTS

Calcium oxide is endowed with properties that, among other things, make it highly suitable for lining ladles employed for steel refining treatments. In fact, the refractories utilized in these ladles must be capable not only of withstanding high temperatures but also of remaining in contact with liquid metal for long periods without any alteration in composition.

The research, performed in collaboration with the Sanac and Dalmine companies, involved laboratory studies on the methods of preparing the refractory material, the production of calcium-oxide bricks and trials on works ladles lined therewith.

3.1. Laboratory experimentation

Experimental bricks were made using as raw material stabilized calcium oxide produced by sintering a mixture of calcium hydrate and chloride. The purpose of the exercise was to ascertain the best pressing

ANNEX

WORKING CONDITIONS FOR THE HANGER TEST APPLIED BY CRM ON SUBMERGED NOZZLES FOR CONTINUOUS CASTING

1. Analysis of steel (in 10^{-3}%)

    C   : 70
    MnO : 510
    Si : 245
    Al : 245
    P   : 21
    S   : 14

2. Powders: see Table A.1.

3. Black nozzle: isostatically pressed nozzle

    Al_2O_3 : 71 – 73%) excluding graphite

    SiO_2 : 32 – 33%) excluding graphite

    Graphite   : 29 – 31%
    Density   : 2.6
    Porosity   : 16%

4. Main test parameters

    Samples   : 15 mm diameter, 150 mm high

    Vertical movement   : amplitude: 10 mm
                         : frequency: up to 60 cycles/min

    Movement of rotation : amplitude : 90°
                           frequency: 20 cycles/min

    Temperature   : 1535°C

    Duration   : from 15 to 30 min. depending on the type of product.

TABLE A.1. Characteristics of commercial powders used to test attacks on nozzles

CHARACTERISTICS POWDERS	1	2	3	4	5	6	7	8	9
Chemical analysis (% w/w)									
CaO	32.8	29.1	34.2	33.4	29.2	28.7	31.5	30.6	30.7
MgO	0.6	0.5	-	-	-	0.9	0.5	0.4	0.4
SiO_2	28.4	30.9	29.1	30.9	29.8	26.1	24.5	32.1	32.6
Al_2O_3	11.8	9.5	5.6	6.1	7.7	15.0	6.1	7.3	6.0
Na_2	2.8	4.7	4.8	2.3	6.7	13.6	5.5	3.8	3.6
K_2O	0.2	2.5	0.3	0.3	0.7	0.3	0.6	0.1	0.5
F	4.4	5.4	1.2	4.4	1.3	6.7	5.5	3.6	9.0
LiO_2	-	-	-	-	-	2.8	-	-	-
B_2O_3	-	-	-	-	-	0.4	-	-	-
Ctot	9.8	8.7	5.6	6.9	9.4	4.6	11.5	9.2	9.1
CO_2	13.0	10.8	8.5	7.1	4.4	0.2	8.9	11.8	11.3
Viscosity at (°C) (Pa x s)									
1 250	1.25	1.11	0.33	0.66	1.41	0.34	0.70	1.20	0.80
1 300	0.80	0.78	0.21	0.42	0.88	0.24	0.31	0.76	0.55
1 350	0.52	0.53	0.15	0.26	0.60	0.17	0.22	0.54	0.40
1 400	0.33	0.36	0.11	0.18	0.40	0.10	0.16	0.40	0.30
1 450	0.20	0.26	0.08	0.12	0.27	0.05	0.13	0.29	0.23
1 500	0.13	0.18	0.60	0.11	0.19	0.01	0.10	0.23	0.16

methods to be adopted and to check on the properties of the ensuing refractory.

The bricks were made by the well-known process of mixing the sinter with tar, followed by pressing and then by treatment at low temperature (300°C) to eliminate volatiles.

Several tests were run in an induction furnace to assess the influence of this experimental material on the metallurgical stability of a steel bath, the results being compared with those obtained using refractories more commonly employed for ladle linings.

Refining treatment was simulated by injecting into steel previously killed with aluminium metal the necessary quantity of an alloy containing Ca and Si. Metallurgical stability of the bath was evaluated by determining the gradual decrease of Ca and Al in solution in the steel after introduction of the refractory sample. The results are illustrated in Figs 2 and 3. They show that the calcium oxide appears to be the most stable refractory, followed by high-alumina and dolomite.

3.2. Works-scale trials

As the aim was to run trials on linings in at least four works ladles, around 50 t of stabilized CaO sinter had to be produced. This was done on an industrial plant complete with a 1.5 m diameter disc pelletizer

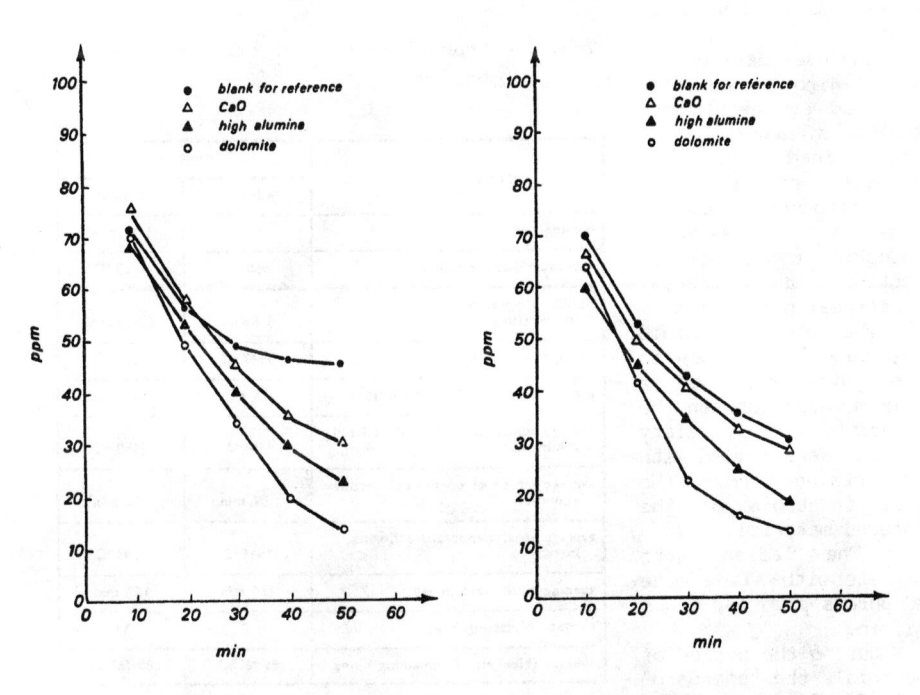

Fig. 2 – Variation in calcium content in solution in steel bath with various types of refractory material

Fig. 3 – Variation in amount of aluminium in solution in steel bath with diverse type of refractory materials

and a 3.5 m long, 1.4 m diameter rotary kiln. The sintered material was graded, packed in hermetically-sealed containers and sent to Nuova Sanac sworks where the bricks were made on the line normally utilized for the production of dolomite bricks. The procedure adopted was that established during the laboratory tests.

All stages of the operation went ahead without any difficulties due to the new type of raw material. Even hydration, much-feared especially during low-temperature tempering, remained within the limite normally encountered with dolomite products.

The same procedures and containers employed for packing dolomite materials prior to despatch were used for the CaO products.

The properties of the bricks produced on the industrial plant bore out the laboratory findings. The relevant figures are given in Table I.

The use-tests on the ladle-lining bricks were performed at the Dalmine Steelworks. Four ladles were lined in two successive phases. These were introduced into the normal steelworks production cycle without adopting any special operational procedures.

The bricks were laid dry, care being taken to keep joint size to a minimum. Gaps between the wearing and safety sections were rammed with CaO obtained from the finer fractions of the sintered material.

The ladles were complete with slide gate and porous plug for argon stirring.

During the period of the trials the "standard" lining of the ladle consisted of alumina refractory bricks.

The first two campaigns were run at Dalmine's N° 1

determination		material		
		green	tempered	coked
bulk density	g/cm³	2.57	2.75	2.40
porosity	%	—	13.1	24.4
crushing strength	N/mm²	—	44.13	42.66
modulus of rupture	N/mm²	—	7.65	10.98
residual carbon	%	—	—	4.3

Tab. 1 – Characteristics of calcium oxide bricks

operating data	campaign	
	first	second
heating rate	1 °C/min	1 °C/min
maximum heating temp.	950 °C	1050 °C
holding time at maximum temperature	5 hours	6.5 hours
average tapping temp.	1647 °C	1656 °C
average length of Finkl treatment	76 min	66 min
average temperature at end of Finkl treatment	1593 °C	1590 °C
average length of continuous casting operation	87 min	80 min
average steel temperature during casting	1544 °C	1548 °C
average time steel in ladle	174 min	167 min
number of casting runs	8	16
average thickness of remaining lining	95 (68%)	88 (63%)

Tab. 2 – Average values concerning use of calcium oxide lined ladle in the first and second campaigns

Steelworks. Two ladles were lined with CaO bricks, there being no problems either in installation or operation. Some characteristics of the two campaigns are indicated in Table 2.

The steelmaking process also included argon-stirring in the ladle, the addition of ferroalloys in the Finkl plant, heating with electrodes to maintain the required temperature, and feeding of Ca-Si cored wire to improve castability.

The average time the steel was in the ladle during each heat was about three hours at a temperature varying between 1650°C and 1550°C.

The first campaign was interrupted after the eighth heat with the refractory lining still in good condition, while the second was continued for sixteen heats. In this case, too, the refractory was still in a good state (see Fig. 4), the average thickness of the remaining lining being 60% of that initially installed.

The influence of the presence of the CaO lining was evaluated in the case of each heat by determining the desulphurization efficiency during treatment in the Finkl plant. The processed data indicate an average improvement of 10% in desulphurization efficiency on passing from an alumina to a lime lining. The sulphur content generally dropped from an original level of 150-200 ppm to 50-100 ppm after treatment (see Fig. 5).

Another two campaigns with more drastic treatment of the steel were performed at Dalmine's N° 1 Steelworks using a smaller ladle than

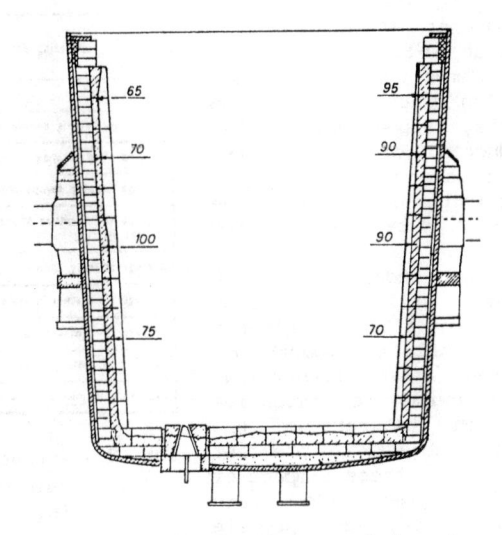

Fig. 4 – Ladle lining remaining at end of second campaign

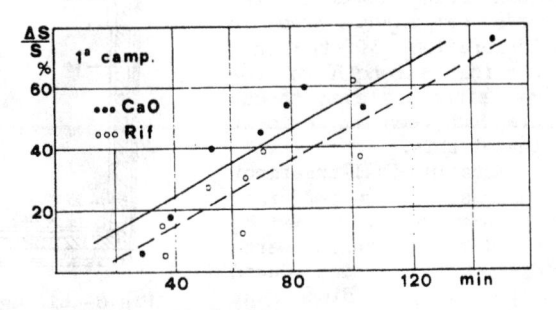

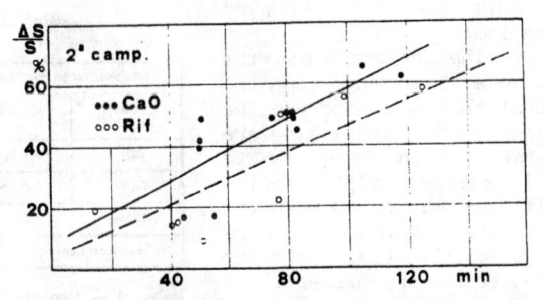

Fig. 5 – Desulphurization efficiency versus Finkl treatment time

in the two previous campaigns: 50 t capacity against 90 t.

Campaigns 3 and 4 were run ten months and eighteen months, respectively, after production of the bricks, which were found to be in excellent condition. The relevant data are reported in Table 3. The third campaign ended with the lining still in good condition after twenty-one heats, while the fourth was stopped after twenty-five heats owing to excessive wear on the part of the lining subject to greatest mechanical stress (impact of steel tapped from arc furnace). The wear profile is illustrated in Fig. 6.

It is considered that the exceptionally long period that the bricks were stored (eighteen months) before being used in the fourth campaignresulted in deterioration of the hot mechanical strength of the lime refractory even though there had been no evident signs of this.

Metallurgical treatment of the steel was performed using a method developed by the CSM, alkaline-earth additives being introduced in special metal dippers at a point near the bottom of the ladle.

The effects of the various treatments on the usual alumina lining and the experimental lime lining are summarized in Table 4. In all cases, desulphurization efficiency improved with the innovatory lining: the more drastic the treatment the greater the improvement.

operating data	campaign	
	third	fourth
heating rate	1 °C/min	1 °C/min
maximum heating temperature	1050 °C	950 °C
holding time at max. temp.	3 hours	1 hour
average tapping temperature	1672 °C	1670 °C
average temperature of steel in ladle after stirring	1614 °C	1612 °C
average stirring time	5 min	6 min
average time steel in ladle	60 min	65 min
average time of casting	52 min	51,5 min
number of casting runs	21	25
average thickness of remaining lining	67%	55%

Tab. 3 – Average values concerning use of calcium oxide lined ladle in the third and fourth campaigns

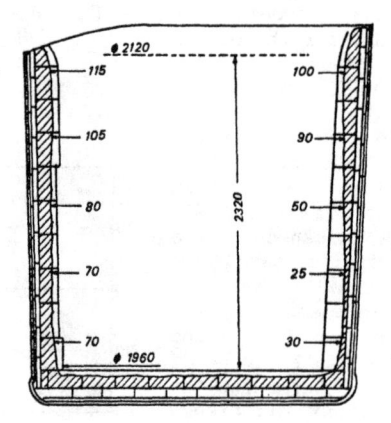

Fig.6– Lining remaining at end of fourth campaign

desulphurization by dipper		CaO ladle		Al₂O₃ ladle
type	quantity	third campaign	fourth campaign	third campaign
CaSi	3,4 kg/t	49,2	—	33,7
CaSi	4,8 »	80,0	66	38,5
CaSiBa	4,3 »	40,0	—	20,0
Without treatment		20,0	—	14,0

Tab. 4 – Desulphurization efficiency $\frac{\Delta S}{S}$ % of dipper-treated heats

4. CALCIUM OXIDE CONTINUOUS CASTING NOZZLES

Difficulties are frequently encountered during the continuous casting of aluminium-killed steels owing to the clogging of the casting nozzle as a result of internal build-up of alumina deposits. This problem is exacerbated in the case of the metering nozzles employed for casting in small moulds.

Several solutions have been proposed to avoid this drawback, for example, argon bubbling into the most critical area of the nozzle, special nozzle shapes, and treatment of the steel with calcium which is introduced near the bottom of the ladle. Unfortunately, howevern the results obtained are not always good enough to quarantee that the heat is successfully cast.

Therefore, to try to find a simple economic solution it was decided to modify the chemical composition of the nozzle. The working hypothesis was that the alumina which causes che clogging would react with calcium oxide. Trials were thus run on nozzles made of CaO refractory.

The research was conducted in collaboration with Vesuvius Italiana. The basic idea explored was that a low-melting calcium-aluminate film would form in the contact zone between the inclusions and the wall of the nozzle and that this would act as a lubricant, favouring the sloughing-off of the deposits that would then be carried away by the flow of liquid steel, thus preventing clogging.

Three types of nozzles were made:
- Monobloc CaO nozzles
- Monolithic $CaO+ZrO_2$
- Double-layer nozzles

The work first involved laboratory development and then direct casting-line tests.

4.1. Monobloc CaO nozzles

The nozzles were made of calcium oxide that had been sintered and stabilized by means of a procedure worked out during previous research.

Moisture-free mineral-oil was used as the binder for forming the nozzles in a 1000 Kg/cm^2 press.

It was found that the best sintering conditions for ensuring a product with the desired characteristics consisted in treatment at 1600'C for three hours in an electric furnace.

No particular difficulties were encountered due to hydration either during fabrication or during storage prior to use. Regarding the subsequent stage, too, by adopting a few simple precautions, it was possible to follow the normal procedures for installing the nozzle in the tundish and preheating prior to casting.

Operational behaviour was assessed on a four-strand continuous billet-caster in an electric steelworks. The experimental nozzles were installed on one of the strands, while the other three were equipped with large-diameter nozzles with stopper-rod for flow control.

The characteristic data of these tests are given in Table 5 whichpresents all the results of the works trials performed during the course of the research.

Seven works trials were run on nozzles made wholly of CaO (these are numbered from 1 to 7 in the Table). Five gave good results, while two, involving the casting of low-C steels, had to be interrupted prematurely owing to excessive nozzle wear.

In the five trials heats that were concluded normally, no problems were encountered with clogging, despite the fact that the steels had an aluminium content that would have made them very difficult to cast using a normal metering nozzle.

On the basis of the results obtained with these tests and by reference also to previous experience, it is evident that calcium-oxide nozzles can be employed to cast Al-killed steel provided this has a carbon content higher than 0.1% and sulphur lower than 0.02%. Another requirement is that the stream between the tundish and the mould should be protected.

There are, however, some use limitations, because the mechanical strength is not such as to guarantee the possibility of casting a sequence of heats nor to permit the casting of particularly aggressive steels.

Test	Nozzle type	Tundish temp.	\multicolumn{4}{c	}{Steel composition}	\multicolumn{2}{c	}{Pouring time}
			C	Mn	S	Al	linea CaO	tot.
1	CaO	1530	0.41	0.84	0.020	0.020	77	77
2	"	1583	0.07	0.27	0.020	0.018	20	67
3	"	n.d.	0.05	0.25	0.025	0.011	18	68
4	"	1546	0.28	0.65	0.028	0.017	75	75
5	"	1546	0.41	0.63	0.021	0.033	70	75
6	"	1543	0.33	0.54	0.025	0.021	70	75
7	"	1554	0.35	0.65	0.021	0.025	71	71
8	CaO/ZrO$_2$	1520	0.30	1.28	0.007	0.056	65	65
9	"	1537	0.39	0.81	0.020	0.018	15	65
10	"	1553	0.11	0.52	0.012	0.021	70	70
11	"	1554	0.11	0.52	0.012	0.021	70	70
12	"	1540	0.23	0.62	0.014	0.037	57	57
13	"	1556	0.21	0.63	0.009	0.022	52	52
13 bis	"	1543	0.21	0.61	0.009	0.020	66	66
14	"	1548	0.19	0.63	0.012	0.035	60	60
15	CaO/ZrO$_2$ mod	1555	0.18	0.89	0.010	0.021	58	58
16	CaO + ZrO$_2$	1547	0.18	1.06	0.014	0.024	62	62
16 bis	CaO/ZrO$_2$ mod	1547	0.18	1.06	0.014	0.024	57	57
17	CaO + ZrO$_2$	1527	0.29	0.64	0.014	0.016	69	69

Tab. 5 — Results of full-scale casting test

4.2. Monolitic CaO+ZrO$_2$ nozzles

To improve the mechanical strength of CaO nozzles, which is not high anough in some cases, the experimentation was extended to include mixtures of lime and zirconia.

So as not to jeopardize the anticlogging properties of the nozzles, attention was concentrated on mixtures containing less than 10% zirconia. It was ascertained that small additions of ZnO$_2$ brought considerable increases in the mechanical strength of sintered samples; e.g. the addition of 5% ensured a 30% improvement in the mechanical strength of a sample sintered at 1550°C.

The forming, sintering and storage methods employed for the all-lime version were adopted to prepare nozzles from a mixture of this composition.

The two works casting trials (16 and 17 in Table 5) demonstrated that the behaviour of the nozzles is very similar to that of the CaO nozzles, whose anticlogging behaviour is retained, despite the ZrO$_2$ addition. The two heats of Al-killed steel were cast regularly without any problems due to adhesion of alumina inclusions.

4.3. Double-layer nozzles

Direct observation and reports published by other authors indicate that the clogging caused by alumina occurs mainly in the upper conical part of the nozzle. It was decided, therefore, to make a two-part nozzle, the upper part formed of CaO and the lower of ZrO$_2$. With this system the anti-clogging effect is guaranteed by the former and the good wear-resistance of the normal refractory by the latter.

When an initial version of the nozzle, shown in Fig. 7a, was subjected to works trials it was found that the lower zirconia part had to be repeatedly cleaned to remove alumina adhesions. The many trials performed (Nos. 8 to 14 in Table 5) were generally brought to a satisfactory conclusion without loss of production, but in nearly every case it was necessary to prod away the adhering alumina to restore optimum casting conditions.

In the light of the results obtained in these initial works trials it was decided to design and build a new, modified nozzle with a smaller ZnO_2 surface area in contact with steel flow. The new nozzle, illustrated in Fig. 7b, has proved successful in two works casting trials (Nos. 15 and 16 bis in Table 5) which were concluded without any problems at all and with the need for only one cleaning operation (in Heat 16 bis).

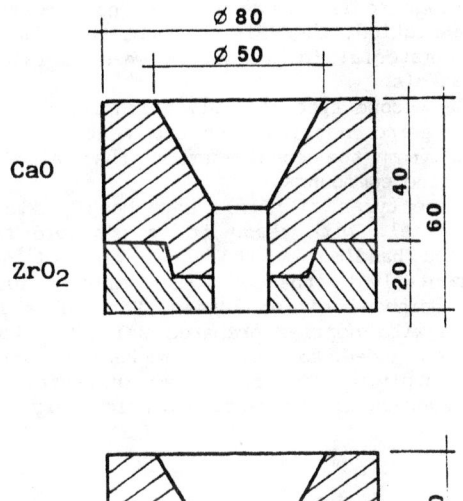

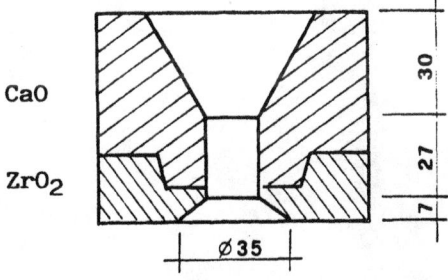

Fig. 7 – Schematics of two-layer nozzle

5. CONCLUSIONS

Two possible applications involving calcium oxide have been examined. Each of these is based on a particular property of this oxide which has most of the attributes of a high-quality refractory, but has not found industrial applications owing to problems connected with its marked tendency to hydrate.

Treatment with additives that promote sintering, followed by high-temperature sintering reduce this drawback and permit the production of CaO components that can be handled and stored without any complicated or costly procedures.

The aim of the initial line of research, which also involved works applications, was to produce calcium-oxide bricks for lining ladles employed for steel-refining treatments. Calcium oxide is considered to be particularly suitable for this kind of application because of its great thermodynamic stability.

Trials have shown that there is no difficulty in producing, storing and utilizing bricks formed of tar-bonded sintered CaO. Ladles lined with these bricks have stood up well to the stresses encountered in steelworks operations, permitting campaigns comparable to those possible with the usual refractories.

It has been found that there is a marked improvement in the desulphurization efficiency of steels treated in ladles lined with CaO bricks.

Though so far there has been no industrial follow-up of this line of experimentation, there is no doubt that due consideration should be given to this material in the future when it is necessary to make particularly clean steels.

The second type of application investigated, namely the production of lime metering nozzles for the continuous-casting of Al-killed steels started from the consideration that alumina reacts with CaO to form low-melting compounds.

Three types of nozzle have been produced using sintered CaO as the raw material. With these it is possible to cast Al-killed steels that cannot be handled by the usual kind of nozzle. The nozzles, produced experimentally by the CSM, were employed successfully for numerous heats on a continuous billet caster. Particularly promising results have been obtained with nozzles prepared using CaO to which small amounts of ZrO_2 have been added to improve mechanical strength, while a double-layer nozzle involving the use of CaO in certain parts and ZrO_2 in others has proved successful for casting particularly aggressive steels.

ESTABLISHMENT OF OPTIMUM HEATING RATES FOR REFRACTORY
STRUCTURES IN THE IRON AND STEEL INDUSTRY

DETLEV WOLTERS
Thyssen Stahl AG, Duisburg, Federal Republic of Germany

Summary

For economic reasons refractory-lined installations must be put back into operation as swiftly as possible after relining and repairs. Pilot tests were carried out to ascertain the maximum rate at which such installations could be heated without the refractory lining cracking or spalling. A furnace designed in 1974, measuring 5 m long by 2.5 m in diameter, fired by natural gas and equipped with devices for measuring expansion and pressure, was fitted with a full-scale lining comprising three layers, the inner layer (working lining) consisting of chamotte, magnesite, silica, mullite or corundum bricks. The heating rates were varied depending on the material used. Various design modifications proved necessary in the course of the 27 tests carried out up to 1982.
The project, which received finance from the ECSC, showed that this design permits all types of brick to be heated more rapidly than in normal practice. The researchers made proposals concerning the heating of hot-blast mains, blast furnace shafts, mixers and heating and heat treatment furnaces.
Finally, a number of comments are made in the light of the current state of the art.

1. AIMS

For economic reasons, the availability of the refractory-lined production units in the iron and steel industry was particularly important at the beginning of the 1970s in view of the healthy state of the steel market. The rate of availability can be increased by, among other things, rapid heating to operating temperature after relining or repairs. However, if heating is too rapid, the refractory material may crack or spall, i.e. the thermomechanical stresses may be too great. In practice, therefore, heating is usually carried out cautiously, which is to say slowly, and the rates determined empirically. If installations are to be heated as quickly as possible so as to reduce down-time and at the same time as slowly as necessary to avoid damage, the thermomechanical processes accompanying heating must be understood. The literature at that time included a large volume of test results for individual bricks. However, the values explained little about what happened in practice and did not give any information on how masonry consisting of such bricks would behave. Thus, full-scale pilot tests were called for. The results of studies of the thermomechanical processes resulting from the heating of new refractory masonry in a furnace similar to those used in practice and determination of the optimum heating conditions would then be transferred to actual operation - for example, hot-blast mains, blast furnace shafts, mixers and heating and heat treatment furnaces. A project (with financial support from the ECSC) was therefore carried out between 1 October 1974 and 30 June 1982 at the refractory materials research centre of Thyssen-Stahl AG.

2. DESIGN AND OPERATION OF THE TEST FURNACE

The main considerations underlying the design of a suitable test furnace were the requirements it should meet and the functions it should perform. The main concern was that it should have full-scale masonry lining, so as to reflect operating conditions accurately. The test furnace therefore had to be similar to actual industrial furnaces as regards size, shape, and operation. It also had to permit certain measurements to be made, so as to determine the thermomechanical behaviour of the lining, with particular reference to the effects of temperature, time and type of brick, on the interaction between the lining and casing. However, the intention was not to test the bricks individually as in the laboratory but rather the operating characteristics of the bricks when built into masonry. A cylindrical design was chosen as being the most common in the iron and steel industry. Figure 1 shows the design of the furnace. The external dimensions were chosen so as to be comparable with operating conditions in practice as regards the stresses involved.

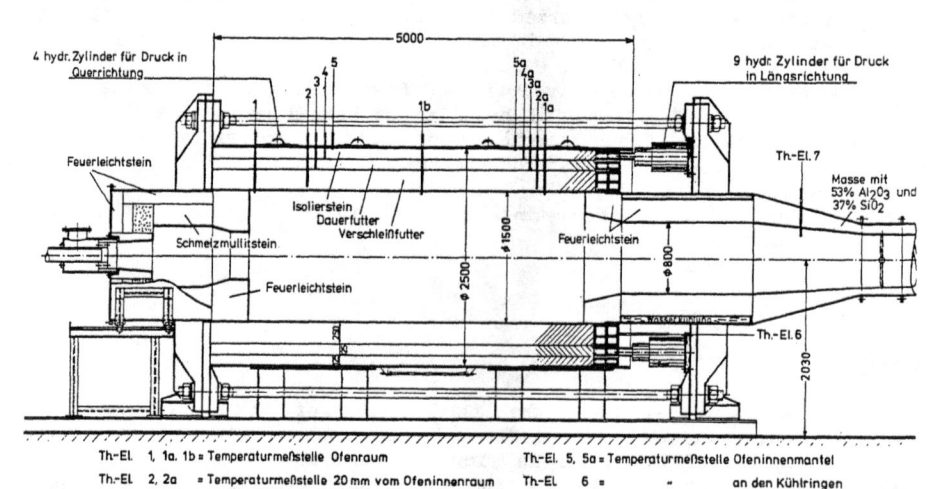

Th.-El. 1, 1a, 1b = Temperaturmeßstelle Ofenraum   Th.-El. 5, 5a = Temperaturmeßstelle Ofeninnenmantel
Th.-El. 2, 2a   = Temperaturmeßstelle 20 mm vom Ofeninnenraum   Th.-El. 6 =   "   an den Kühlringen
Th.-El. 3, 3a   =   "   250 mm   "   Th.-El. 7 =   "   Rauchgasabzug
Th.-El. 4, 4a   =   "   375 mm   "

Glossary:

4 hydr. Zylinder für Druck im Querrichtung = 4 hydraulic cylinders for transverse pressure
9 hydr. Zylinder für Druck in Longsrichtnung = 9 hydraulic cylinders for longitudinal pressure
Feuerleichtstein = refractory brick   Schmelzmullitstein = fused mullite brick
Isolierstein = insulating brick   Dauerfutter = safety lining
Verschleissfutter = working lining   Masse mit = baffle containing
Temperaturmessstelle = temperature measuring point
Ofenraum = furnace interior   vom Ofeninnenraum = from furnace interior
Ofeninnenmantel = inner shell   an den Kühlringen = at cooling rings
Rauchgasabzug = waste gas outlet

Fig. 1. ECSC test furnace

It was decided to use a three-layer refractory lining in order to follow normal industrial practice while at the same time to examine if possible both the behaviour of the individual layers and their interaction. In addition, we wished to take account of the expansion behaviour of the entire furnace resulting from the varying heat gradient from the interior to the casing caused by the three layers. The layers comprised a 250 mm thick working lining consisting of the type of brick under study, a 125 mm thick safety lining consisting of AIII chamotte brick (A30) or lightweight refractory bricks, and an insulating layer applied directly to the casing and also 125 mm thick.

The casing consisted of a 15 mm thick sheet of St37 steel with a slit in the top. The total length was 5 m and the internal diameter 2.5 m. The furnace was fired by a natural gas burner, adjustable over a range of 1:10 and with a maximum output of 62.5 Nm^3/h. The system was designed to permit furnace temperatures of 1 700°C with the addition of oxygen. The furnace temperature and the temperatures in the various brick layers were measured and monitored continuously by means of thermocouples.

The tensions in the refractory linings and the compressive stresses on the casings arising during heating and operation were to be measured by pressure meters and possibly controlled. Each layer of masonry had to have its own system for measuring and regulating the longitudinal stresses. Each layer was expected to exhibit different expansion and fracture behaviour and we also wished to study the interaction between the various layers. The pressure units for measuring the axial pressures were mounted on the offgas side with the burner side as skew-back. Thus three water-cooled compression rings - one for each layer - were mounted on the offgas side, each with a pressure cylinder with pressure gauges set at an angle of 120° to each other. Nine pressure cylinders were used for measuring and regulating the axial pressures, each able to produce or absorb a pressure of 400 kM per pressure unit. As mentioned above, the cylindrical furnace had a slit in the top for measuring the tangential forces. This was fitted with four pressure cylinders (500 kM each for tangential displacement or force regulation).

All the 13 pressure cylinders were controlled electronically, for both displacement and force. 'Displacement regulation' is the term given to the immediate application by the pressure cylinder of a pressure equal to that resulting from the tendency of the refractory masonry to expand on heating, so that no expansion can in fact take place. 'Force regulation' means that the expansion of the refractory masonry on heating can be measured with a specific counterpressure applied by the pressure cylinder. The regulation system was from the outset designed in such a way that it was possible to switch over from displacement to force regulation and vice versa in the course of a test. All 13 pressure cylinders could be controlled individually.

3. TYPES OF BRICK AND LINING

The types of brick for the working layer and the body of the masonry and type of lining were selected on the basis of three criteria:

a) popular types of trade bricks should be used;
b) brick types with characteristically different expansion behaviour should be used for the working layer;
c) the linings should be designed in such a way as to make the measurement data readily transferrable to industrial practice.

The bricks, cement and type of lining were selected on the basis of 1970s technology. There have been many changes since. The types of brick

TABLE 1. Chemical and physical properties of the types of brick

Tests	Insulating Lining Lightweight Refractory Brick	Safety Lining AIII Chamotte (A30)	Lightweight Refractory Brick.	AO Chamotte (A40)	Working Lining Magnesite	Silica	Mullite	Corundum
SiO$_2$ %	46.1	57.2	33.6	49.7	1.30	94.7	24.6	0.60
Al$_2$O$_3$ %	37.4	35.6	64.8	45.5	0.64	0.88	74.2	98.0
TiO$_2$ %	0.90	1.81	0.43	1.64	0.04	0.68	0.03	<0.03
Fe$_2$O$_3$ %	0.83	1.95	0.59	1.33	0.27	0.23	<0.10	<0.10
CaO %	13.2	0.30	0.15	0.30	2.20	2.75	0.15	0.15
MgO %	<0.10	0.30	<0.10	0.20	95.0	0.10	0.10	0.10
Bulk density g/cm^3	0.52	2.09	1.00	2.34	2.88	1.80	2.66	3.24
Apparent porosity %	80	13.8	68	15.0	17.9	21.8	14.1	17.4
Cold crushing strength N/mm^2	0.64	54	2.8	60	48	38	66	66
Thermal expansion at 1 000°C %	0.56 by 700°C	0.41	0.54	0.55	1.36	1.21	0.52	0.82

used are shown in Table 1 together with a number of characeristic chemical and physical properties.

Dry-pressed AO type (nowadays (A40) chamotte bricks were used for the working lining in the first series of tests. They exhibit low thermal expansion and used to be widely used in the upper and middle shafts of blast furnaces. In the test furnace, the chamotte full arches were bonded with ceramic-setting chamotte cement (type CO) with joints 2 mm thick. The CO cement was also used for the joints between the working layer and the safety lining of AIII chamotte bricks (nowadays A30). The AIII bricks and the joints with the insulating layer were bonded with a ceramic-setting CII cement. A chemical-setting cement containing approximately 40% Al_2O_3 was used for bonding the bricks used for the insulation layer (ASTM group 23 lightweight refractory bricks with approximately 37% Al_2O_3) and for bonding this layer to the casing.

Side arches were used for the insulating and safety linings in all cases, and the same type of lightweight refractory brick was used for the insulating layer in all six series of test.

For the second series of tests magnesite bricks containing approximately 95% MgO were selected for the working lining. These were expected to behave very differently from the chamotte bricks in view of the high coefficient of thermal expansion. Bricks of this kind used to be widely used in hot-metal mixers and convertors. Magnesite cupola bricks were bonded with chrome cement (approximately 55% Cr_2O_3 and 44% Al_2O_3). The safety and insulating linings were the same as for the tests with chamotte bricks. Silica bricks containing approximately 95% SiO_2 were used for the working lining in the third and sixth series of tests. These share a high thermal expansion coefficient with magnesite bricks, but a spontaneous reaction is likely on heating as a result of phase changes.

Silica bricks were also chosen because they are one of the materials used for the lining of hot-blast mains, parts of hot-blast stoves and coking furnaces. Silica end arches were used with a ceramic-setting silica cement containing approximately 92% SiO_2. For the sixth series of tests, additional axial and radial expansion joints were included in the masonry. From the third series onwards, the safety lining was changed. Instead of AIII chamotte bricks, ASTM group 30 lightweight refractory bricks containing approximately 65% Al_2O_3 were used, bonded with a chemical-setting corundum cement (containing approximately 93% Al_2O_3). This was following traditional practice in the lining of hot-blast mains.

For the fourth series of tests, mullite end arches containing approximately 74% Al_2O_3 bonded with ceramic-setting cement containing approximately 70% Al_2O_3 were used for the working lining.

The range of high-alumina products was rounded off in the fifth series, in which corundum bricks containing approximately 98% Al_2O_3 were used for the working lining. End arches were bonded with a chemical/ceramic-setting cement containing approximately 85% Al_2O_3. Corundum bonders were used in the working linings in all cases. In the fifth series of tests, three skintled radial joints, each 15 mm wide, were included at intervals of approximately 1.4 m.

4. TEST CONDITIONS AND PROCEDURE

Different conditions were adopted for the various linings in the light of the stresses expected in the different types of materials. The following parameters were varied: heating rate, final temperature, holding time and pressures applied by the pressure units. The test parameters of the 27 tests are set out in Table 2.

TABLE 2. Summary of conditions for the various test series

Brick type used in working lining	Test No.	Heating rate K/min	Furnace temperature °C	Duration of test h	Working lining ring At 20°C	Working lining ring At operating temperature	Safety lining ring At 20°C	Safety lining ring At operating temperature	Insulation lining ring At 20°C	Insulation lining ring At operating temperature	Tangential pressure units (4 units) At 20°C	Tangential pressure units At operating temperature
							daN					
Chamotte bricks	1	1	530	50	3.000	120.000	1.500	rising & variable	–	–	2.000	rising & variable
	2	1	530	50	3.000	120.000	1.500	rising & variable	1.500	unchanged	2.000	rising & variable
	3	10	1.365	50	1.500	120.000	1.500	120.000	–	–	2.000	rising & variable
	4	10	1.365	50	120.000	120.000	120.000	120.000	–	–	2.000	rising & variable
	5	10	1.365	50	1.500	1.500	1.500	1.500	–	–	2.000	rising & variable
Magnesite bricks	6	0.5	530	72	120.000	120.000	120.000	120.000	3.000	3.000	4.000	rising until shutdown
	7	1	1.360	72	120.000	120.000	120.000	120.000	3.000	3.000	4.000	
	8	1	1.460	50	3.000	3.000	3.000	3.000	900	900	Furnace casing welded	
	9	1	1.460	50	3.000	3.000	3.000	3.000	900	900		
	10	3.5	1.460	50	3.000	3.000	3.000	3.000	900	900		
Silica bricks	11	1	1.315	50	1.500	1.500	1.500	1.500	1.500	1.500	longitudinally, tangential pressure units removed	
	12	1	1.300	50	1.500	1.500	1.500	1.500	1.500	1.500		
	13	0.3	1.300	90	1.500	1.500	1.500	1.500	1.500	1.500		
	14	9	1.300	50	1.500	1.500	1.500	1.500	1.500	1.500		
Mullite bricks	15	1	1.400	50	500	1.500	1.500	1.500	1.500	1.500	As in previous series of tests	
	16	1	1.400	50	500	1.500	1.500	1.500	1.500	1.500		
	17	2	1.400	50	500	1.500	1.500	1.500	1.500	1.500		
	18	5	1.400	50	500	1.500	1.500	1.500	1.500	1.500		
Corundum bricks	19	0.5	1.400	76	500	120.000	500	111.000	500	102.000	As in previous series of tests	
	20	0.5	1.400	76	40.000	40.000	500	108.000	500	54.000		
	21	1.0	1.400	66	500	40.000	500	500	500	500		
	22	4.0	1.400	50	500	40.000	500	500	500	500		
Silica bricks	23	0.05	1.300	826	30.000	90.000	500	500	500	500	As in previous series of tests	
	24	0.05	up to 800 800–1300	660	30.000	90.000	500	500	500	500		
	25	0.05 0.10 0.33	up to 500 500–900 900–1300	520	30.000	90.000	500	500	500	500		
	26	0.08 0.17 0.33	up to 500 500–900	340	30.000	90.000	500	500	500	500		
	27	1	up to 1300 1.300	50	30.000	90.000	500	500	500	500		

The heating rates were increased in the course of each test series as far as was considered relevant or until the limits of the equipment were reached - as in the case of the magnesite, mullite and corundum brick linings, since bricks of these materials have such a high heat storage capacity that the burner output was not adequate to increase the heating rate still further. The lower heating rate limit was originally 0.3 K/min.

This was substantially lowered for the second series of experiments with silica bricks (tests Nos.23-27) by modifying the burner.

The final temperatures in the furnace were determined on the basis of operating and material-specific data. For example, in the case of silica bricks, which achieve maximum expansion at 800 - 1 000°C, a furnace temperature of 1 300°C was regarded as an adequate maximum.

A maximum duration of 50 hours was originally planned for all brick types, since it was thought that this would be adequate for the temperature to stabilise over the entire thickness of the furnace wall, so that the expansion behaviour of the entire refractory masonry and the casing could be assessed on the basis of a definitive final state. However, this did not prove to be the case with any of the bricks or in any of the test series, even if the duration of the test was extended to 90 hours.

The cooling of the furnace after each test was monitored. The linings were examined visually between the tests in order to obtain a picture of any cracking, spalling or changes in the joints.

The 13 pressure measurement units used to determine tensile and compressive stresses had also been intended to regulate displacement and force. It was originally planned first of all to set all the pressure units for displacement regulation so as to be able to absorb and record all the axial and tangential forces occurring during heating of the masonry. Right from the design stage for the steel components, it was assumed that the pressure resulting from the heating of the masonry would largely be absorbed both axially and radially by the cemented joints and that the resistance of the pressure cylinders would be adequate to permit either displacement regulation or force regulation at will. It was intended, therefore, to measure the expansion of the masonry when force regulation was being applied. However, it emerged in the very first tests with chamotte bricks that the axial stresses developed in the working lining were so great that the force-absorption capacity of the pressure cylinders (1 200 kN) was not adequate to maintain the displacement-regulation function. In some cases, therefore, it proved necessary to switch over to force regulation right from the outset.

In the course of the second series of tests, with magnesite bricks, unexpectedly strong tangential forces occurred which forced the slit in the steel casing to open and tore the tangential pressure units out of their mountings. This necessitated substantial changes to the design of the furnace and the measurement devices.

The steel casing had to be welded together, since the design did not permit the installation of new pressure units with a higher capacity. The only substitute for the tangential pressure measurement systems with the new version of the casing was to measure the expansion of the casing resulting from the expansion of the masonry. Three measurement bands were therefore placed around the casing to measure the increase in its circumference. The radial pressure was calculated as a function of the expansion of the circumference and the geometrical dimensions of the casing. It was found to be approximately 0.8 N/mm^2, or 16 times higher than the tangential pressure units could cope with, in the tests with magnesite bricks.

In spite of certain limitations in the operation and performance of the test furnace, certain important results were obtained, from the practical viewpoint, and these have led to an improved understanding.

5. TEST RESULTS

The test results have been described in detail, represented graphically and evaluated in the final report on the research project and a dissertation (Refs.1 and 2). This paper deals only with the most important points.

As Table 2 shows, different parameters were set for the various tests with the result that each test proceeded differently. The table does not indicate the reduced heating rates after relining or the holding times in the lower temperature range to guarantee complete drying.

The example from the test series with magnesite bricks shown in Figure 2 is typical of all the tests. In this case the heating rate was 1 K/min. The figure shows the average furnace temperature, the average brick temperatures and the expansion behaviour of the individual brick layers (Figure 2a to 2c) measured via the force regulation by the pressure units.

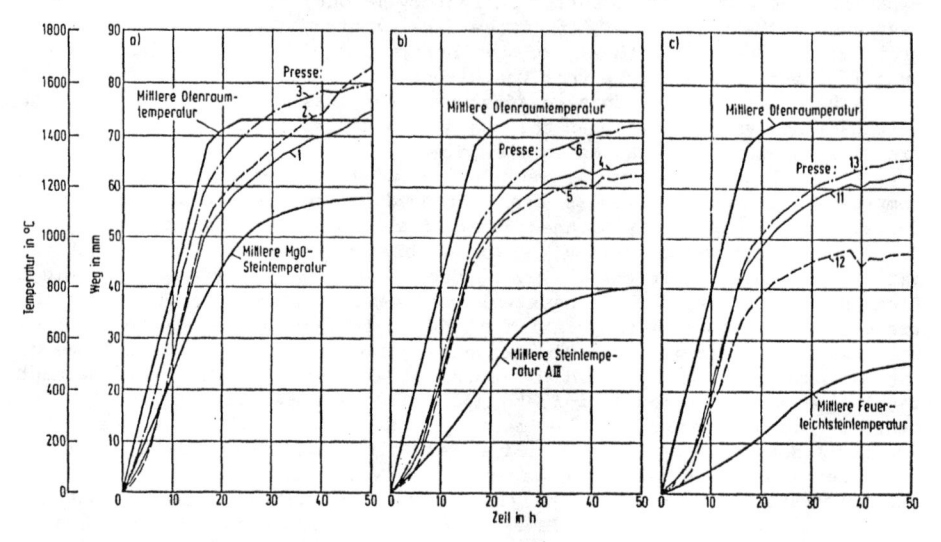

Glossary:

Temperatur = temperature
Mittlere Ofenraumtemperatur = average furnace temperature
Presse = compression unit
Mittlere MgO Steintemperatur = average magnesite brick temperature
Mittlere Steintemperatur AIII = average AIII brick temperature
Mittlere Feuerleichtsteintemperatur = average lightweight refractory brick temperature

Fig. 2. Test 8 with axial expansion of a) magnesite masonry, b) AIII chamotte masonry and c) lightweight refractory brick masonry at constant force

The furnace tests were supplemented by the following laboratory tests for certain types of brick used for the working lining:

a) softening under load of bricks as received with loads of up to 0.5 N/mm^2 and heating rates of 2, 5 and 10 K/min. and cooling rates of

0.5, 2, 5 and 10 K/min.; microscopic assessment of test samples for changes in texture and cracking;
b) comparison of physical properties before and after a test series.

The results of the additional mineralogical tests on the chamotte, magnesite and silica test samples used for the softening-under-load tests can be summarised as follows.

Virtually no differences could be detected in the AO chamotte bricks heated to 1 000°C: when received they already had microscopic cracks in the coarse grain, which remained unchanged on heating and cooling. However, the magnesite bricks heated to 1 450°C exhibited increased cracking and loosening of the matrix with increasing heating rates and loads. The same is true of the silica bricks heated to 1 350°C. The results of these tests facilitated the systematic choice of heating rates to be applied in the test furnace. Following the tests in the furnace, the physical properties of the chamotte bricks had not been significantly changed, i.e. the stresses had not resulted in any damage. While no cracks were visible in the magnesite bricks, the cold crushing strength and hot transverse strength were substantially lower than as received. The weakening resulted from microscropic cracks which formed during the tests. The silica bricks (tests 11-14) were largely destroyed, so that no physical data could be determined from the bricks used. However, at the microscopic level it was found that the texture had been compacted and the structure destroyed.

Significant differences in the axial and radial expansion behaviour of the individual types of bricks and the effect of this on the safety and insulation lining were determined in the various test series. Basically, the expansion behaviour of masonry constructed from all the brick types can be calculated from the values determined for individual bricks in the laboratory. However, the resultant pressures can only be determined in pilot tests, such as those conducted in this project.

As regards the connection between the increase in the heating rate within a given test series and the characteristics of the bricks, it may be noted that no difficulties were encountered with AO chamotte bricks. Even at the fastest heating rate (10 K/min) the test results obtained with the whole range of axial and radial pressures suggested that bricks of this type could be used without difficulty in similar industrial installations. The tests resulted in no cracking or spalling, and even the safety lining remained in good condition.

By contrast, the high thermal expansion coefficient of magnesite bricks meant that even at low heating rates (0.5 K/min) the radial expansion was so great that it was impossible to prevent the casing from deforming. When the longitudinal slit in the casing was welded shut, the entire radial pressure came to bear on the body of the masonry and destroyed the insulating layer. In the first test with the silica bricks, the lowest possible heating rate of 0.3 K/min was found to be too high, since it led to cracking and spalling. After this series of tests both the silica bricks and the entire body of the masonry had been destroyed. The burner was therefore modified to permit much lower heating rates (down to 0.05 K/min $\triangleq$ 3 K/h) and the furnace was completely relined. Axial and radial expansion joints were incorporated in the light of the theoretical expansion. Nevertheless, spalling was observed at isolated points on the edges (tests 23-27). The insulating layer was also damaged by excessive radial expansion with the mullite brick lining.

Four tests were carried out with a completely new corundum brick lining. Three expansion joints were incorporated on the basis of the theoretical axial expansion. These joints did not close completely and the bricks were undamaged at the end of the series of tests.

If the three-layer masonry is taken as a whole, it can be seen that the expansion behaviour of any given layer was extremely dependent on the others. In the tests with chamotte bricks the axial expansion of the AO chamotte layer was initially reduced because of the layers being cemented one to the other. However, as the tests proceeded this bond was broken, thus cancelling out this effect. The opposite effect was observed with the magnesite and silica linings, since their high coefficients of expansion and the strong radial bracing of the masonry which this necessitates meant that the safety and insulation linings were also forced to expand axially according to the expansion values of the working lining. Contrary to expectations, the cement joints between the bricks did not appear to absorb even part of the axial pressure, although the type and thickness of the cement layer must be taken into account.

6. CONCLUSIONS

Certain inadequacies in the pilot system became apparent in the course of the tests so that it was not possible to investigate all the intended aspects of the thermomechanical behaviour of the different types of masonry at different heating rates. Following the tests, the authors deduced from the measurement data and observations that all types of brick could be heated faster than is normal in practice (Refs.1 and 2). The following conclusions were drawn in 1982 with certain reservations concerning the comparison of pilot and industrial conditions:

- in hot-blast mains with chamotte and high-alumina bricks in the inner ring, the heating rate could be reduced from 10 to 12 days to six to seven days provided the requisite holding times were observed and assuming an operating temperature of 1 150°C;
- under the same conditions, silica brick lined hot-blast mains could be heated in two to three weeks rather than four;
- the usual drying and heating period of up to 12 days for the upper section of a blast furnace shaft lined with chamotte or high-alumina bricks could be substantially reduced, since the refractory masonry would permit heating rates of 400 to 500 K/h after the drying phase. However, the authors doubted whether heating rates as high as this would be feasible;
- a heating rate of up to 50 K/h could be the aim for hot-metal mixers lined with working linings consisting of magnesite bricks without any risk to the masonry. A total heating period of 10 to 12 days could be reduced to three to five days. However, there are doubts as to whether sufficient energy could be delivered to the mixer in such a short period;
- heating rates of up to 100 K/h would be possible for heating or heat treatment furnaces.

Axial and radial expansion joints would be required in all cases.

7. FURTHER CONSIDERATIONS

Even with the present state of the art the higher heating rates proposed following this ECSC project would not, as pointed out at the time, be immediately feasible. When refractory-lined installations are being heated as rapidly as possible, more parameters must be taken into account than was possible in the test furnace.

For example, the shape of the bricks has a great influence in that the larger and more complex they are the more slowly the installation may be heated. If the linings also comprise refractory baffles, the drying properties and related holding times also influence the heating rate. Neither is it a straightforward matter to extrapolate from the results obtained with the pilot furnace to the situation with large installations such as mixers or blast furnace shafts with in some cases substantially different geometries. Generally speaking, the large heating capacity required for rapid heating can only be obtained with large burners, which on the one hand are expensive and, on the other, impractical for use in some cases. This means that one must use the burners specific to the installations in question, which are not always ideal. Designers also point out that, for purposes of heating, the refractory lined installation must be regarded as a whole. Above all, no excessive temperature gradients must be allowed to develop, since the relative movements between the individual masonry layers and particularly between the lining and the metal casing should be kept as small as possible. The development of brick types with higher resistance to temperature change would also be useful with an eye to increased heating rates. However, the main advantages lie in the greater resistance to temperature shocks in the higher temperature range (such as when clinker coatings become dislodged from the walls of the blast furnace shaft) or the possibility of cooling the installation more rapidly.

Generally speaking, designers and manufacturers advocate slow heating - unlike the operators, who would be in favour of the most rapid heating possible. However, they all want to avoid damage.

The designers and manufacturers constantly refer to their years of experience with trouble-free heating. Nevertheless, things could also be done differently - as can be seen from the example of a steelworks which specifies the heating rate for its 2 000 t mixer and has for a long time now been heating it in six to seven days, i.e. at least 30% more quickly than the normal rate. The aim should be for all involved to apply theoretical findings and other people's experience as well as their own in order to achieve more rapid and hence more economic but nonetheless risk-free heating.

REFERENCES

1. SCHWEINSBERG, H., Festlegung der schnellstmöglichen Aufheizung von feuerfest zugestellten Aggregaten in der Eisen- und Stahlindustrie. ECSC technical research. Final report - contract 6210-CA/1/106 and 7210-CA/129, February 1983.
2. FLOSDORF, F.J. Beitrag zur betriebsnahen Ermittlung der thermomechanischen Vorgänge und der optimalen Bedingungen beim Aufheizen von neuerstelltem feuerfestem Mauerwerk in Aggregaten der Eisen- und Stahlindustrie.
Dissertation submitted to the RWTH, Aachen, July 1982, Faculty of Mining and Metallurgy.
3. Feuerfestbau. Vulkan-Verlag Essen 1987, pp.183-198.

INSTALLATION OF INSULATING PROTECTION LININGS FOR CONTINUOUS CASTING TUNDISHES BY FLAME GUNNING

JACQUES PIRET
CRM, Abbaye du Val-Benoît, B-4000 Liège, Belgium

Summary

Gunning in an oxygen/natural gas flame is proposed as a technique for applying insulating protection linings to tundishes. This method has the following advantages over existing processes involving cold tundishes with internal insulating protection (boards or slurry casting):

- formation of a purely ceramic-setting insulating lining, i.e. with a completely hydrogen-free bond;
- reduction of thermal losses at the beginning of casting, since not only are such linings insulating but they also have a high heat content because of the way they are applied;
- the possibility of selecting the density and thickness of the protection lining as a function of the stresses in the various zones of the tundish.

1. INTRODUCTION - AIMS

Tundish linings usually comprise three layers:

- a safety layer (4 to 7 cm thick) applied directly to the metal casing. This often consists of chamotte but sometimes a lighter material is used in order to provide insulation and hence protect the casing, as well as performing the primary function, i.e. safety;
- a permanent layer (15 to 20 cm thick) which is also the main layer, consisting of bricks or monolithic high density concrete;
- a protection layer (approximately 5 cm thick) the main function of which is to permit easy deskulling without damage to the permanent layer.

This protection layer is particularly important since it is directly in contact with the steel and can affect purity from the point of view of inclusions and hydrogen content. Also for reasons of quality, the aim is to keep superheating to a minimum at the tundish stage, which means that the protection lining must act as an effective insulator right from the start of casting if this is to be possible without the risk of blockages at the start-up phase.

At the beginning of this study, two main types of protection lining were being used, i.e. dense basic coating applied manually or by gunning, and prefabricated insulating boards.

The dense basic coating provides no insulation and makes deskulling fairly difficult, with damage to the main layer. Apart from being expensive to buy and install, prefabricated insulating boards cause considerable amounts of hydrogen to be absorbed at the beginning of casting because of the organic bonds they contain (Figure 1) - even if they are fairly substantially preheated, as is usually the case. More recently, insulating bodies very similar in composition to the boards, but placed by gunning techniques have been proposed. Our intention therefore, was to develop, with financial aid from the European Community, a new type of insulating protection lining which would not spoil the quality of the steel, particularly by releasing hydrogen.

We therefore adapted our flame gunning technique (Refs.1 and 2), which permits ceramic bonding by spraying a mixture of refractory powder through an oxygen/natural gas flame, thus obviating the need for organic or hydraulic bonds, which are hydrogen sources (Ref.3).

Furthermore, the other refractory materials currently used for tundish protection linings are not burnt, and there is therefore a delicate transition between the disappearance of their low temperature bond (hydraulic or organic) and the formation of the ceramic bond (Figure 2). This fragility is an advantage from the point of view of easy deskulling, but also means that the steel comes into contact with an easily erodable material which may cause impurities.

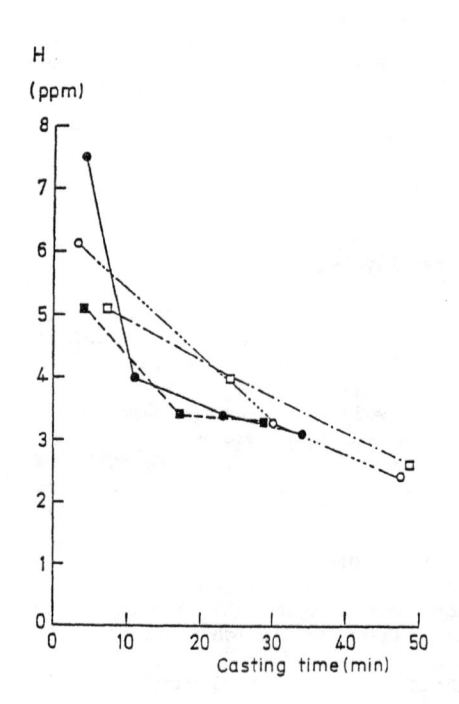

Fig. 1.

Evolution of the hydrogen content of liquid steel for a ladle in first position in the CC sequence trials with various qualities of insulating boards

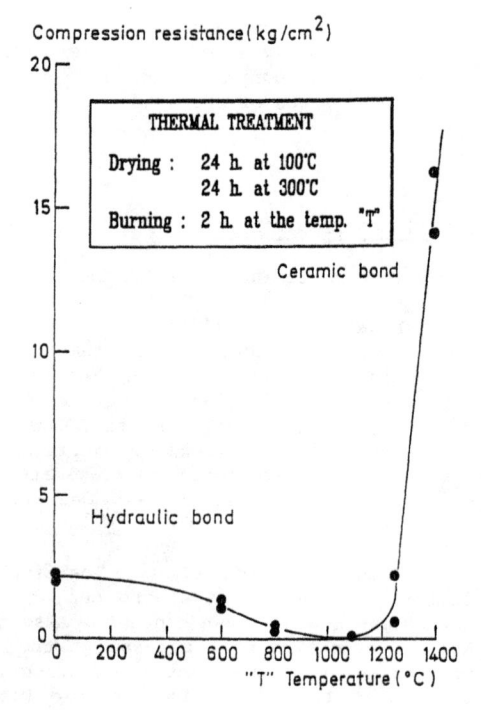

Fig. 2.

Evolution of the compression resistance of a 'slurry' gunning material with the burning temperature

We hope, by means of flame gunning, to maintain the ease of deskulling while at the same time improving the mechanical properties of the material in contact with the steel.

A further objective of this new gunning method is to improve working conditions and speed in the tundish relining shop, since workers will no longer have to enter the tundish in order to renew the protection lining, and the rotation rate can be increased by keeping the cooling period to a minimum, i.e. until the skull has solidified.

2. DESCRIPTION OF GUNNING SYSTEM

The equipment, which is in three sections, is shown in Figure 3:

- gas-feed system, with flow meters and pressure regulators;
- powder-feed unit;
- gunning lance.

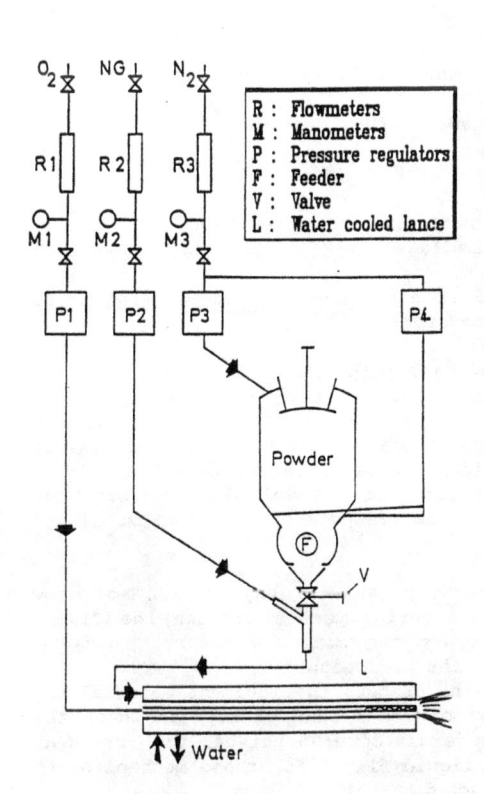

R : Flowmeters
M : Manometers
P : Pressure regulators
F : Feeder
V : Valve
L : Water cooled lance

1 _ Tundish translation
2 _ Lance horizontal rotation axis
3 _ Lance vertical rotation axis
4 _ Lance translation

Fig. 3.
Sketch of the CRM flame gunning device

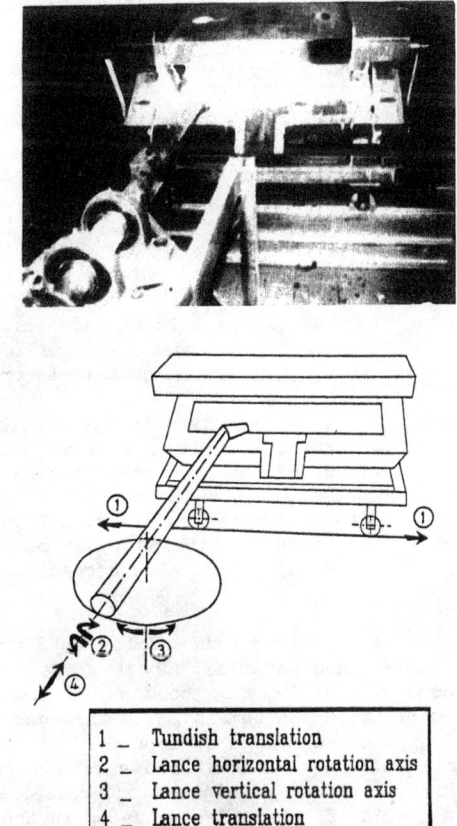

Fig. 4.
Decomposition of the movements in the tundish flame gunning system

The oxygen passes directly from the pressure regulator P_1 to the central pipe of the lance. The natural gas collects the powder to be gunned and transports it to the outer pipe of the lance. The valve V remains closed throughout the preheating phase, to avoid any powder being released prematurely. The refractory powder reservoir is kept under pressure with nitrogen at a higher pressure than that of the natural gas so as to avoid any natural gas entering the container. The container is also fitted with a fluidizing device.

After a few preliminary trials with a small fixed casing, we constructed a gunning device comprising the following elements (64):

- a receptacle shaped like a tundish (length: 1.2 m; width: 0.45 m; depth: 0.32 m) with an adjustable lid and mounted on a carriage permitting automatic lateral movement;
- the lance was mounted on a device permitting translational movements, rotation around a vertical axis and rotation around its longitudinal axis;
- the 'tundish' was placed in an enclosure fitted with fume extraction equipment.

3. PHYSICAL INTERPRETATION OF THE PHENOMENA OCCURRING DURING FLAME GUNNING

Flame gunning of linings involves the following three steps:

- preheating the material in the flame of the burner;
- impact on the hot wall;
- fixing on the wall.

Temperature is a predominant factor in the operation, but other properties of the material must also be taken into account if the various mechanisms at play are to be understood.

Two conditions must be fulfilled if a grain of gunned material is to adhere to the wall onto which it is projected:

- its kinetic impact energy must be dissipated in a different form, as otherwise the shock will be elastic and the particle will rebound by virtue of the principle of conservation of energy. This transformation of the kinetic energy cannot take place unless friction and plastic deformation are possible. A liquid phase is therefore essential;
- the grain of material must be retained on the wall by the forces of cohesion, which also necessitates the presence of an effective liquid phase.

We propose the following mechanism as an explanation of how flame-gunned material adheres. As the material passes through the flame, some of the ceramic bond melts and coats the more refractory grains of material, which remain solid throughout the operation.

At the moment of impact, the grains of both the incident material and the wall are covered with a fine layer of the liquid phase, in which the kinetic energy dissipates and cohesive forces develop between the grains as a result of the surface tension in the liquid film. For these mechanisms to be effective, the following conditions must be met:

- the grain size of the bond must be sufficiently fine for it to melt well before the point of impact;
- the bond must perfectly coat the refractory phase, which means that the liquid phase generated by the bond must be able to deform quickly

- in order to assume the precise shape of the refractory grains. A liquid phase of low viscosity and low surface tension is best for this purpose;
- the grain size of the refractory phase must also be fine in order to:
  . permit adequate heating of the particles;
  . prevent particles rebounding as a result of excess kinetic energy due to their mass.

The presence of a surfactant would therefore be useful and, as will be seen below, chromic oxide, which, according to the literature (Ref.4), reduces the surface tension of liquid silicic materials, increases the efficiency of gunning.

4. RESULTS OF GUNNING TESTS

4.1 Effects of Heating

The results obtained in our tests reflect the relative importance of the various factors mentioned in the previous paragraph.

The main parameter is the application of heat during the operation, since this determines not only efficiency, i.e. the amount of material that adheres to the wall, but also the density of the layer, which determines its insulating action.

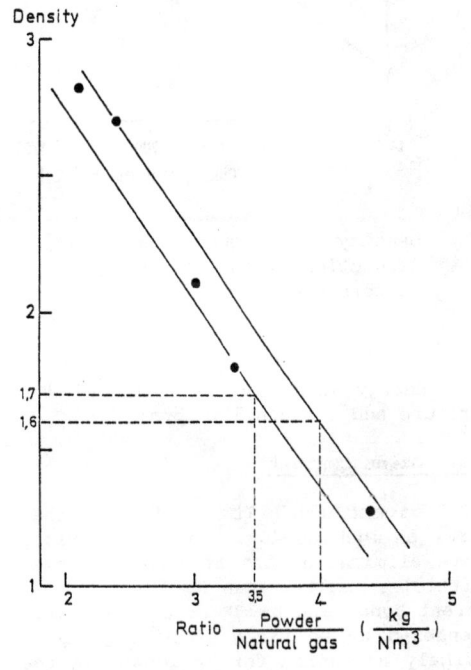

Fig. 5.
Evolution of the density of the gunned refractory as a function of the ratio powder amount/ natural gas volume (preliminary trials on a small vessel)

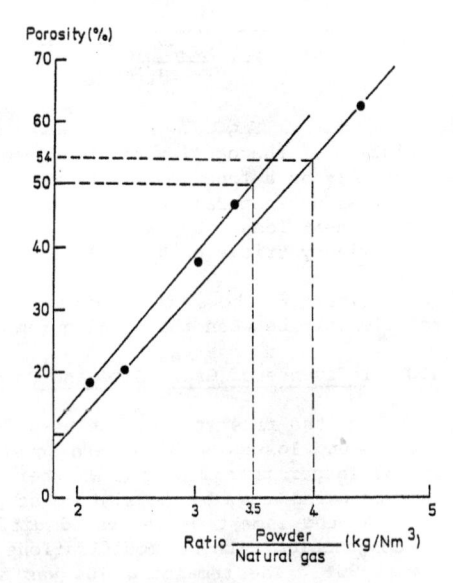

Fig. 6.
Evolution of the porosity of the gunned refractory as a function of the ratio powder amount/ natural gas volume (preliminary trials on a small vessel)

The influence of the powder/gas ratio on the density and porosity of the gunned material became apparent even in the preliminary tests (with a vessel measuring 0.4 x 0.4 x 0.3 m) (Figures 5 and 6). This influence was equally apparent in the tests using the much larger vessel similar to a tundish. It was also observed that the level of thermal loss resulting from the opening of the lid had a great influence on energy consumption at constant density of gunned product.

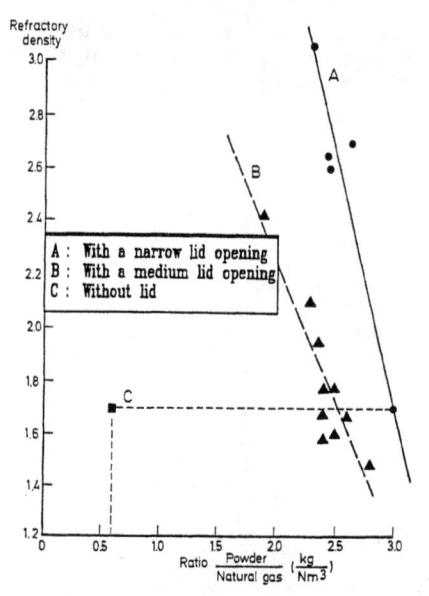

Fig. 7.
Evolution of the density of the gunned refractory as a function of the ratio powder amount/natural gas volume with various heat losses by radiation (preliminary trials on a small housing)

Fig. 8.
Density of the gunned layer as a function of the working temperature

Figure 8 shows the influence of energy delivery in terms of the relationship between the gunning temperature and the density obtained.

4.2 Influence of Grain Size and Chromic Oxide Content

For the first tests, we used 0-0.5 mm mixtures. It was observed that the rebound losses, which were sometimes as much as 40%, mainly comprised grains larger than 200 μm. We therefore eliminated this fraction and even took to using a powder consisting of particles smaller than 100 μm.

At the same time we tested different bonds and a 13% Cr_2O_3 granulate. In conjunction, these modifications enabled us to bring efficiency up to almost 90%. The remaining 10% was largely accounted for by losses in the fumes. A dust-trapping system is therefore essential.

5. PROPERTIES OF GUNNED MIXTURES

Table 1 shows the density, porosity and cold mechanical properties of the gunned materials as a function of the gunning temperature. If a density (or porosity) similar to that of boards or light gunned materials is aimed at - and this was in fact our objective with a view to ensuring that the protection lining performs an insulating function - it is found that flame gunning produces a material with substantially higher cold crushing strength and transverse strength than those in current use.

TABLE 1. Comparison of the properties of the linings obtained by flame gunning and conventional insulating protection layers

	Density	Apparent porosity	Cold crushing strength	Cold transfer strength
Flame gunning				
to 1400°C	1.2	63	1	≈ 0
to 1500°C	1.6	53	15	4
to 1600°C	2.2	32	43	9
to 1700°C	3.0	11	65	37
Basic boards	1.6	53	4.3	3.3
Humid insulating body	1.4	54	0.2	≈ 0

The micrographs in Figures 9.1 and 9.3 show examples of the microscopic structure. In 9.1, which shows an excessively dense material (magnification 200 x), the various phases can be distinguished:

- phase A: MgO grains:
- phase B: chromite grains:
- binder phase C: magnesium silicate:
- binder phase D: aluminium-calcium silicate.

The micrograph in Figure 9.3 shows the structure of a material of the same composition but substantially higher porosity.

Finally, Figure 10 shows the tundish after application of the gunned lining and Figure 11 shows a section through the insulating layer obtained by flame gunning.

6. SIMULATION OF THERMAL LOSSES OF STEEL IN TUNDISH

In order to simulate the thermal phenomena accompanying the cooling of steel in the tundish, the CRM adapted its model for calculating thermal losses for ladles.

Figures 12 and 13 show the main results of calculations for a small capacity tundish (6 t). These calculations were made first for two standard cases:

- a) the conventional technique, i.e. a dense basic coating preheated intensely up to a temperature of almost 1 300°C (consumption: 480 Nm^3 natural gas or 18 GJ per preheating);
- b) a lining consisting of insulating boards permitting surface heating up to approximately 1 000°C (consuming 40 Nm^3 gas or 1.5 GJ per preheating);

and for comparison, for the same tundish with a lining applied by flame gunning (336 Nm^3 natural gas or 12 GJ plus 300 Nm^3 oxygen per run).

9.1

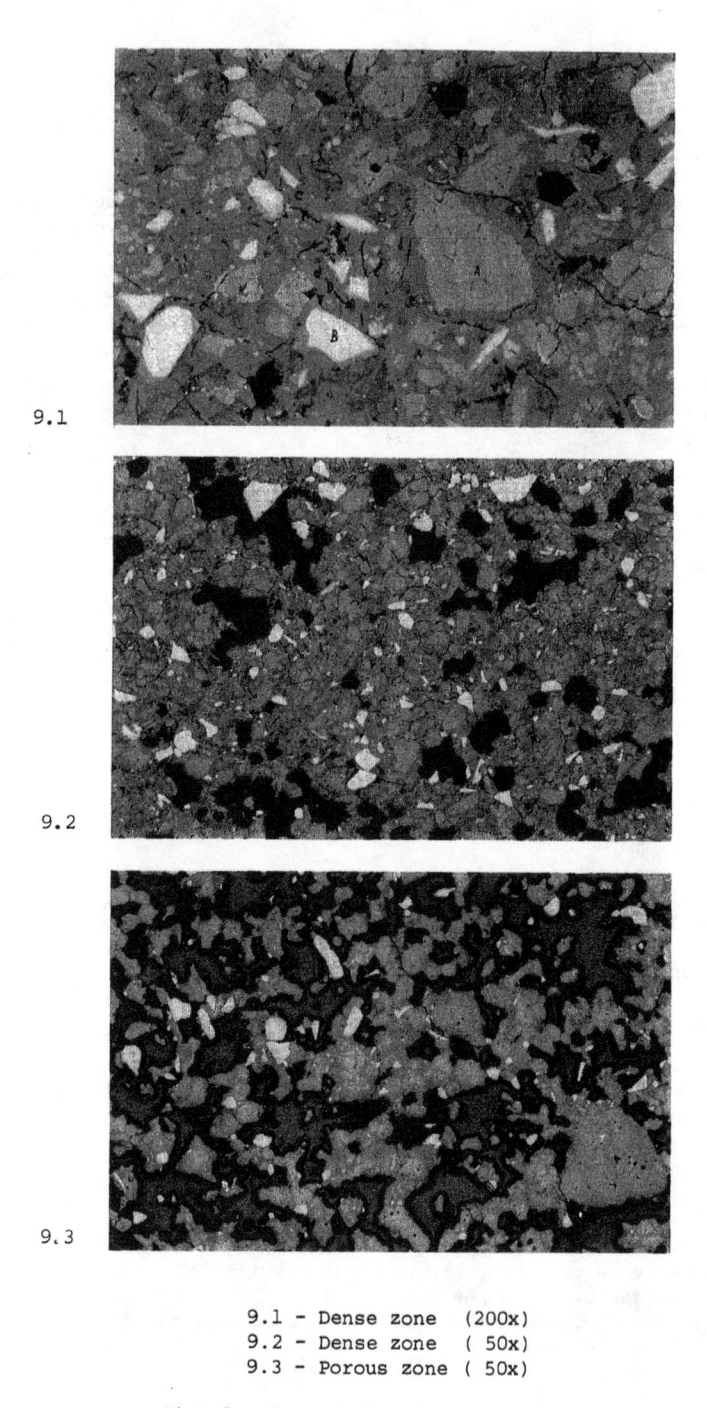

9.2

9.3

9.1 - Dense zone (200x)
9.2 - Dense zone (50x)
9.3 - Porous zone (50x)

Fig. 9. Micrographs of gunned material

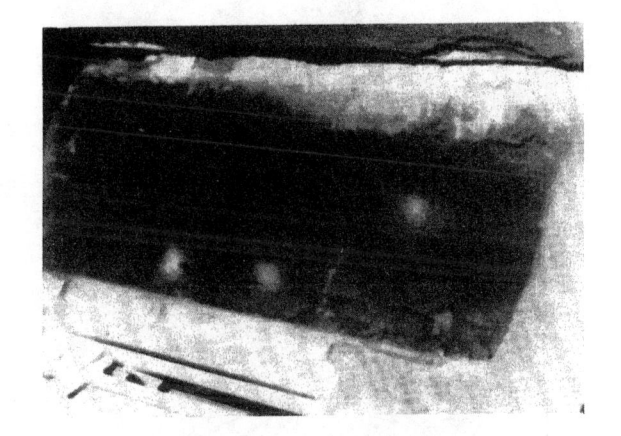

Fig. 10.
Aspect of the half-scale tundish
after applying a flame gunning

Fig. 11.
Section in the insulating layer
obtained by the flame gunning

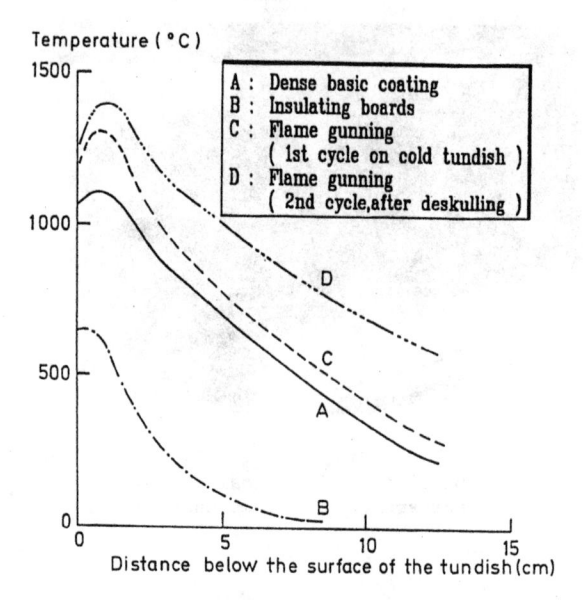

Fig. 12. Thermal profile of the tundish lining before casting (after 2 min. of handling) for three protective layers (6 t tundish)

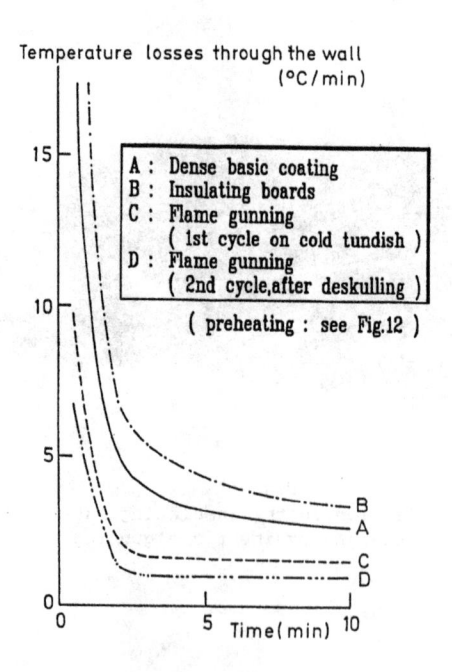

Fig. 13. Temperature losses of the steel bath in the tundish during the beginning of casting for three protective layers

The high temperatures essential for gunning (approximately 1 500°C) produce very effective preheating, giving the wall of the tundish high enthalpy (Figure 12). Together with the low conductivity of the gunned layer, this means that the thermal losses from the bath are substantially reduced, particularly at the beginning of casting (Figure 13).

7. CONCLUSIONS

Flame gunning is a highly innovative way of applying protection linings to tundishes. This study is a first step which demonstrates that it is possible to use this technique to apply a lining with both insulating properties and high mechanical resistance. Other noteworthy points are as follows:

- guaranteed absence of hydrogen;
- possibility of casting with less superheating;
- reduction of risk of impurities from erosion of the lining;
- flexibility, which means that the thickness and mechanical properties of the protection lining can be varied from one zone of the tundish to another to take account of the various stresses encountered.

Finally, the costs involved would, we think, be similar to those of other processes and automation would be at least equally justifiable as for the competing techniques.

8. REFERENCES

1. J. PIRET and B. MAIRY. Détermination des caractéristiques optimales des matériaux réfractaires en fonction de leurs conditions d'utilisation. CEE/CRM No.6210 CA/2/201.
2. J. PIRET and D. NAUDTS. Pyrogun Technique for Hot Repairs of Furnace Linings. Interceram, Special Issue Aachen 1981, Vol.31 (1982) (pp.350-353).
3. Europäische Tagung über Feuerfeste Stoffe im Stranggiessbereich, 10 and 11 March 1982. Minutes drawn up by the VDEh.
4. Traité de céramique et materiaux minéraux. C.A. Jouenne, Edition Septima 1979.
5. J. PIRET and J. MIGNON. Pre-heating of the Steel Ladle. CBM Metallurgical Reports No.37, December 1973 (p.3-11) or Préchauffage d'une poche d'aciérie. CIT of CDS No.9, 1974 (pp.2037-2050).

LIST OF REFRACTORY RESEARCH SUPPORTED ON THE ECSC STEEL RESEARCH PROGRAMME

Final reports are published in Luxembourg and can be ordered under the following address:

> Directorate General XIII
> Information Market and Innovation
> Bâtiment Jean Monnet
> L - 2920 LUXEMBOURG

(F = in French, DE = in German, EN = in English, IT = in Italian)

Year of publication	Title/ Beneficiary	Publication No (Contract No)
1981	Détermination des charactéristiques optimales des materiaux réfractaires en fonction de leurs conditions d'utilisation. Determination of the optimal properties of refractory materials as a function of their working conditions. (CRM, Liège)	EUR 7293 F
1982	Etude de l'usure du revètement réfractaire des convertisseurs à l'oxygène à l'aide de traceurs radioactifs. Study of refractory lining wear in oxygen converters using radioactive tracers. (ARBED, Luxembourg)	EUR 7727 F
1982	Réparation par projection des fours d'aciérie. Rapair of steelworks furnaces by spraying procedures. (IRSID, Saint-Germain-en-Laye)	EUR 7582 F
1983	Determination of the optimal properties of refractories as a function of their service conditions. (CRM, Liège)	EUR 7293 EN
1984	Festlegung der schnellstmöglichen Aufheizung von feuerfest zugestellten Aggregaten der Eisen- und Stahlindustrie. Determination of the fastest rate of heating for refractory lined units. (Thyssen/VDEh)	EUR 8568 DE

Year of publication	Title/ Beneficiary	Publication No (Contract No)
1984	Amélioration des briques réfractaires cuites. Improvement of burnt refractory bricks. (CRM, Liège)	EUR 8868 F
1985	Impiego dell'ossido di calcio come materiali refrattario per il rivestimento delle civiere per l'affinazione fuori forno dell'acciaio. Use of calcium oxide as refractory material for lining ladles employed for steel refining treatments. (CSM, Roma)	EUR 10089 IT
1985	Application of thin basic facings to steel ladles. (BSC, Rotherham)	EUR 10067 EN
1985	Mathematical model for the determination of thermal spalling in refractory material on basis of the practical relationship of the appearance of rupture, physical properties and physical conditions. (Hoogovens, Ijmuiden)	EUR 10068 EN
1987	Energieeinsparung durch Einsatz ungebrannter feuerfester Baustoffe. Energy-savings by using unfired refractories. (Hoesch, Dortmund)	EUR 1165 DE
1987	Réparation par projection des fonds de convertisseurs à l'oxygène a soufflage par le fond. Spraying repairs of the bottom of bottom-blowing oxygen converters. (IRSID, Saint-Germain-en-Laye)	EUR 10310 FR
1989	Scariatori contente ossido di calcio come agente anticlogging per il collagio degli acciai calmati all'aluminio. Nozzles containing CaO for aluminium-killed steel casting. (CSM Roma/Deltasider Milano)	EUR 12169 IT
1989	Réfractaires de poches à acier. Steel ladle refractories. (IRSID, Saint-Germain-en-Laye)	EUR 12137 FR

RESEARCH NOT YET COMPLETED

Year of publication	Title/ Beneficiary	Publication No (Contract No)
expected in 1989	Revêtement de paniers répartiteurs par gunitage à la flamme. Lining of distributor vessels by flame guniting. (CRM, Liège)	7210.CE/201 (to be published soon)
expected in 1991	Einsatz plasmabeschichteter feuerfester Werkstoffe. Improvement of steel quality through the use of plasma coated refractory materials. (VDEh, Düsseldorf)	7210.CE/103
expected in 1991/92	Erhöhung der Haltbarkeit von Stahlgusspfannen durch Einsatz von neuem und verbessertem Feuerfestmaterial. The use of new and improved refractories to increase the life of steel casting ladles. (Hoesch, Dortmund)	7210.CE/102
expected in 1991/92	Optimisation des réfractaires de coulée continue d'aciers inoxydables. Optimization of refractories for the continuous casting of stainless steels. (Ugine, Savoie)	7210.CE/303
expected in 1991/92	Inserti speciali per l'impiego in zone sollecitate dal flusso di metallo liquido in colata continua. Special inserts for use in zones affected by hot metal flow in continuous casting. (CSM, Roma)	7210.CE/403
expected in 1991/92	Refrattari per la fabricazione di acciai a bassissimo contenuto di impurezze. Refractories for the making of steel with very low impurity contents. (CSM, Roma)	7210.CE/404

Session II

Chairman: M. KOLTERMANN (*Hoesch Stahl AG, Dortmund, Federal Republic of Germany*)

THE EUROPEAN FEDERATION OF MANUFACTURERS OF REFRACTORY PRODUCTS (PRE):
ITS TECHNICAL ACTIVITY AND ITS ROLE IN INTERNATIONAL STANDARDIZATION

MONIQUE LEFEBVRE
Technical Secretary
European Federation of Manufacturers of Refractory Products
44 rue Copernic, 75116 Paris, France

Summary

Since its institution in 1953, the European Federation of Manufacturers of Refractory Products (PRE), whose members include producers from 14 countries of Western Europe, has aimed to facilitate trade between manufacturers and users by standardizing and harmonizing the methods used to determine the properties of refractories. The 49 Recommendations on classifications, sampling, sizes and test methods resulting from its technical activities are used both by ISO Technical Committee TC 33 on Refractory Materials as a basis for standardization and by the standards institutes of PRE member states for revising outdated documents or formulating new standards.
The aim of the PRE's current technical work is to improve the old methods devised for traditional materials by adapting them to new products, taking into account users' latest quality requirements.
In the early 1950s, a number of branches of industry in Europe felt the need to join forces and form associations to facilitate commercial, economic, technical and scientific exchanges.
Even before the Treaty of Rome was signed, therefore, the European Federation of Manufacturers of Refractory Products was founded in Venice in 1953, shortly after the European Coal and Steel Community (ECSC) in 1951 and in parallel with other associations in the ceramics sector, such as the European Federation of Tile and Brick Manufacturers (TBE) or the European Federation of Ceramic Sanitary-ware Manufacturers (FECS).
From its original 11 member countries, the Federation has expanded to encompass refractories manufacturers from 14 countries in Western Europe: Austria, Belgium, Denmark, France, the Federal Republic of Germany, Greece, Italy, the Netherlands, Norway, Portugal, Spain, Sweden, Switzerland and the United Kingdom. In 1988 their total production amounted to some five million tonnes of refractories, well above that of the United States or Japan.

1. PRE ACTIVITIES AND ORGANIZATION

The work of the PRE, which began by exchanging statistics, has expanded constantly in line with the growth in trade in refractory products between European countries since the birth of the modern steel industry in the late 1950s.
Investment in this new steel industry founded on innovative principles was enormous. The construction of large-scale plant and the desire to improve performance led manufacturers of refractory products to develop new materials and diversify production.

As a response to the new demands of steel producers, a large European market evolved bit by bit. Since trade between European countries never ceases to grow, it has become vital to find ways of determining the intrinsic properties of the products on offer which are clear and intelligible to both manufacturer and user. The importance of the PRE's technical work has increased accordingly.

Currently, under the guidance of an Executive Committee which takes the policy decisions, the PRE carries out its work through the following bodies:

- an **Economic Committee** for regular exchanges of views on market conditions and for information on the production, consumption, import and export statistics of PRE member states;
- a **Technical Committee** whose aim is to unify the methods for determining properties of refractory products.

The **technical work** is done by subcommittees for general topics and by working groups for specific topics, which meet periodically at least once each year. This work, which often involves inter-laboratory comparative tests, culminates in the publication of Recommendations approved by the delegations of all member states.

Although these publications have since the 1960s made a major contribution to the establishment of a common language between suppliers and users, the PRE organization felt the need to improve its dialogue with users in the steel industry and to involve them in its work by setting up joint working groups together with the steel industry in 1969, designated as SIPRE working groups. There are currently five of these: Test Methods, Blast Furnaces, BOF Electric Arc Furnaces, Continuous Casting and Secondary Steelmaking. They usually meet once every two years, the first two groups one year and the other three the following year.

For the working group on Test Methods, the meetings enabled certain test methods proposed in the Recommendations to be discussed jointly or formulated after trials by joint manufacturer/user groups. Currently, for example, tests on the oxidation resistance of non-oxide-containing materials are being conducted in 15 laboratories belonging to the two professions.

The meetings of the other working groups: Blast Furnaces and Steelworks, enable steel producers to explain trends in steel processes and their latest requirements in refractory products, and manufacturers to propose solutions involving research and development of materials to comply with these demands.

2. PRE TECHNICAL WORK:

- **Compilation of Recommendations**
- **Liaison with the ISO**

In the absence of international standards on refractory products, the technical work carried out by the PRE since it was founded aimed first to produce a set of 'Recommendations' on methods of determining a wide range of properties, enabling valid comparisons to be made between the test results obtained by laboratories of different suppliers and users in different European countries. The guarantees provided on the most straightforward and usual of these properties have been used for many years as a basis for specifications for the delivery of refractory materials.

The next objective of the PRE's Technical Committee was to ensure that these documents were recognized and converted to ISO standards by submitting them to ISO Technical Committee TC 33 on Refractory Materials, which came into being at a later date. PRE results are thus at the root of most ISO standards published to this day.

Lastly, PRE texts, which are usually sent to the standards Institutes in the various member states, are used for revising old standards or formulating new ones.

The PRE Recommendations, currently 49, are collected in a booklet published in its third edition in 1978 and revised in 1985. They cover a very wide range of methods of determining the properties of refractories, under the following headings:

- classification;
- standardization of dimensions;
- sampling;
- test methods.

The annexed table summarizes the PRE recommendations and ISO standards at 31 July 1989, specifying the dates on which the corresponding documents were last revised. It should be noted that:

- most methods for determining properties of traditional refractory products (dense and insulating shaped products, standard castables) are already included in the booklet published in 1978;
- the ISO standards on the same topics were published several years later (ten years or more in some cases);
- PRE work since 1980 has concentrated chiefly on perfecting specific tests or adapting methods devised for traditional materials to new products;
- PRE is to carry out the preliminary study on the new topics approved by ISO Committee TC 33.

3. STANDARDIZATION WORK IN PROGRESS (see table)

To keep pace with changes in the types and properties of refractory products supplied to users, the current work of PRE in the various areas mentioned above aims to improve traditional methods and to adapt them to the new products, methods and quality requirements.

3.1 Classification

The general standard classification of refractory products (ISO 1109) is based on chemical analysis of the materials and distinguishes the following three families: silica-alumina products, basic products and special products not belonging to the first two groups.

It was decided to revise this standard in order to take into account the wide range of products available today. The PRE did this by considering each family in turn:

- basic products: Recommendation PRE/R48, compiled in 1988 and now under consideration by the ISO, applies to refractories containing less than 7% residual carbon and classifies groups of products as a function of their main constituent content (MgO, CaO, Cr_2O_3). It also provides information on the type and state of the raw materials and on the type of bonding of the material. The PRE also intends to consider the classification of basic products with high carbon content;
- for products in the silica-alumina series, a document based on criteria analogous with those defined for basic products is under preparation;

- for special products, a distinction is to be made between oxide-based products such as zircon, zirconia, etc. and non-oxide products such as carbides, nitrides, and so on.

The supplementary classifications define sub-groups of products according to the change in the value of a property with temperature.

For insulating and fibre products, the PRE Recommendations opted for the permanent linear change as the criterion for classification.

For unshaped materials, the same property was selected for the PRE Recommendations compiled in 1978. This one criterion is, however, inadequate for assessing the behaviour of some of these materials under temperature. Attempts are being made to find at least one more appropriate criterion for classification.

In the absence of standardized test methods for unshaped materials and fibre products, ISO is considering only the classification of insulating materials.

3.2 Dimensions

PRE work here led to ISO standardization of the most common shapes. Current work involves studying size measurement methods and assessing defects in dense shaped products (see paragraph 3.4.6).

3.3 Sampling

To ensure that the value of a property determined by a specific test method is meaningful, sampling must produce an item representative of the batch being tested.

The PRE began looking at the application of the statistical principles of sampling shaped refractory products in 1964. The resulting ISO standard 5022 was published in 1979 and subsequently adopted by many countries as their national standard. It defines inspection plans, based on both producer's and user's risks, for achieving a precise assessment of the quality of the supply from as small a sample as possible by monitoring both attributes (size, appearance, defects, etc.) and measurements (determination of properties).

The user's risk is defined as the probability of accepting a batch in which the proportion of defective items is equal to the limit quality (LQ).

The producer's risk is defined as the probability of rejecting a batch in which the proportion of defective items is equal to the acceptable quality level (AQL).

For raw and unshaped materials, the recent ISO standard 8656 defines sampling plans based on the minimum density of the intended sample as a function of maximum grain size.

3.4 Test Methods

A distinction is drawn here between general test methods applicable to all categories of usual products and the specific tests developed more recently, either for determining properties of new families of refractory products or for meeting users' latest quality requirements.

3.4.1 Chemical analyses

The chemical analysis of a material is one of its most important properties for the following reasons:

- the general classification standards for refractory products are based on their main element content, which is thus one of the properties most often sought for quality inspection purposes;
- impurities content is also frequently determined as an indicator of the quality and refractoriness of the material.

Following the development of methods of chemical analysis by wet process, suitable for silica-alumina and basic refractory products, joint ISO/PRE working groups currently deal with:

- standardization of physical methods of analysis (X-ray fluorescence, atomic absorption, etc.) which are now replacing the traditional chemical methods. Work began with X-ray fluorescence. A document specifying the general principles of the method, i.e. bead preparation, flux type, interaction corrections, calibration, etc., is under preparation;
- analysis of refractory products containing non-oxides: work began with material containing silicon carbide and led to the development of reference materials; the analysis of products containing silicon nitride is also to be studied.

3.4.2 Studies of texture

The studies involve the measurement of true and bulk densities, apparent and true porosities of raw materials, granular material and shaped products, and the determination of properties related to pore size distribution.

The bulk density of a granular material is a very important property, forming the basis for trade between manufacturers and their raw materials suppliers. There are several possible methods, and test results may differ according to the type of material tested; the method used must therefore be specified together with the value obtained.

Like PRE Recommendation R30, ISO standard 8840 of 1987 specifies the mercury vacuum and arrested water absorption methods; the mercury method, which is more easily reproducible, is taken as the reference method. Because of the hazards of using mercury, the second method is recommended for routine work.

The bulk density and porosity of shaped dense and insulating products are frequently determined, being a simple way of assessing the quality consistency of the batches and hence frequently requested by users as guaranteed values for specifications.

Permeability to gases and acid resistance are important properties for some specific applications; gas permeability is an interesting feature, being related to pore size distribution; acid resistance indicates the behaviour of materials for installation in industrial chimneys.

3.4.3 Properties under load

The properties of refractory products under load when cold are measured frequently; cold crushing strength has long been used in quality inspection.

The cold crushing strength test for dense shaped products, which has been standardized in the various countries for many years, involves operations which may or may not require packing between the test piece and the press. The results obtained differ according to the method used, and it has not yet been possible to draw up an ISO standard in spite of the work in this field.

Since it is impossible to standardize packing, the ISO recently decided to standardize the non-bedding method, which is to be regarded as the reference and known as the 'Cold Compressive Strength' method.

The former Recommendation R14 of 1967 is currently being revised by the PRE and is to be in two parts:

- one describing a method identical to the current ISO draft;
- the other describing a method which allows packing to be used and which can be applied for routine purposes; this will still be known as the 'Determination of the cold crushing strength'.

Among properties under load at high temperatures, refractoriness under load is the one most often sought by users as an indicator of the refractoriness of the product and its behaviour as a function of temperature.

The creep test, which requires the same apparatus as the refractoriness under load test, is specific to certain applications. It investigates the behaviour of materials under load at a specified temperature over time.

The hot modules of rupture test assesses the behaviour of refractory products subjected to mechanical stresses at high temperature. For products containing little or no oxidizable material, the test conducted in air is reproducible and has been standardized.

When materials higher in carbon are involved and the test has to be conducted in a neutral or reducing atmosphere, it has not yet been possible to establish a method yielding reproducible inter-laboratory results.

3.4.4 Thermal and refractory properties

The determination of the pyrometric cone equivalent is one of the oldest tests used to define the refractoriness of materials, and involves comparisons with standardized reference pyrometric cones.

Thermal conductivity is a property needed for calculating heat losses through walls of refractory linings. There are several methods of determining this property, based on different principles. PRE and subsequently ISO preferred to standardize the hot wire (cross-array and parallel) method, which measures the property at the true temperature rather than at the average temperature of the wall over a temperature gradient. It should be noted that the parallel hot wire method, currently an ISO draft standard, was perfected by a SIPRE working group.

The permanent linear change of dense and insulating refractory products is an important property, since supplementary classifications of unshaped materials, insulating and fibre products are based on it. It is also frequently required by users for assessing the shrinkage behaviour of materials inside brickwork and hence the stability of the structure.

Corrosion and thermal shock resistance are properties directly connected with the conditions in which the materials are used, and are difficult to simulate in the laboratory.

The old PRE Recommendations specify tests which are simple but not particularly meaningful. ISO is currently investigating the 'Ribbon Test' for thermal shock resistance, which simulates actual working conditions more closely by subjecting a sample of refractory material placed in a wall to thermal cycles. Even though this test is useful as a research tool, however, it may be difficult to standardize, since its inter-laboratory results are still scarcely reproducible.

3.4.5 Specific tests

Most PRE documents on tests designed for specific families of products such as unshaped or fibre materials or non-oxide products have been revised recently in view of developments and improvements in the quality of the materials over 20 years.

3.4.5.1 Unshaped materials

The Recommendations on specific tests of unshaped materials compiled since 1975 were concerned with quality inspection, test piece preparation and the determination of properties using test pieces unfired, during firing and after firing. Between 1984 and 1986 these Recommendations were revised to cater for the new materials developed since the 1970s, particularly low-cement castables. The current aim of the PRE is to perfect specific tests on unshaped basic products; two new Recommendations are to be issued shortly.

Because of the fundamental discrepancies between the PRE Recommendations and the ASTM standards, a working group has been set up in the framework of ISO to seek a compromise between the European and North American methods of determining properties.

3.4.5.2 Ceramic fibre products

Recommendation PRE/R41 of 1977 concerned methods of determining permanent linear change in dimensions, the thermal conductivity coefficient and tensile strength. The current PRE study, which should shortly produce a revised Recommendation, covers not only improved procedures for determining these three characteristics but also ways of identifying four additional properties: thickness, bulk density, compressibility and shot content. The new PRE draft is to be a study document for the next ISO meeting.

3.4.5.3 Products containing non-oxide substances

Recommendation PRE/R35 of 1977, issued by a SIPRE working group, originally applied to tar-bonded, fired and tar-impregnated basic refractories and was revised in 1985 to include resin-bonded products. It concerns the determination of certain properties at room temperature on test pieces as delivered, after coking by a well-defined method and after complete elimination of carbon, and is applied to products with a maximum residual carbon content of 7% after coking. For products with higher carbon content which may also contain graphite and certain anti-oxidants, a supplement indicating precautions to be taken in determining certain properties was added in 1987.

Recommendation PRE/R35 is now being used as a basis for ISO work on this subject.

No standards or recommendations yet exist on methods of determining the high temperature properties of non-oxide products. Following a survey by a SIPRE working party, it was decided to conduct inter-laboratory tests on the oxidation resistance of these products. This study is currently in progress.

3.4.6 General topics related with quality requirements

Ever since its foundation and the compilation of Recommendations, the PRE Technical Committee has felt the need to allocate a field of validity to the various methods proposed. However, at this time, it was only possible to issue general and sometimes imprecise documents such as:

- **Documentation sheet FD 5 (1975)**

Inter-laboratory reproducibility of test methods, explaining the differences which may be observed between the values of a given property as measured by a specified method;

- **Recommendation PRE/R23 (1976)**

Dimensional tolerances of dense and insulating shaped products, giving tolerance limits according to the type of product and details on warpage and taper control;

- **Recommendation PRE/R31 (1977)**

Recommendation for examining the technical clauses of contracts specifying means to control the conformity of a delivery.

Since about 1980, ISO work on quality control has produced ISO general standards 9000 to 9004, adopted by the European Communities as standards EN 29000 to 29004. These texts are now used by suppliers and users in most European countries as reference documents for deliveries of refractory products.

The PRE texts named above do not therefore meet the new quality requirements. At the request of Sweden and the United Kingdom, ISO has included the following new subjects in its programme:

- **size measurement of refractory bricks;**
- **appearance and defects of refractory bricks;**
- **reproducibility and repeatability of test methods.**

For the first two, an ISO working party was set up and is in close collaboration with the corresponding PRE working group; a PRE document has been submitted to ISO TC 33 for consideration.

ISO has asked the PRE Technical Committee to undertake a preliminary study of the third subject. A survey is being conducted to define this very wide-ranging problem more closely and to find a way of solving it at minimal cost.

CONCLUSION

The technical work of the European Federation of Manufacturers of Refractory Products, which began in an attempt to standardize the methods for determining properties of refractory products and to facilitate exchanges between the various countries of Europe, has been adopted on an increasingly international scale by being used as a basis for ISO standardization. Although initially limited to traditional products, PRE work currently aims to improve existing methods or to develop new ones to cater for:

- new products, by developing tests specific to the new refractories, which have proliferated over the past 15 years;
- new methods which have emerged in parallel with new techniques, such as computerization, for improving the performance of measuring equipment;

- the new quality requirements of the various sectors of industry.

REFERENCES

1. HALM, L. (1958), 'The European Federation of Refractories Producers and its Technical Activity', **The American Ceramic Society Bulletin**, Vol.37, No.12, pp.501-506.
2. Recommendations de la Fédération Européenne des Fabricants de Produits Réfractaires (PRE) (July-September 1969). **Bulletin de la Société Française de Ceramique** No.84, pp.3-34.
3. PRE Booklet of Recommendations (1978), revised in 1985.
4. PADGETT, G.C. (November 1987), 'International and National Standardization: A European Viewpoint', **2nd International Conference on Refractories**, Tokyo.
5. PADGETT, G.C. (1988), **Review of the National and International Standardization of Refractories**, Symposium, China.

PRE RECOMMENDATIONS AND ISO STANDARDS
31/07/89

Subject	Recommendation	ISO Standard	Observations
CLASSIFICATION			
- Dense shaped products	Under revision	1109 (1975)	Revision of ISO 1109: new subject
- Nomenclature of manufacturing processes		2246 (1972)	
- Basic products containing less than 7% C	R48 (1988)	DP 10081	
- Products in the alumina - silica series	Under investigation.		
- Unshaped materials (dense and insulating)	R42 (1978)	1927 (1984)	
SUPPLEMENTARY CLASSIFICATIONS			
- Castables	R43 (1978)		
- Ramming mixes and mouldables	R44 (1978)		
- Shaped insulating products	R39 (1971)	DP 2245	
- Dense shaped acid-resisting refractory products	R40 (1972)	DP 10080	
- Fibre products	R49 (1989)		
VOCABULARY FOR THE REFRACTORY INDUSTRY		836 (1968)	Revision of ISO 836 in progress
DIMENSIONS			
- Rectangular bricks	) R3 (1977)	5019/1 (1984)	
- End arches and side arche	) R36 (1977)	5019/2 (1984)	
- Large end arches, end arch bonder bricks			
- Checkers for glass furnaces	R20 (1977)	5019/3 (1984)	
- Bricks for electric arc-furnace roofs		5019/4 (1984)	
- Skewbacks	R37 (1977)	5019/5 (1984)	
- Basic bricks for oxygen converters		5019/6 (1984)	
- Refractory bricks for rotary kilns	R38 (1977)	5417 (1986)	
. Dimensions			
. Hot-face identification marking:			
- by notch	R46 (1985)	9205 (1988)	
- by colour	R47 (1985)		
SAMPLING			
- Sampling and acceptance control of shaped refractory products	R7 (1977)	5022 (1979)	
- Raw materials and unshaped products		8656 (1988)	

TEST METHODS

Chemical Analysis

- Loss on ignition — R24 (1976)
- Silica, alumina and fireclay products — R33 (1977) + supplement (1984)
- Basic products — under investigation — DP 10058
- Chemical analysis by instrumental methods — under investigation — new subject) subjects being
- Non-oxide containing refractories — new subject) studied by joint
) ISO/PRE working group
- Moisture content of a raw material or an unshaped product — R11 (1966)

Texture

- True density of raw materials and refractory products — R8 (1978) — 5018 (1983)
- Bulk density of granular materials — R30 (1977) — 8840 (1987)
- Bulk density apparent and true porosity of dense shaped refractory products — R9 (1976) — 5017 (1988)
- Bulk density and true porosity of shaped insulating refractory products — R10 (1976) — 5016 (1986)
- Permeability to gases — R16 (1972) — DP 8841
- Resistance to acids — R22 (1975) — 8890 (1988)

Under-load properties

- Cold crushing strength of insulating shaped products — R15 (1968) — 8895 (1986)
- Cold crushing strength of dense shaped products — R14 (1967) — DP 10059   Revision of PRE/R14 in progress
- Cold modulus of rupture of dense and insulating shaped products — R21 (1978) — 5014 (1986)
- Hot modulus of rupture — R18 (1978) — 5013 (1985)
- Refractories under-load — R4 (1956) — 1893 (1989)
- Creep in compression — R6 (1978) — 3187 (1989)

Thermal properties

- Pyrometric reference cones — 1146 (1988)
- Refractoriness — 528 (1983)
- Thermal conductivity:
  . cross-array hot-wire method — R32 (1977) — 8894-1 (1987)) method devised by
  . parallel hot-wire method — DIS 8894-2 (1988)) STPRE working group
- Permanent change in dimensions of dense shaped products — R19 (1978) — 2478 (1987)

- Permanent change in dimensions of insulating shaped products | R13 (1978)
- Resistance to corrosion | R34 (1972)
- Resistance to thermal shock | R5 (1977) | 2477 (1987) under investigation | ISO document in preparation based on different principle from that of PRE Recommendations

Specific tests

Unshaped materials

- Control as received
  . basic castables | R25 (1984)
- Preparation of test pieces
  . basic castables | R25-2 (draft) R26 (1984) R26-2 (draft)
- Determination of characteristics of unfired test pieces | R27 (1986)
- Determination of characteristic values during and after firing | R28 (1986)
- Cements and grouts | R45 (1984) | new subject

Ceramic fibre products

- Permanent linear change in dimensions
- Thermal conductivity (panel method)
- Tensile strength
- Thickness
- Bulk density
- Compressibility
- Shot content |) R41 (1977))))))) |) Revised PRE/R41) scheduled for late) 1989))))

Products containing non-oxide substances

- Physical methods for tar bonded and tempered, resin bonded, fired and tar impregnated basic refractories | R35 (1985) + supplement 1987
- Resistance to oxidation | under investigation | Method devised by SIPRE working group

inter-laboratory tests under control of SIPRE working group

General matters related to quality requirements

- Recommendations for the examination of the technical conditions of special contracts | PRE/R31 (1977)
- Brick size measurements | under investigation | new subject
- Measurement of defects | under investigation | new subject
- Reproducibility and repeatability of test methods | under investigation | new subject

UNSHAPED REFRACTORIES
AN EXAMINATION OF ASSESSMENT AND QUALITY CONTROL METHODS

B. CLAVAUD
Director for Research and Development
Lafarge Réfractaires Monolithiques

M. JACQUEMIER
Head of Applied Research
Lafarge Réfractaires Monolithiques

H. LE DOUSSAL
Head of the Refractory Service and Technical Ceramics
Société Française de Céramique
France

Summary

As a European producer of high performance unshaped refractories and a licenser throughout the world, we are keen to find the most appropriate way of combining the necessary laboratory assessment, realistic analysis of these characteristics once installed on-site and the procedure to be adopted for checking and defining the quality level of production lots. Indeed, our methods (ISO, PRE, ASTM) have often proved unable to reconcile these three requirements in a faithful and cost-effective way.

1. INTRODUCTION

Throughout the world, refractory products are undergoing rapid change, the inevitable evolution of an industry which has to adapt to progress: new constraints, economy, speed of execution and flexibility. Unshaped refractories offer advantages in all these respects and at the same time have undergone obvious technological improvement.

Japan, which passed through these changes earlier than Europe and the United States, can be taken as an example of the following trends:

- an increase in the proportion of unshaped refractories (Figure 1)
- rapid growth of concrete-like and gunning materials (Figure 2): the present figures are:

Unshaped refractories - (60-65% concrete (including gunning)
(10-15% ramming materials and plastics)
(10-15% jointing materials)
(10-15% miscellaneous)

This period of technological change does, however, raise a number of issues:

- how can unshaped refractories be realistically assessed and ranked, in quality terms, with equivalent bricks;

- how can the issue of the quality of the installation work be tackled, since this is often the vital determinant of performance (II).

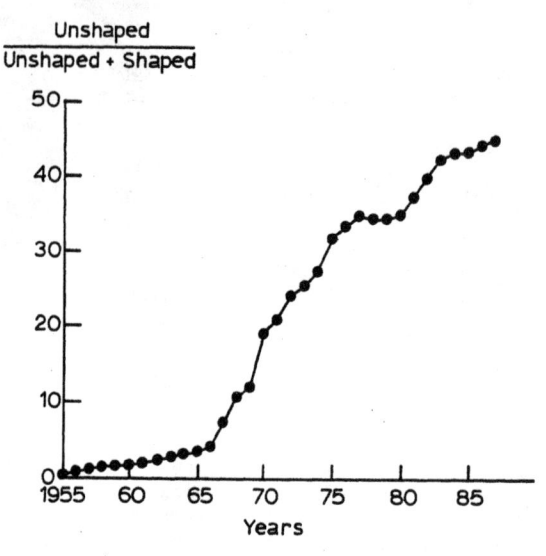

Fig. 1. Conventional refractories production over time in Japan (Ref.1.)

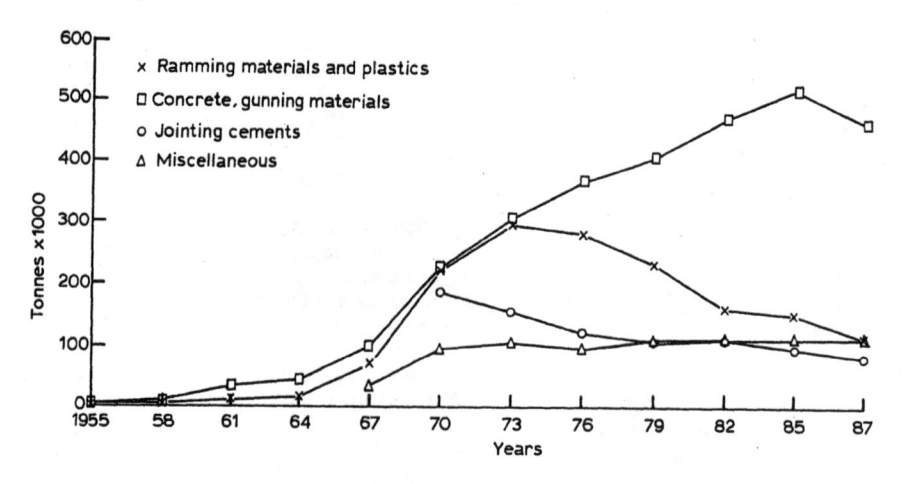

Fig. 2. Trends in unshaped refractories in Japan
Source: MASUMI TODA

Even a competent user cannot at present adequately answer these questions - whereas others are often completely at sea. The resulting feeling of helplessness in dealing with problems and suppliers makes it difficult to improve performance. Nevertheless, this is a prerequisite for more widespread progress and an advancing economy.

Now we come to standards! These exist for unshaped refractories but they unfortunately comprise rather a patchwork in that:

- they are usually modelled on shaped products and are therefore only marginally relevant;
- they are sometimes very specific (petrochemicals);
- they completely or largely ignore the issue of 'industrial on-site installation'.

At the present time, the PRE recommendations do comprise a whole, even if they do not cover the 'installation' aspect. It is revealing to examine the relevant ISO standards: two standards (ISO 1927 - classification, ISO 8656 - sampling) concern unshaped refractories as against 27 covering shaped ones. **The issue still requires a great deal of international examination and our initial conclusion is that the difficulty of the subject has sidetracked attempts at standardization and has confused laboratories not fully competent in the field.** Our aim here is therefore to throw out a few ideas acquired over 15 years of practical experience at our Research Centre, in factory laboratories and in the field.

2. ANALYSIS OF THE PROBLEM

With this conclusion in mind, and faced with managing a technology used in every continent, we have studied this entire problem. We will therefore examine in turn each of the essential features set out in the summary.

2.1 Methods

As we see it, assessment and quality control methods generally comprise written procedures adapted to, and applied in, specific applications. They may be restricted to a given company, profession, country or group of countries. Where they are recognised at each of these different levels, they may be chosen as references and therefore become standards.

These reference documents form a common technical language and they are drawn up in the light of acquired know-how. They may sometimes be initially restricted to one country (petrochemicals). In applying them, cross-checks will always be needed, given that it is impossible to set out in the text - and we see this for example with unshaped products - all the details of what often proved to be vital descriptions of the material. This is true of the current problems associated with the recommendation PRE/R26 - consistency test using the cup method - in which the key factor is the attachment to the vibrating table.

2.2 Refractory Materials

Methodological considerations are rather complicated by the fact that temperature rises and the holding time modify the material's properties both temporarily and definitively (continued sintering of highly porous materials). A minor truncation of the method can thus generate significant differences (e.g. softening under load can be affected by the rapidity with which the temperature is increased).

Certifying products on the basis of their applications, as proposed by certain bodies, is quite unrealistic. After all, the product/process/man relationship is a strikingly interdependent one. The same product used for the same purpose can vary widely in performance. This is because specific features of the process can result in significant differences. Such is the case with the tiles on which the ladle lands:

in steelworks X: they last 25 castings
in steelworks Y: they last 39 castings.

Differences between the steelworks include the temperature and grade of the steel, the height of the fall, the cycle...

To sum up this section, let us examine shaped materials stabilized by firing in the factory. The quality of this stabilization is indistinguishable from the product. The product is thus 'almost' completely defined: its shaping as such (the brick, its shape) and its level of heat treatment. Indeed, the recognised assessment and quality control methods adequately cover these materials.

2.3 Unshaped Refractories

As noted above, these are more difficult to deal with than shaped refractories:

- they have to be installed (the key stage);
- they are fired 'in-situ' at temperatures usually close to the average operational temperature. Their 'modification' gradient is therefore linked to the thermal gradient. The result is a whole range of values for the same parameter from one side of the wall to the other. Taking the example of standard concretes (cement content in excess of 15%):

20° — 500°C	— 800°C —	1200°C ....
hydraulic bonding		ceramic bonding
	intermediate temperature lowest mechanical strength	

In a nutshell, unshaped products resemble a kit when they leave the factory, whereas shaped products are in their finished form. The issues that then arise are 'instalment skill' and 'installation instructions'.

Without wishing to go too much into details, unshaped refractories are also 'perishables' (for example, water vapour may reduce the activity of cement in concrete while ramming materials and plastics may become less workable). The equation must therefore contain a time element:

* quality at t = 0 = quality of material
   (production   +
                   quality of installation

* quality at t = x = quality of the material itself
                               +
                     quality of installation

We must therefore consider the various stages of the process and the repercussions of each (Figure 3). The importance of the installation phase cannot be overlooked. It is therefore, above all, that the faults of preceding stages can become cumulative.

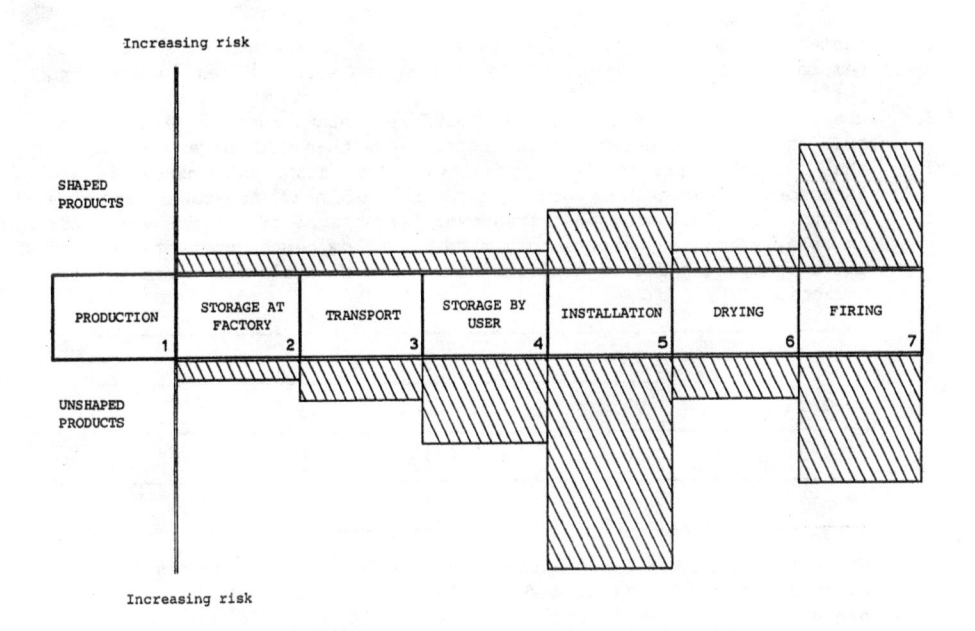

Fig. 3. Shaped and unshaped products
Arbitrary scale of vulnerability to the stages between
leaving the factory and first use

In conclusion

Compared with shaped refractories, there are two key aspects:

- how to determine the preliminary firing temperature before assessment and the length of this firing;
- how to estimate and reproduce quality during the industrial installation phase.

Indeed, in spite of the intensive efforts of PRE and ASTM, it is necessary to rethink the quality control set up for unshaped refractories. Predominantly for historical reasons, there is in effect a tendency to monitor quality as if they were fired shaped materials - although they are fundamentally different.

2.4 Assessment

It is only when this process is completed that a product, completed in accordance with accepted techniques, achieves real stature. The difficulty of comparing the sets of values obtained using divergent methods means that it is important to choose appropriate systems and units. Finally, assessment can be limited to basic values (statistical monitoring) or may involve a virtually unlimited number of values (Youngs module, Poissons coefficient...).

Even today, there are three snags:

1. these results are artificial in that they are drawn from tests on samples installed in a laboratory (hence often maximizing performance);
2. the continued use of several very different standards (for example the ASTM and PRE procedures for concrete) means that values vary widely;
3. the results reflect the 'condition' of the test piece before assessment - in other words the degree of heat treatment and its length. The following post-treatment percentages of mullite were, for example, found in a refractory with ultra-low cement content (0.3% of CaO), containing 20% of a silica binder and fine aluminium (as mullite stoichiometry):

	2h	5h	25h
1250°C	0	0	0
1300°C	trace	3	5
1400°C	7	10	13

While the first two difficulties can be sidestepped, the significance of problem 3 cannot be overlooked. After all, the constraints on the materials are:

- the maximum temperature achieved with this application
- the lifetime of the material:
  - oxygen-blowing lances   : 60/600 minutes
  - blast furnace spouts   : 20/50 days
  - channel induction furnace   : 300/600 days
  - blast furnace   : 5-10 years

Better simulation of tests, and a more informed choice of materials, would therefore be assisted by a table contrasting heat treatment values with varying product lifetimes.

3. POSSIBLE PROPOSALS

1. Initially, a **single international system** should be adopted so that what is becoming a shared technology can benefit from exchanges across the board. This would appear possible for unshaped refractories because **ISO standards** are virtually non-existent.

2. Methods should be structured to include the advanced parameters now responsible for the real inadequacies of our standards. This is set out in Figure 4, which introduces five new key ideas:

- C (4) - the possibility of working on industrial samples;
- C (5) - the possibility of comparing these industrial samples with in-situ samples, leading on to the additional concept of a difficulty rating for the tool being lined;
- D (6) 'a la carte' definitions of treatment procedures;
- E (8) the concept of measures specific to unshaped refractories, e.g. thermal conductivity;
- F (9) as a corollary to the four preceding comments, the need to have access to new documents covering the presentation and definition of these products.

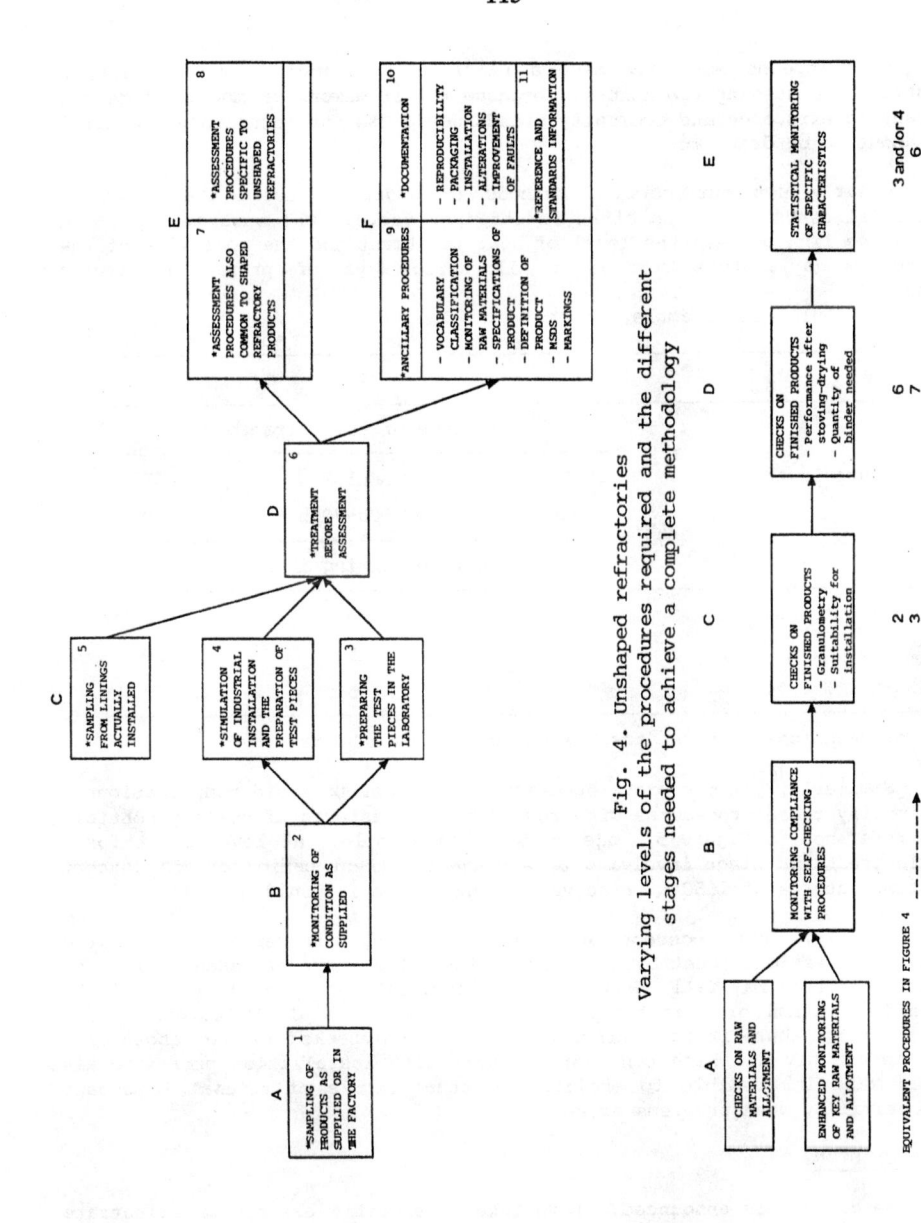

Fig. 4. Unshaped refractories
Varying levels of the procedures required and the different stages needed to achieve a complete methodology

Fig. 5. Unshaped refractories
Stages in the quality control process

We Europeans must now make a start on this work, predominantly a question of assessing the issues, because PRE framework recommendations are already in existence and Community standards (a CEN for refractory products) are about to be drawn up.

3. **Clearly set out certain methods** in order to avoid the snare of artificial assessment, so often rejected by users. One such method would retain the link between the level of heat treatment and the real life of the refractory (a feature common to all three ways of preparing samples (Figure 4).

One could, for example, adopt:

	1	2	3
		Lifetime of the refractory	
Temperature level	SHORT 50h	MEDIUM 50h-700h	LONG 700h
		Length of firing in hours	
800 - 1200°C ...	-	50	100
1200 - 1600°C ...	5	25	50
1650°C (*)	5	5	-

(*) For technical reasons, the maximum period is 5 hours.

Similarly, it may prove necessary - in order to avoid complications - to simplify these procedures with regard to some aspects of quality control. With reference to Figure 5, one could, for example, at level E opt for a single treatment stage involving an average treatment period of 25h (except for temperatures of 1650°C or above, in which case it would be 5h).

4. **Finally, this concept of overall quality, as represented by the installed material, must be integrated with other forms of assessment.** The extent that this may fall short of the optimal laboratory situation could be defined in terms of variation. For example: C (3)-C (4) compared with C (3)-C (5), where it is known that C (4) can be checked in the laboratory. This is one way in which considerable work with installation professionals should make it possible to abolish the usual circle of refusal to accept responsibility when problems arise.

4. CONCLUSIONS

We could, as announced, have taken a specific example to illustrate our argument - as with the work done to promote the classification of unshaped refractories in terms of their behaviour at limit temperatures. May we suggest that you consult the text set down as the draft standard AFNOR NF B 40-380, which we hope will one day become a European standard.

Rather than conclude by quoting a major philosopher "science without conscience merely ruins the soul", we prefer "without a sound basis, methodology merely hinders".

Our methods, shaped into standards, form a language whose words must reflect reality.

REFERENCES

1. Evolution and present situation of unshaped refractories in Japan-MASUMI TODA-SHINAGAWA, TECHNICAL REPORT No.32, 1989.
2. Les Non-Faconnes réfractaires à haute performance en Acierie, B. CLAVAUD, P. MEUNIER and J.P. TARGE, Colloque SFC, 24/25 April 1989, Paris.
3. Draft standard AFNOR NF B 40-380.

DATABASE ON REFRACTORY MATERIALS FOR IRON AND STEELMAKING

EMILIO CRIADO*, ANDRES PASTOR*, ROSA SANCHO**
*Instituto de Cerámica y Vidrio,
CSIC, Arganda del Rey, Madrid, Spain
**Instituto de Información y Documentación en Ciencia y Technologia,
CSIC, Madrid, Spain

Summary

A general view of the scientific and technical information provided by the most important international databases related to the properties and performance of refractory materials for steelmaking processes was made.
The following databases have been analyzed:

CERAB (USA), CHEMABS (USA), COMPENDEX (USA), METADEX (USA-UK), NTIS (USA), PASCAL (France).

The information has been reviewed for the period 1980-1987, taking into account the notable technological changes in refractory materials that occurred in the late 1970s.
The 2464 references studied have been classified as follows: Papers 1452 (59%), Patents 817 (33%), Congresses 160 (6.5%), Others 35 (1.5%). The references have been published according to both the type of refractory materials and the steelmaking process in which they were used. A specific index containing more than 400 entries facilitates access to any reference required.

INTRODUCTION

The iron and steel industry remains the most important consumer of refractory materials (60-65%) in the developed countries, despite the large scale introduction of new processing techniques, in particular new smelting and casting processes. Progress in the refractory industry in new materials technology during the past 20 years has been due largely to the drive of the steel industry and the strong international competition among steel producers. Their need to produce steel faster, more cheaply and more cleanly has forced refractory producers to develop materials capable of keeping furnaces in operation for longer periods. Demand has led to the increased use of unshaped refractory materials as compared to shaped pieces, particularly derived from the high price ratio, 1-5/10, existing between monolithic and shaped materials. The share of unshaped refractory materials used in the iron and steel industry and foundries now accounts for 80-85% of the total production of such materials.
Although this development has been due primarily to the efforts to increase the life cycles of equipment for steel furnaces, it was also caused by the ability of all refractory users to perform lining maintenance on kilns and custom shape production using new refractory products.

One of the biggest changes has effectively gone unnoticed, overshadowed by other ceramics developments, that is, the transition of the industry from tonnage production to one that specialises in the large scale custom fabrication of high technology products.

Most of the high technology materials going into refractory products include SiC, Si_3N_4, Sialon, BN, AlN, particulate composites including graphite, fibrous composites and cementitious materials. A large proportion of these high technology materials is already being used by the iron and steel industry, which in this way helps test their reliability under actual operating conditions.

1. OUTLOOK FOR THE REFRACTORIES INDUSTRY

The development of the refractories industry will remain tied to the steelmaking industry, despite the other major emergent or conventional consumers (cement, petrochemical, energy, etc.). Expected worldwide steel production and other major factors affecting it are shown in Figure 1.

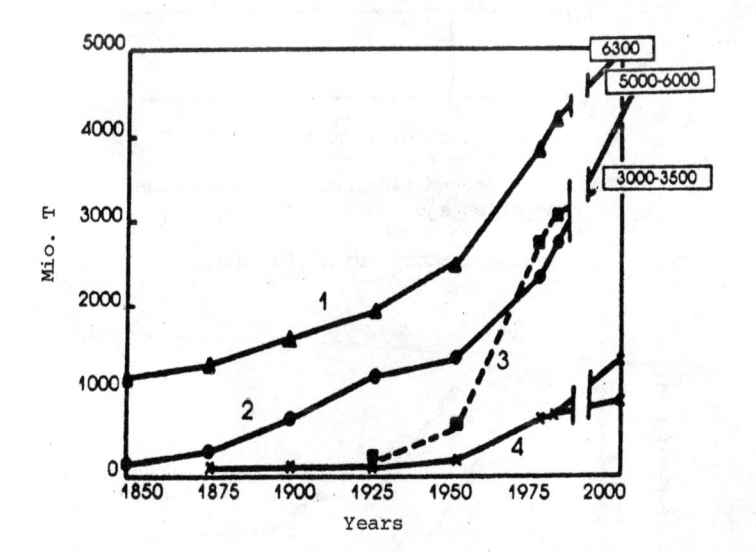

1. World Population
2. Coal Production
3. Oil Production
4. Crude Steel

Fig. 1. Forecast for some significant magnitudes

Source: II International Congress on Refractories, Japan 1987

European studies on the perspectives for a worldwide crude steel production of 750-800 million tons, forecast a refractory specific consumption and production as shown in Figure 2. As we approach the end of the 1980s the figures are close to the 15 kg per ton crude steel, i.e. about 9 kg brick and 6 kg monolithic.

A near levelling off in the decrease in specific consumption in the near future according to Japanese sources is reflected in Figure 3. In other words, the reduction of refractories unit consumption in the 1970s was brought about by changes in steelmaking processes, especially by the increased percentage of continuous casting. The reduction in the 1980s is mainly the result of adopting magnesia carbon bricks in converters. However, the rate of reduction was slower in the 1980s, that is 0.4 kg/t/yr as opposed to 1.5 kg/t/yr in the 1970s.

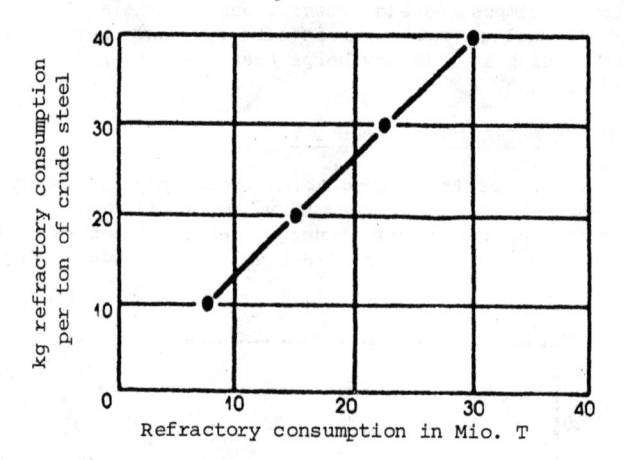

Fig. 2. Expected tendency in refractories production and specific consumption

Source: II International Congress on Refractories, Japan 1987

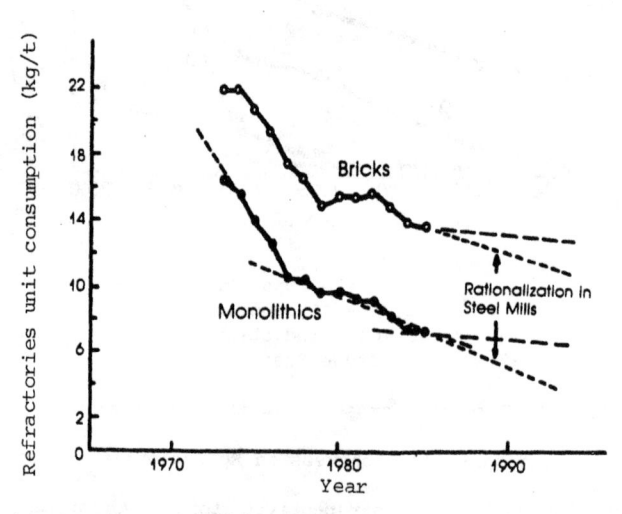

Fig. 3. Forecast for specific refractories consumption

Source: II International Congress on Refractories, Japan 1987

From the late 1980s to the 1990s it is expected that the refractory brick unit consumption will fall only 0.1 kg/t/yr, while the monolithic consumption will maintain its present levels. But if rationalization and concentration of the steel industry continues, the specific consumption of refractories will decrease by more than 0.4 kg/t/yr.

In worldwide terms the refractory industry is fairly small when compared to other industries, particularly the chemical and automotive industries. Refractory production plants are usually small or medium sized enterprises. There are about 300 refractory firms in the European Community, with approximately 35 000 people employed, about 14% fewer in 1985 than in 1980. The USA refractory industry comprised more than 150 companies with 13 000 employees. Nearly 45 plants have been closed during the past 10 years and overall industry employment has dropped drastically from 22 000 workers in 1980. In Japan the number of companies in the past 10 years has dropped from 95 to 79, and employment has fallen from 17 000 to 12 000 workers in the same period.

To face this situation there is a general agreement on the actions taken by the most developed countries:

- A process of concentration through a series of mergers and take-overs has made the refractory industry structure highly concentrated with larger diversified refractory producers dominating a major percentage of their domestic markets.
- The great majority of the bigger companies are enhancing the effectiveness of their research and development tasks and since the possibilities of rationalizing refractory and steelmaking production operations are feasible to a large extent, there is a clearly increasing tendency to high performance refractories.
- The majority of the small companies concentrate their efforts towards specific market sectors.

At this stage, the technical support which can be given by a higher level of information exchange should not be overlooked. The aim of this report is to make a contribution in this area.

2. PURPOSE OF THIS STUDY

The scope of this work was to produce a comprehensive collection of references classified by refractory material type or by steelmaking process covering the most usual worldwide information sources about this subject. This work was also intended as an evaluation of the existing databases concerned with this particular field of technology.

In this way, a general review of the scientific and technical information published on the research and performance of refractory materials employed in steel and steelmaking processes during 1980-1987 was carried out. The time frame was chosen on account of the important changes taking place in specific consumption of refractories in the steel industry, as shown in Figure 3.

As a first step, all bibliographic references to papers published in periodical or non-periodical format were retrieved by access to the appropriate databases. More than 3300 references were retrieved.

3. DATABASES USED

The subject matter comprised two technological branches, iron and steelmaking and refractory materials. There was no specific database comprising both subjects, so it was necessary to employ six databases closely related to the topics, namely:

CERAB (Ceramic Abstracts) USA 1980*
CA (Chemical Abstracts) USA 1967*
COMPENDEX (Engineering Index) UK 1969*
METADEX (Metals Abstracts) USA-UK 1969*
NTIS (National Technical Information Service) USA 1962*
PASCAL (Bulletin Signalétique) France 1973*

* Available from

These databases contained information furnished by the following sources:

- Scientific and Technical Journals;
- Meetings, Workshops and Seminars;
- Patents;
- Technical Reports, Books and PhD Thesis.

Searching was performed through the ESA/IRS host in Frascati (Italy), except for CERAB, which was searched through the Pergamon Infoline host in London.

3.1 Bibliographic Searching Strategy

Three sets of key words were combined:

A) Refractory Material Types
B) Refractory Materials Branch
C) Iron and Steelmaking Processes.

The key words employed are listed in Table 1.

TABLE 1. Search Strategy

MATERIAL	BRANCH	PROCESS
ALUMINA	CASTABLE*	BLAST FURNACE
ALUMINOSILICATE	CEMENT*	COWPER
BASIC	CERAMIC*	MIXER
BAUXITE	FIBER*	STEELMAKING
CHROMITE	MONOLITHIC*	COKE
DOLOMITE	REFRACTOR*	CONVERTER
GRAPHITE		FOUNDRY
MAGNESITE		LADLE
MULLITE		FURNACE
SIALON		TUNDISH
SILICON CARBIDE		CONTINUOUS CASTING
SILICON NITRIDE		SLAG
ZIRCON		HEAT TREATMENT
ZIRCONIA		

* Marked key words were truncated to recover singular, plurals and other derivative forms.

3.2 Information Retrieved

According to this searching strategy, a total of 8838 references were published during the whole period covered by the databases (1962-1987). This number was reduced to 5772 references for the selected period, 1980-1987. It is striking to note that this figure represents more than 65% of the total information contained in the databases.

TABLE 2. Retrieved references distribution by databases

Document Type	CERAB	%	CA	%	COMPENDEX	%	METADEX	%	NTIS	%	PASCAL	%	Total	%
PAPERS	139	84.7	957	46.8	238	92.9	181	73.6			568	93.4	2083	62.27
PATENTS	23	14.0	957	46.8			4	1.6	1	4			985	29.44
CONGRESSES	2	1.2	98	4.7	16	6.2	51	20.7	1	4	31	5.1	199	5.95
TECH. REPORTS			33	1.6	2	0.8	8	3.2	23	92	3	0.5	69	2.06
DISSERTATIONS			1	0.04							5	0.8	6	0.18
BOOKS							2	0.8			1	0.1	3	0.09
Total	164	4.90	2046	61.17	256	7.65	246	7.35	25	0.75	608	18.17	3345	

TABLE 3. Chronological distribution of retrieved references

Document Type	72	73	74	75	76	77	78	79	80	81	82	83	84	85	86	87	Total	%
PAPERS					1		13	94	185	184	167	169	155	201	201	82	1452	58.9
PATENTS				2	1	1	8	37	82	67	91	107	116	123	103	79	817	33.1
CONGRESSES	1		1	1	1	1	13	9	35	26	14	15	29	3	11		160	6.5
TECH. REPORTS			1				2	4	2	3	5	4	2	5	2		30	1.2
DISSERTATIONS								1					2				3	0.2
BOOKS									1					1			2	0.1
Total	1		2	3	3	2	36	145	305	280	277	295	304	333	317	161	2464	

In order to compile only the most specific and relevant references, only those containing key words in titles and descriptors, avoiding abstracts, were selected. The final number of references chosen was 3345.

The distribution of the different types of documents in each of the databases is shown in Table 2.

Chemical Abstracts published more than 60% of all the references; the other databases ranged between 1% (NTIS) and 18% (PASCAL).

After eliminating repetition between databases, the final number of references came to 2464. The chronological distribution by document type is set out in Table 3.

During the study period, 1980-1987, the number of relevant references published per year (300) climbed slowly. The low figures for 1979 and 1987 were due to delays in updating the databases.

Almost 60% of all the references were papers in scientific journals. More than 33% were patents, the number of which grew faster than the other document types.

The 1452 papers were published in 19 languages in 294 different journals in 32 countries. Only 14 journals published more than 15 papers; nevertheless two of these, Taikabutsu (Japan) and Ogneupory (USSR), published more than 200 papers in the period considered. The most productive journals are listed in Table 4.

TABLE 4. Most productive journals

Number of references 1980-1987	JOURNAL NAME, COUNTRY
215	TAIKABUTSU, Japan
205	OGNEUPORY, USSR
99	REFRACTORIES, USA
26	TAIKABUTSU OVERSEAS, Japan
21	FACHBERICHTE HUETTENPRAXIS METALWEITERVERAR-BEITUNG, West Germany
20	AMERICAN CERAMIC SOCIETY BULLETIN, USA
19	METALLURG, USSR
17	KOMPLEKSNOE ISPOLZOVANIE MINERALOGO SYRYA, USSR
17	STAHL EISEN, West Germany
16	RADEX RUNDSCHAU, Austria
16	SPRECHSAAL, West Germany
15	IRON STEELMAKER, USA
15	JOURNAL AMERICAN CERAMIC SOCIETY, USA
15	MATERIALE OGNIOTRWALE, Poland

On the basis of the 1452 papers, the most widely used language was English (32.5%), followed by Russian (25.8%), Japanese (18.6%) and German (8.5%). However, the high use of English was due in part to the cover to cover translations of Russian and Japanese journals.

Most of the retrieved patents were Japanese in origin (52.4%), other important producers were USA and USSR (6%). The journals, meetings, patents, technical reports, books and dissertations, as well as authors and institutions, are listed in the final indexes.

4. CLASSIFICATION OF THE INFORMATION RETRIEVED

The 2464 retrieved references were subsequently indexed by assigning to each the most suitable key words from the lists of refractory materials (Table 5) and iron and steelmaking processes (Table 6).

TABLE 5. Refractory materials classification

GENERALITIES
REFRACTORY MATERIALS HIGH ALUMINA,
FIRECLAY, SILICEOUS AND SILICA

High Alumina, Group I
High Alumina, Group II
Fireclay and Low Alumina Fireclay
Siliceous
Silica

BASIC REFRACTORY MATERIALS

Magnesite
Magnesite-Chrome
Chrome-Magnesite
Chromite
Forsterite and Olivine
Cordierite
Dolomite
 Calcia

SPECIAL REFRACTORY MATERIALS

Materials Based on Carbon and
Graphite
Zircon
Zirconia
Silicon Carbide
Carbides (Other than Silicon Carbide)
Nitrides and Borides
Spinels (Other than Chromites)
Materials Mixtures with
Oxides and Carbon MgO-C
  Tar, Pitch impregnated
  Polymers
  Metallic Additives
    Aluminium
    Magnesium
    Silicon
  Phosphate Bonded
Al_2O_3-C
  Tar, Pitch impregnated
  Polymers
  Metallic Additives
    Aluminium
    Magnesium
    Silicon
  Phosphate Bonded
Fusion-Cast Refractory
Sialons
Polymers
Phosphate Bonded
Tar, Pitch impregnated

Metallic Additives
Composites with fibers
  Ceramic Fiber in Ceramic Matrix
  Metallic Fiber in Ceramic Matrix
  Organic Fiber in Ceramic Matrix
  Ceramic Fiber in Metallic Matrix
Refractory Fibers
  Aluminosilicate Fibers
  High Alumina Fibers
  Others: Silica, Zirconia and Carbon
Barium and Titanium Compounds

UNSHAPED REFRACTORY MATERIALS

Joining Products
  Adhesives, Binders
  Refractory Cement
Materials for Monolithic Construction
and Patching
  Castable Refractory
    Calcium Aluminate
    Basic
    Others
  Ramming and Plastic Materials
    Ramming
  Refractory Materials for House
  Casting
    Tuyeres, Tapping Holes
    Casting Runners and Channels
  Materials for Gunning and Slinging
  Plasma Spraying Materials
  Coating Materials

BY-PRODUCTS FROM IRON AND STEEL-
MAKING WASTES

Blast Furnace Slag
  Cement
  Glass-Ceramics
  Building Materials
  Glass Fiber
Converter Slag and Others
  Cement
  Glass-Ceramics
  Building Materials
Fly Ash, Silica Fume
  Cement
  Glass-Ceramics
  Building Materials
Others: Refractories, Bauxite,
Carbon, Ferroalloys, etc.
  Cement
  Glass-Ceramics
  Building Materials

TABLE 6. **Iron and steelmaking processes**

    Blast Furnace
    Runner and Tapping Holes
    Cowper
    Mixer
    Open-Hearth Furnace
    Converter
     Porous Plug
     Lance
    Electric Arc Furnace
     Plasma Furnace
    Transport Ladle
     Torpedo Ladle
    Casting Ladle
     Closing System, Nozzle, Slide Gate
     Secondary Metallurgy
    Continuous Casting
     Tundish
     Casting System, Submerged Nozzles, Shroud
    Foundry
     Sand Molds
     Induction Furnace
     Cupola Furnace
    Mineral Ores Treatment
     Sintering, Pelletizing
     Direct Reduction
    Heat-Treatment Furnaces
    Coke-Ovens
     Walls
     Doors
    Slag
     Chemical Composition Slag
     Metallurgical Slag Process
     Slag-Refractory Reaction
    Nonferrous Metallurgy
     Ferroalloys
     Nickel Industry
     Copper Industry
     Zinc Industry
     Phosphorus Industry
     Aluminium Industry
     Lead Industry
     Other Industries

The classifications followed both the European Refractories Association Classification (PRE) and the Refractories in Iron and Steelmaking Processes Recommendations (ISO 836/68). However, the European Ceramic Documentation and other modifications have also been used for greater precision. This applies specifically to monolithic and special refractory materials, where specific items were included for mixtures of oxides and graphite, refractory materials for casting, binders and metallic additives, by-products from the steelmaking industry, etc.

General breakdowns for the references by refractory material type and iron and steelmaking process are displayed in Tables 7 and 8. High alumina and magnesite were refractory types receiving the most attention (19% and

12% respectively) with production, properties, converters and transport ladles the most frequent aspects considered (13% and 12% respectively).

TABLE 7. Classification of references by refractory material types

	No. of References	%
Generalities	97	4.2
High Alumina	442	19.1
Fireclay	21	0.9
Silica	126	5.4
Magnesite	285	12.3
Magnesite-Chrome	175	7.6
Forsterite and Olivine	17	0.6
Dolomite	93	4.0
Carbon and Graphite	95	4.1
Zircon, Zirconia	111	4.8
Silicon Carbide	143	6.2
MgO-C, Al_2O_3-C	212	9.2
Additives, Binders	89	3.8
Refractory Fibers, Composites	118	5.1
TOTAL SHAPED	**2024**	**87.5**
Unshaped	289	12.5
TOTAL	**2313**	
By-products from Iron and Steelmaking Wastes	151	6.1
TOTAL REFERENCES	**2464**	

TABLE 8. Classification of references by iron and steelmaking processes

	No. of References	%
Production and Properties	595	24.1
Generalities on Use	112	4.5
Blast Furnace, Cowper and Casting House	225	9.1
Mixer	3	0.1
Open-Hearth Furnace	32	1.3
Converter	339	13.7
Electric Arc Furnace	128	5.2
Transport Ladle	84	3.4
Casting Ladle and Closing System	294	11.9
Secondary Metallurgy	83	3.4
Continuous Casting and Tundish	167	6.8
Foundry	133	5.4
Heat Treatment Furnaces	87	3.5
Coke Ovens	38	1.5
Slags	55	2.2
Nonferrous Metallurgy	89	3.6
TOTAL REFERENCES	**2464**	

4.1 Classification of References by Refractory Material Type

Comparing the percentage distribution of published works with some representative production figures for refractory materials brings to light some interesting trends (Table 9).

TABLE 9. Classification of refractory production (x 10^3 t.)

REFRACTORY MATERIAL	JAPAN 1985	%	CEE 1985	%	USA 1985	%
HIGH ALUMINA	169	14.8	500	15.3	317	17.0
FIRECLAY	446	39.1	1014	30.9	699	37.6
SILICA	14	1.2	92	2.8	20	1.1
BASIC	270	23.7	1165	35.6	663	35.7
SPECIAL	241	21.1	502	15.3	160	8.6
Shaped Subtotal	1140	55.9	3273	65.4	1859	68.2
MONOLITHIC	898	44.0	1728	34.5	865	31.7
Total	2038		5001		2724	

Source: Japan, The Refractories Association of Japan
CEE, European Refractories Producers Association
USA, Refractories Division of American Ceramic Society

a) Refractory materials with low industrial production figures but large numbers of published works. This points to those materials involved in rapid technological change. For instance, silicon carbide, zircon-zirconia, oxides and graphite mixtures, etc. A similar market trend was exhibited by high alumina refractories.
b) Basic and silica refractory materials which displayed a more even balance between industrial production and volume of bibliographic references. This indicates technologically more mature materials.
c) Fireclay materials, which yielded fewer references because of a sharp fall in industrial production.
d) Monolithic materials, a very special case, since the low publication levels stood in contrast to a sharp rise in industrial production. This can be ascribed to competition among producers, giving rise to a certain level of secrecy within the industry.

4.2 Classification of references by iron and steelmaking processes

The distribution of references by iron and steelmaking processes is shown in Table 8.

The large number of references not attributable to a specific process may be due to the inadequacies in the indexing methods employed by the databases. Otherwise the most innovative steelmaking processes were the converter, casting ladle, closing ladle system and continuous casting. The lower number of specific references to blast furnaces is probably due to the level of development achieved by this technology.

5. CONCLUSIONS

The main American and European databases do not currently offer access to databases in the USSR, the world's leading producer of steel and refractories. The same holds true for databases in Japan, the country generating the most inventions in the field of refractory materials, reflected by the high figure (53%) of patents made in Japan.

Of the six readily accessible databases consulted, Chemical Abstracts (USA), followed by PASCAL (France), were the most productive in response to searches for information on refractory materials employed in steelmaking.

Most (62%) of the publications were articles in journals, followed by patents (30%). Papers submitted to meetings accounted for only 6%, technical reports for just 2%. The shares of doctoral dissertations and books were practically negligible.

Taikabutsu (Japan), with 215 papers, and Ogneupory (USSR) with 205, were the most prolific journals during the period considered.

The USSR ranked first in the number of journals and papers published. The Federal Republic of Germany, USA and Japan all published similar numbers of journals dealing with refractories, with Japan the second ranking producer of papers.

Reference to 87 different meetings held during the period were uncovered; a total of 160 papers were presented to the meetings. The most papers were submitted to:

- International Colloquium of Refractories (FRG)
- Steelmaking Conference (USA)
- National Open Hearth and Basic Oxygen Steel Conference (USA).

The progressive increase in the volume of information and the growing technological interest in refractory materials employed in steelmaking, combined with the widely scattered sources of the information available, are grounds for establishing a single interdisciplinary database for the compilation of all information published worldwide in this strategic field.

Session III

Chairman: G. C. PADGETT (*British Ceramic Research Ltd (BCRL), Stoke on Trent, UK*)

REQUIREMENTS FOR REFRACTORY MATERIALS IN MODERN STEEL PRODUCTION

GERHARD KLAGES
Refractory Technology Centre, Thyssen Stahl AG, Duisburg,
Federal Republic of Germany

HEINZ SPERL
Steelworks Committee and Refractories Technology
Association of German Metallurgists, Düsseldorf,
Federal Republic of Germany

Summary

Refractory materials for lining metallurgical vessels have always played an important part in raw steel production. Although their function used to be almost exclusively one of protecting the steel casings of the vessels, they have to be seen today as an intrinsic component of metallurgical processes - not least as a cost factor. The development of refractory products has been correspondingly intensive, one of the aims being to allow for the constant improvements and changes taking place simultaneously in the steel industry.

1. RAW STEEL PRODUCTION AND THE CONSUMPTION OF REFRACTORIES

As shown in Figure 1, the raw steel production of the Federal Republic of Germany has varied between 35 and 45 million tonnes since 1975. Continuous casting increased its share from 25% to 89% during this period. Open-hearth steel production fell from 6.7 million tonnes (= 16.9% of total production) in 1975 to zero in 1983.

Consumption of refractory materials showed the same trend: total consumption fell by nearly 45% from approximately 1 275 000 tonnes to 700 000 tonnes. Specific consumption declined continuously, also by about 45%, from around 32 to 17.5 kg.

The decisive factors in this drop in the consumption of refractories were better refractory properties and greatly improved iron and steel production at all levels from pig-iron plants to rolling mills. The continuous casting process played an important part in this, as did the replacement of open-hearth by electric and converter steel.

Figure 2 shows the trend in consumption of the various refractory materials for steel production in the Federal Republic of Germany. Distinctions are made between shaped and unshaped and between basic, acid and high-alumina materials.

Comparison of shaped and unshaped products shows a drop of some 45% in the consumption of shaped products and approximately 60% in unshaped.

Basic materials form the majority in both groups. Production of shaped basic items has remained at about 200 000 tonnes since 1975. Consumption of unshaped basic grouts and cements has fallen almost directly in parallel with total production.

The reasons for the drastic fall-off in refractory consumption for steel production have already been discussed in numerous publications, and

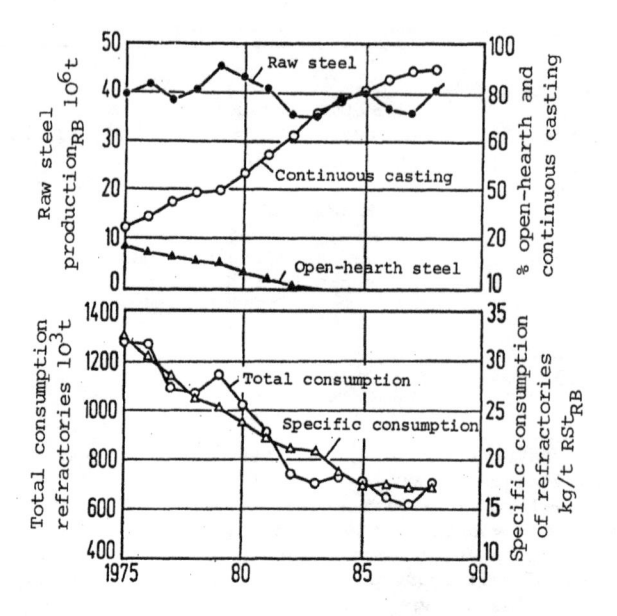

Fig. 1. Raw steel production and consumption of refractories, Federal Republic of Germany (Federal Statistical Office data)

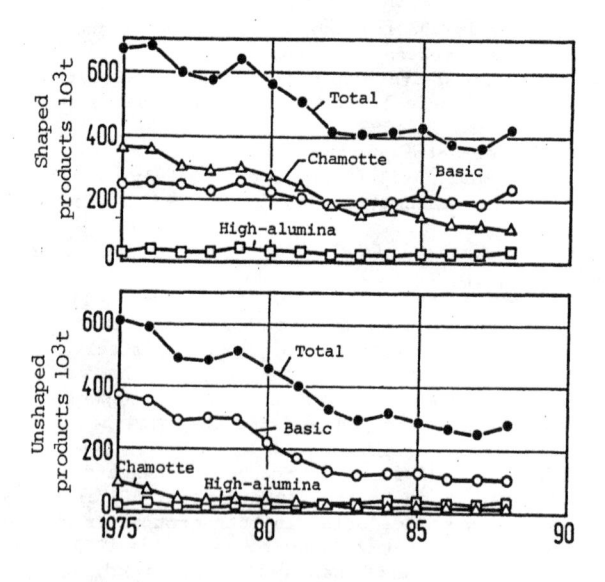

Fig. 2. Consumption of refractories for raw steel production, Federal Republic of Germany (Federal Statistical Office data)

we do not intend to examine them here. There is, however, a significant levelling out of the downward consumption curve, as may be expected, but a further drop will be almost inevitable as a result of new and improved processes and working methods and the necessary demand for greater availability of refractory-lined installations.

Until recently, linings were developed and improved empirically to meet requirements as closely as possible on the basis of the extensive practical experience of the refractory and steel industries. It is, however, becoming increasingly difficult to adapt lining techniques to new refractory products and changed metallurgical processes at short notice and without risk. One possibility seems to be a comprehensive analysis of the behaviour of installations, enabling the appropriate refractory properties, i.e. the ideal refractory products, to be earmarked. Examples are given below of proven ways of developing optimal linings by investigating stress parameters. For the sake of brevity, we have restricted this to the linings for an AOD converter and a blast furnace.

2. REFRACTORY LINING OF AN AOD CONVERTER

This first example is a refractory lining for an 80-tonne AOD converter producing only stainless steels.

Figure 3 shows a longitudinal section through the converter. The four lateral tuyeres near the base of the vessel are located at the height of the fourth ring of the wall lining, 250 mm above the floor. The free volume of the converter after relining is 33 m^3, and the bath height at 75 tonnes is 2 180 mm. The bath level is 500-600 mm higher during blowing, and sinks by approximately 800 mm from the first to the 200th casting.

At high temperatures, the dolomite refractory lining of the converter is subjected to various levels of stress from the changing oxidizing and reducing blowing conditions at high temperatures. W. Rubens (Ref.1) examined all relevant phases of the process for his dissertation, with the aim of minimizing slag attack on the refractory lining.

The refractory materials are one of the chief cost factors in the AOD process. An IISI study (Ref.2) compiled the specific consumption of refractories for AOD converters in various regions of the world in 1980 and 1982 (Figure 4). It can be seen here that more than 90% of linings in Europe consisted of dolomite with an MgO content of less than 45%, whereas significant proportions of magnesite, magnesite-chrome and chrome-magnesite bricks were used in North America and Japan. Specific consumption in North America and Japan is clearly lower than in Europe, but the lower price of dolomite neutralizes the cost difference for European steelworks.

Figure 5 shows a developed view of the refractory lining of the 80-tonne AOD converter and the levels of stress on the various zones, which are lined with refractory materials and brick lengths appropriate to the causes and degrees of wear. Wear mechanisms differ in the following two zones:

- zone 1: tuyeres
- zone 2: wall area (slag line, trunnion zone)

Even the starting conditions of the process affect refractory durability (Ref.3). Relevant factors here are the volume of blown metal slag, which should be as small as possible, its chemical composition and the Si content of the blown metal. The following measures improve the durability of the working lining:

- elimination of acid components from the slag;

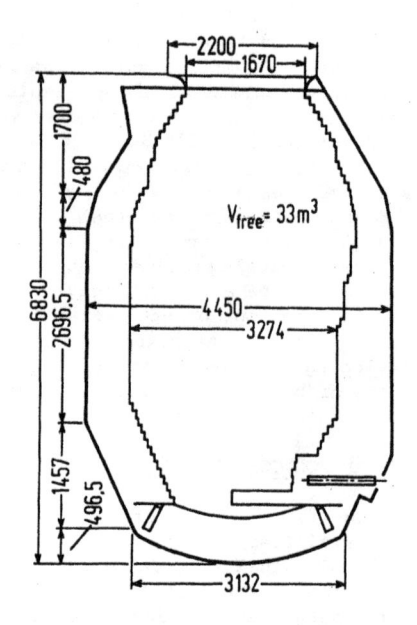

Fig. 3. Longitudinal section through the 80-t AOD Converter after relining (Ref.1)

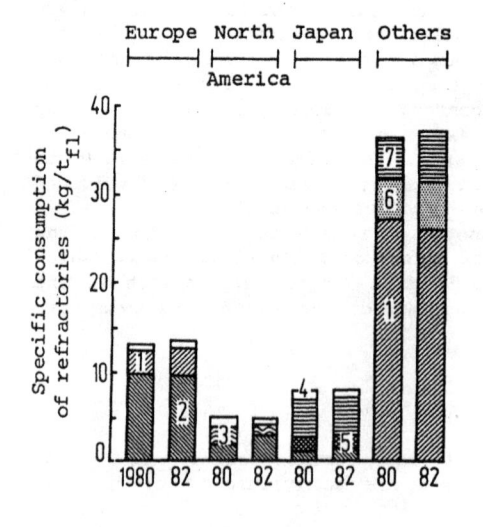

1. Pitch-bonded Dolomite brick (M_gO < 45%)

2. Fired Dolomite brick (M_gO < 45%)

3. Ceramic brick from fluxing fines

4. Others

5. Fired Dolomite-Magnesite brick (45% ≤ M_gO ≤ 90%)

6. Pitch-bonded Dolomite-Magnesite brick (45% ≤ M_gO ≤ 90%)

7. Fired Magnesite-Chrome and Chrome-Magnesite brick

Fig. 4. Specific consumption of refractories for AOD Converter by region (Ref.2)

I   Pitch-bonded Dolomite brick
II Direct-bonded Dolomite brick, fired
III Direct-bonded Dolomite-Magnesite brick, fired
IV As III, fluxing fines
V   Refractory CaO brick, fluxing fines

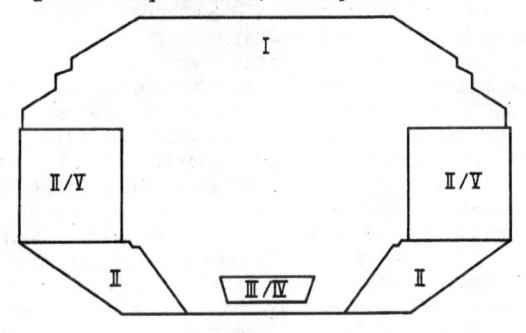

Fig. 5. Developed view of refractory lining of an 80-tonne AOD Converter (Ref.1)

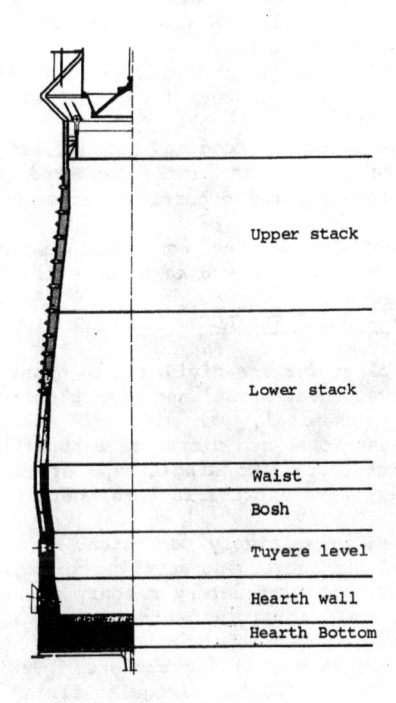

Fig. 6. Diagram of blast furnace zones (Ref.6)

- binding of the iron oxide with reactive slag of sufficient basicity;
- avoidance of excessive melt temperatures during refining;
- adjustment to high slag basicity at the end of the refining period.

The main areas of wear are the tuyere zone and the various parts of the slag zone, particularly near the trunnions, where values of up to 10 mm/melt have been known (Refs.3 and 4).

In view of the various relevant factors, a direct-bonded fired dolomite-magnesite brick is installed in the tuyere zone, where the main antagonist is slag rich in iron and chrome oxides which attack the CaO components of the brick. Tests (Ref.1) have revealed metal inclusions in the sintered dolomite approximately 5 mm beneath the hot surface of the dolomite-magnesite brick. Calcium ferrite forms in contact with the refining slag, while the periclase (MgO) is removed from the slag.

Decomposition is slower when dolomite-magnesite brick from fluxing fines is used, since the larger crystallite diameter of CaO and MgO restricts the infiltration of slag and metal.

The wall area is affected mainly by silicate slagging: slag with low CaO and high SiO_2 content infiltrates the dolomite brick. The reaction with the CaO component of the brick quickly produces C_2S and later C_3S. Infiltration does, however, cease after a certain time, since the high melting points of C_2S and C_3S increase viscosity. The infiltrated surfaces of the brick are then worn away by the movement of the bath. Chemical corrosion is followed by mechanical erosion, the attrition rate increasing with stronger bath movements, particularly turbulence during the refining phase of the next casting.

In the trunnion zone, which is prone to heavier wear, bricks of CaO fluxing fines were tried out. Silicate slag slows down decomposition because of the large size of the crystallite.

While pitch-bonded dolomite brick is used in the back wall area, ceramic-bonded dolomite brick is essential for the belly area of the wall so as to prevent recarburization of the melt.

The remaining zones, such as hood and bottom, are not problem areas in the AOD process and are lined with simple tempered pitch-bonded dolomite brick appropriate to the maximum occurring temperature of approximately 1 720°C.

The average durability of the converter is around 600 castings; specific consumption is approximately 6 kg/t raw steel.

3. REFRACTORY LINING OF A BLAST FURNACE

Figure 6 shows a blast furnace divided into upper stack, lower stack, waist, bosh, tuyere level, hearth wall and hearth bottom (Ref.5). The three main wear mechanisms - chemical, thermal and mechanical stress - are extremely complex in these zones and differ as a function of the position in the blast furnace. Furthermore, no single type of stress is ever the sole cause of wear: there are always other factors contributing to the abrasion process.

In the **upper stack**, the largely mechanical abrasion is caused by the impact of the charge and its subsequent settling in layers.

In the **lower stack**, the refractory masonry is affected mainly by the rising hot gases. The dust particles entrained with these gases have an additional sand-blasting effect on the lining. As in the upper stack, there is also mechanical stress from settling charges. Periodic heat shock and changes of temperature, e.g. from an irregular firing schedule, place the refractory material under extreme stress. There is also wear-inducing chemical attack by alkalies, lead and zinc, the circulatory materials of blast furnaces.

The stresses on the refractory material in the **waist** are similar to those in the bosh and lower stack; the waist can thus be regarded as a transition zone.

The **bosh** lining is in direct contact with liquid phases. There is mechanical stress from dropping charges, which mount up against the walls. Chemical attack occurs when the refractory material reacts with pig iron, slag, alkalies, zinc and lead. Temperature and temperature changes can be counted as thermal effects.

At **tuyere level**, oxidation by the hot blast is a further wear parameter. The thermal shock of temperature changes, which are at a maximum here, places still greater stress on the refractory lining when combined with water intrusion.

In the **hearth** there is chemical attack by pig iron, slag and the circulatory materials plus thermal shock from changes of temperature, which can alter the distribution of stresses inside the lining and cause the 'brittle zone' to expand. Flows of erosive debris, affected by changes in the state of the coke zone and by clinker coatings, e.g. of oldhamite (CaS), constitute a mechanical stress. Advancing attrition zones or 'mushrunning' are noticeable particularly in the hearth bottom/hearth wall transition area.

Current knowledge of temperature zones and reaction sequences and mechanisms enabled us to install grades of refractory brick of highly disparate composition in specific zones across the section of the blast furnace, the aim being to match types of refractory material to types of stress.

With the exception of the upper stack, where a hard chamotte grade gives sufficient abrasion resistance, two lines of approach were followed, each aiming to reduce wear: that of the refractory manufacturer and that of the blast furnace operator.

Brick grades with high heat conduction properties and a low minimum critical reaction temperature constitute the **thermal solution**, in which a sufficiently low temperature at the surface of attack inhibits wear. By contrast, the **refractory solution** involves brick linings with low heat conduction and high minimum critical reaction temperature, the aim being to minimize continuous wear.

The example we have taken is the wear pattern in Thyssen Stahl AG's blast furnace (HO) No.1 in Schwelgern (Figure 7), presented at the XXVth Refractory Symposium in Aachen (Ref.6). The main wear zones are still substantially the same today. From an adequate residual brick thickness in the upper stack, wear increases considerably towards the tuyere level and the residual brick thickness diminishes. At tuyere level, temperature changes and abrasion mean heavy wear. In the hearth, the bulging in the transition zone is clearly visible.

Many different types of brick are currently available on the refractories market. Down to tuyere level, there are graphite, semi-graphite and SiC bricks for the thermal solution. SiC brick is bonded with direct bonding components in the -SiC and/or -SiC modification, or with indirect bonding systems such as silicon oxynitride, silicon nitride and/or SiAlONs. For the refractory solution, chrome-doped mullite brick and corundum grades with the non-oxide SiAlON bonding have been developed.

At tuyere level, carbon brick and semi-graphite are installed for the thermal solution, and prefabricated components of a low-cement concrete with black corundum fluxing fines with 10% added Cr_2O_3 for the refractory solution.

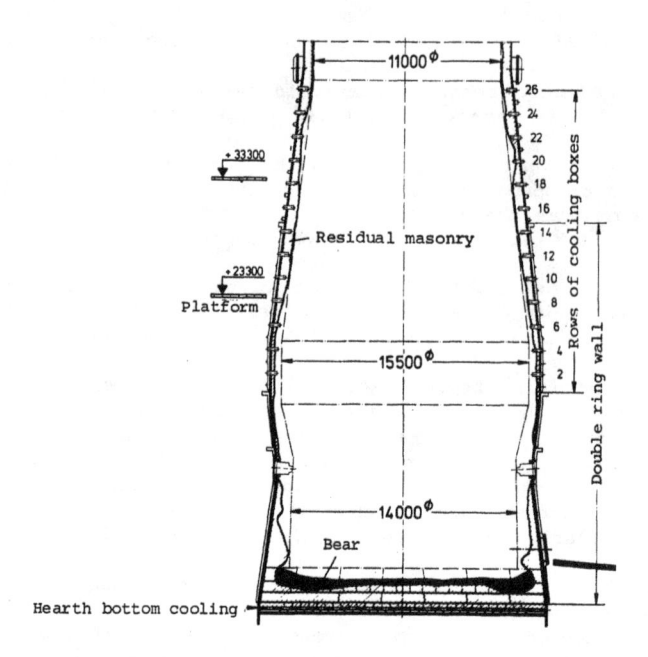

Fig. 7. Wear profile - Blast furnace I/Schwelgern (Ref.6) 12/75

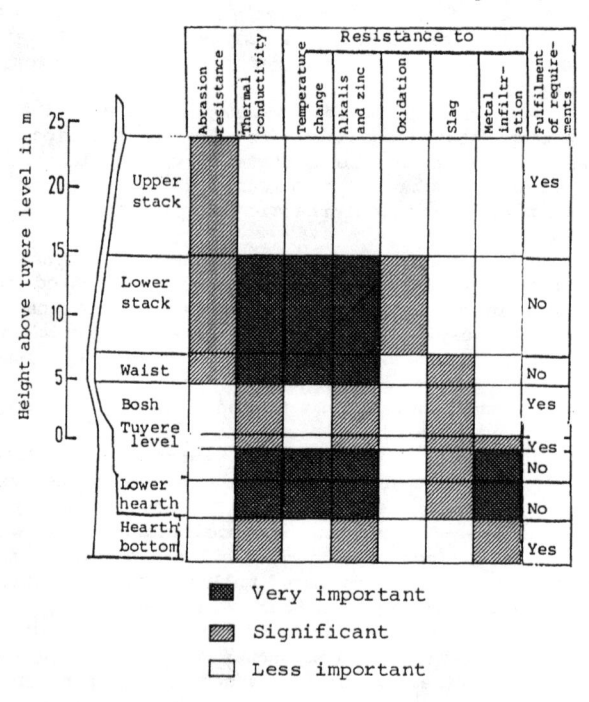

Fig. 8. Required characteristics of refractory linings in the various zones of a blast furnace (Ref.5)

The side walls of the hearth are lined either for the thermal solution, e.g. with microporous carbon brick, or for a combination of the thermal and refractory solutions.

For the combined method, a 'ceramic cup' of low-cement prefabricated components is masoned over the carbon brick. For the hearth bottom, it has proved useful to protect the usual carbon brick with a ceramic-bonded mullite layer and to raise the pool depth.

Despite these efforts, and even with the grades of brick used and improved cooling, the degree of wear in the various zones is still very variable. The aim of achieving regular wear across the entire lining of a metallurgical vessel has still not been met. The statement made by G. Winzer et al. in Aachen as early as 1982 (Figure 8) that the grades of refractories available for zone lining of blast furnaces did not yet meet users' requirements still holds good.

The combined effects of several wear parameters - chemical, thermal and/or mechanical - make it extremely difficult for the supplier to achieve all the desired ratings in a single refractory product.

REFERENCES

1. RUBENS, W. Dissertation, Techn. Univ. Clausthal, 1989.
2. IISI-Studie. Refractory Materials for Steelmaking, Brussels, 1985.
3. ARDITO, V.P. Second International AOD Conference, San Francisco, 1980.
4. ISHIDA, J., YAJIMA, T. and WARITA, T. 3rd Int. Iron and Steel Congress, Chicago, 1978.
5. WINZER, G., HÜSIG, K-R. and BESEOGLU, M.K. XXVth Int. Refractories Symposium Aachen, 1982.
6. ESCHENBERG, R., KEYK, W., KLAGES, G., KOWALSKI, W. and SMEETS, L. XXVth Int. Refractories Symposium Aachen, 1982.
7. FUSENIG, R. and STILL, G. 4th German/Japanese Refractories Seminar, Düsseldorf, 1989.

FUTURE STEELMAKING REQUIREMENTS AND IMPLICATIONS ON REFRACTORIES

DR R BAKER
British Steel Technical, Swinden Laboratories
Moorgate, Rotherham, UK

Summary

Demands continue to be placed on the steelmaker to produce a product which is cleaner, more precise in composition, more consistent in physical properties and at a cost which is competitive both with that of other steelmakers and also with that of competing materials.
In order to meet these challenges a number of developments have been introduced by the steelmaker which aim to improve control of composition and quality, increase through-process yield and reduce energy costs. The implications of the developments currently being pursued and others being contemplated on the refractory supplier are detailed in the following areas.

a) multistage operations to effect closer control of sulphur, phosphorus, oxygen, hydrogen, carbon and nitrogen levels;
b) developments in the main primary processes consistent with the objectives of tighter control of composition but also further reduction of costs;
c) developments at the continuous casting plant to reduce re-oxidation, avoid re-introduction of dissolved gases, improve flow control and save yield by increasing the level of sequence casting;
d) developments on the casting machine and between the caster and the mill to reduce the intensity of cooling, reduce temperature loss in the cast semi and increase the proportion of throughput which is hot connected to the rolling mill.

1. INTRODUCTION

Steel is a collective word for a range of materials which has traditional uses in transport, packaging, construction, consumer durables and general engineering. Steel is also expected to perform in a wide range of service conditions and environments, involving high or low temperatures, aggressive liquids or gases, and often under high and fluctuating loadings. Moreover, many of these applications demand that steel products must be aesthetically pleasing over long periods of service. Such requirements place ever increasing demands in terms of strength, toughness, corrosion resistance, fatigue resistance, strength to weight ratio, surface properties and cost if steel is to survive the challenge of other metals, ceramics and polymers in its traditional markets and also find uses in new market areas.
This challenge is being met by continual development of new steels with improved dimensional tolerance and surface properties which lend

themselves to the application of durable protective or decorative coatings and ease of fabrication by the manufacturer or user, often using increasing mechanised procedures. In order to guarantee satisfaction customers are demanding demonstration of strict quality control and assurance procedures. These demands have challenged the steelmaker to tighten his procedures, sophisticate his process route, which is now extremely complex and, in turn seek better and more assured quality from his suppliers, a major one being the refractories industry.

2. QUALITY DEMANDS ON STEEL AND IMPLICATIONS ON PROCESS ROUTE

Targets in terms of future chemical composition demands and consistency have been discussed and documented previously (Refs.1 and 2). For example, a medium term target for a pipe plate steel to resist corrosive gas conditions was quoted as S 10ppm, P 15ppm, N 35ppm, O 25ppm, H_2 2ppm with a longer term aggregate of these elements of 45ppm. Moreover, any residual oxides or sulphides would be modified in shape to reduce stress raising potential. Similarly specifications for many strip grade products have very tight limits on carbon, oxygen, nitrogen, aluminium and particularly steel cleanliness. The latter is especially important if the sheet is to be deep drawn and formed in thin section such as in DWI tinplate cans.

Such tight chemical specifications have required installation of multistage processing facilities in order to accommodate different process chemistry procedures, the use of specially prepared fluxes, insulating powders or covers and increased use of stirring or reactant gases injected through porous elements, submerged lances or via super surface jets; some of these gases also act as carriers for fluxes, alloys or solid fuels. As most of these reactions are carried out with iron or steel in the liquid state, the problem of containment of slag and steel at temperatures as high as 1750° is a demanding one. Ideally the reactor or container should be inert to the slag-metal reactions being practised. Failing this the refractory material should be stable, have a long life, have a composition which is complementary to the objectives of the slag metal reactions and not introduce detrimental species such as dissolved gases, O,N,H or harmful inclusions.

The attention focussed on the process stages to remove sulphur and phosphorus in the liquid hot metal and remove carbon, sulphur, and phosphorus in liquid steel has resulted in the pioneering of procedures such as:

(a) injection of iron oxide with or without oxygen in the BF runner to desiliconize iron to <0.2% Si, an essential precursor to simultaneous desulphurization and dephosphorization;
(b) injection of iron oxide/lime/fluorspar mixtures into the torpedo ladle, transfer ladle or spare converter to desiliconize and dephosphorize simultaneously;
(c) synthetic slag treatment of liquid steel as an alternative or supplement to iron treatment in the secondary steelmaking ladle to remove phosphorus, using lime/spar/iron oxide followed by sulphur using lime/spar (Figure 1).

These slags and the less used but more effective alternatives which contain soda ash are very aggressive to ladle refractories; particularly at the slag line and optimum procedures, refractory materials or reactor design must be developed if steelmaking costs are not to become excessive.

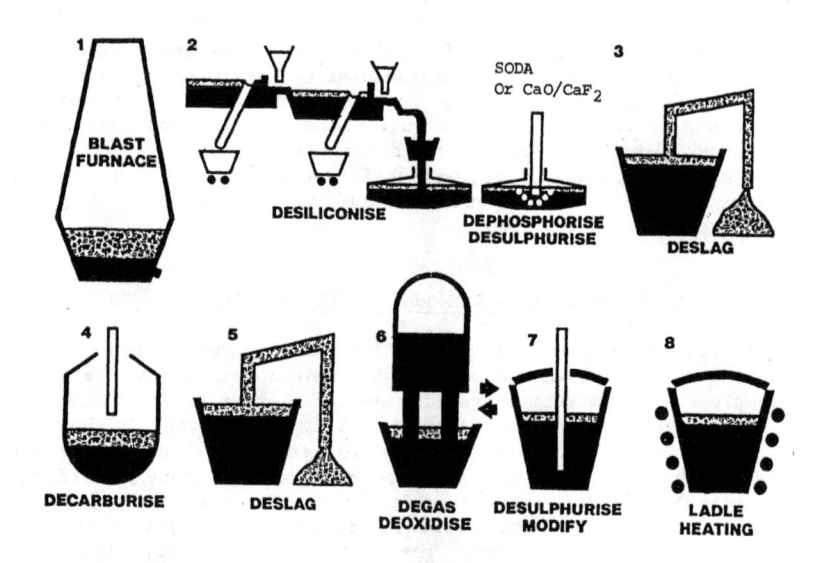

Fig. 1. Procedures for use in processing high purity steels from blast furnace iron

Figure 2 illustrates the type of wear suffered in the shoulder of a torpedo ladle in a plant where hot metal desulphurization is practised. The use of more aggressive slag mixtures will require further protection of this zone.

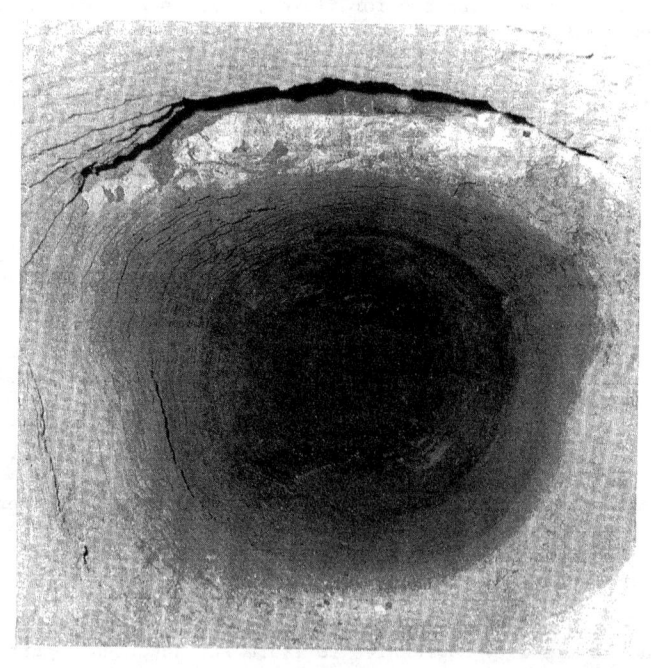

Fig. 2. Slag washing of torpedo ladle mouth

Procedures in the main oxygen steelmaking vessel itself have also been changed to accommodate changes in hot metal composition and requirements for tighter control of carbon, nitrogen and phosphorus. If pre-refining for removal of silicon, phosphorus and sulphur has been practised, the vessel contains reduced quantities of slag and the aim will be to tap as soon as end point carbon and temperature are achieved with minimum after-blowing to avoid nitrogen increase. Alternatives are to tap at a slightly higher carbon content and complete decarburization in the degasser or introduce argon through porous elements in the vessel base to reduce carbon to low levels without excessive overblowing or nitrogen pick-up. These reduced slag procedures and processing conditions, where excessive build-up of slag iron oxide is avoided, should lead to reduced chemical attack on the oxygen steelmaking vessel. Vessel lives consistently in excess of 2000 heats should be the target. However, it is important that bottom stirring elements are carefully designed, installed and preheated to ensure they achieve a full campaign life. It is also known that the increased agitation in the melt can lead to cleaner vessel bottom and scouring in the knuckle area. Careful refractory design and installation is essential. Figure 3 shows areas of particular concern in the BOS vessel, namely charge pad, tapping breast and bubbling elements.

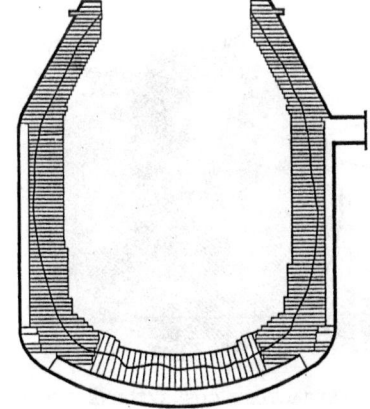

Fig. 3. Wear areas in oxygen steelmaking vessel

A particular concern in the multistage treatment of liquid metal is the prevention of slag carryover between process stages. Systems have already been developed for the holding back of slag in the steelmaking vessel at the start of tapping and detection of slag at the end of tapping so that carryover can be minimized. Such systems are vital if the efficiency of stage refining is not to be jeopardised. These methods will be mentioned in the paper by Mr Hardy and Dr Whiteley.

Currently over 50% of the refractory consumption and costs is expended in the areas of hot metal treatment and transport, the steelmaking vessel, tapping ladles and secondary treatment areas (Figure 4).

The procedures outlined above for improving steel quality will change the balance of this consumption and cost in terms of reductions in the main steelmaking vessel with problems shifting to the treatment of hot metal and the multifarious stirring, injection, degassing and heating processes post-steelmaking. The main target area will be the slag line and there is a job to be done by the steelmaker, refractory technologist and refractory designer and installer to ensure steel enhances its cost and quality competitiveness.

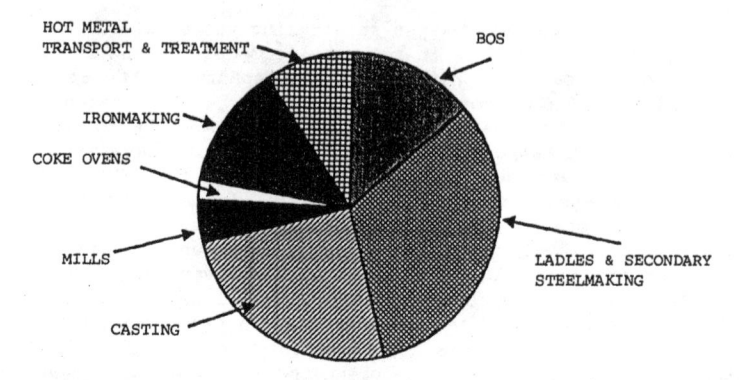

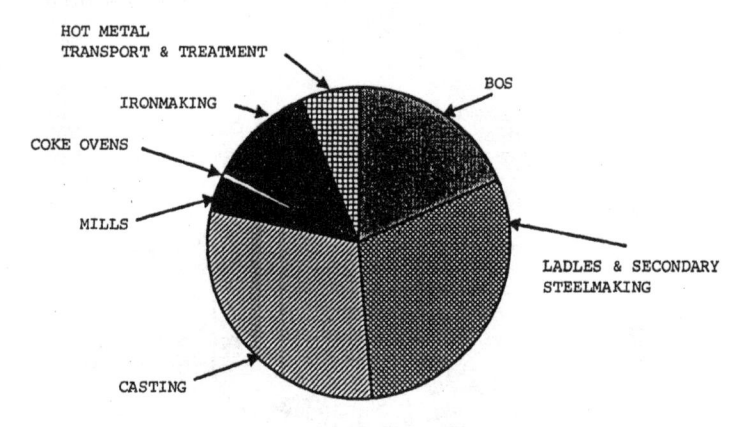

Fig. 4. Refractories consumption and costs

Figure 5 indicates the type of wear profile in a steel ladle mainly at the slag line and impact area.

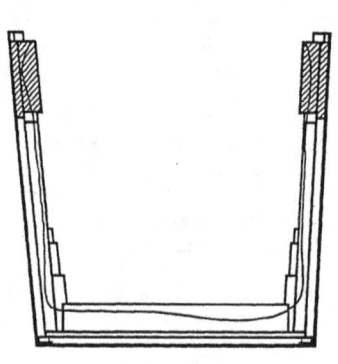

Fig. 5. Wear profile in steel ladles

The current move towards arc heating of steel in secondary steelmaking use of reactive slags and longer residence times will pose an increasing problem for the steelmaker and refractory technologist. Currently synthetic lime:alumina:silica slags are the basis of the treatment but, as lower phosphorus levels are demanded, additions of sodium silicate or spar may be required.

Once liquid steel reaches the casting machine there is little further benefit to be gained in composition improvement other than removal of inclusions and reaction products in the continuous casting tundish, although the possibility of fine trimming in the, now, large tundishes is under consideration. Tundish designs are becoming more sophisticated to ensure equalization of temperature between strands, flotation of inclusions for collection in the tundish cover powders and tundish filtration members are being developed further to improve cleanness. These components can be designed and located as transmission filters or surface collection filters and there is on-going collaboration between the steelmaker and refractory supplier to design systems which are effective in inclusion collection without causing operational problems. The materials used must also have a useful life compatible with that of the tundish.

As the proportion of steel continuously cast increases so the consumption of refractories at the caster has risen to some 25% of the total shown in Figure 4 but, as many of the components are special shapes of high grade material for flow control, e.g. sliding gates, rotary valves and nozzles, or holloware such as SENs and shroud tubes, the cost proportion is even higher at 30%. Figure 6 is a newly developed rotary valve to control metal flow, avoid air entrainment and blockages through alumina build up and metal freezing.

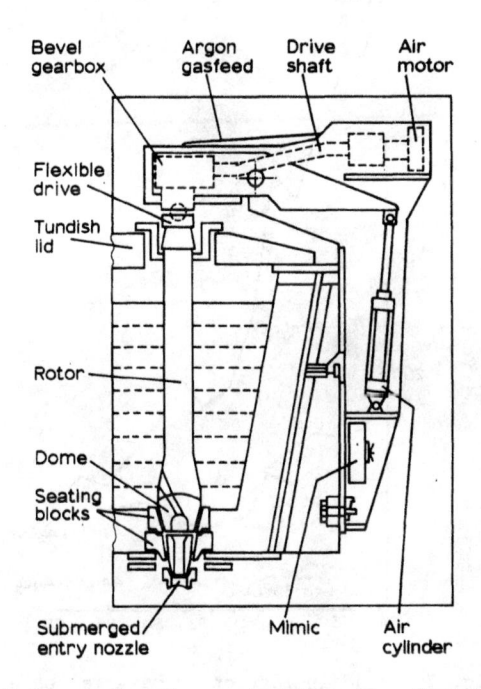

Fig. 6. Rotary valve for controlling flow from tundishes

The importance to the steelmaker of high integrity materials in the casting area is paramount not only for containment but also for lack of degradation, which would introduce exogenous inclusions into the steel, and long life, because of the importance of sequence casting as many ladles as grade commonality will allow. Long sequences save steel yield and time and allow maintenance of a plant tempo which has benefits throughout the process chain, not least being in terms of refractory consumption in the multiple stages.

In this area the tundish itself is the largest consumer of refractories in weight terms. As the working lining is renewed each sequence, every effort must be made to ensure that this is not an unacceptable source of hydrogen. The special tubes and shrouds are changed several times during long sequences using mechanical means and it is important that correct seating is achieved to avoid air leakage and reintroduction of oxygen or nitrogen into the steel. As the sulphur content of liquid steel is reduced, its propensity for picking up nitrogen is increased and the tolerability of solidified product to hydrogen is reduced. Thus the steelmaker's difficulties and his requirements for care with refractory selection and preparation are enhanced. As with ladles, the slag line area is the one where wear is most severe and as steel level tends to be maintained constant to retain the head for flow control, reinforcement of the slag line of teeming shrouds and nozzles is necessary.

In general as steelmaking has faced the problem of cost and quality competitiveness refractory consumption has followed the same trend as energy consumption and yield losses, i.e. consistently downwards (Figure 7).

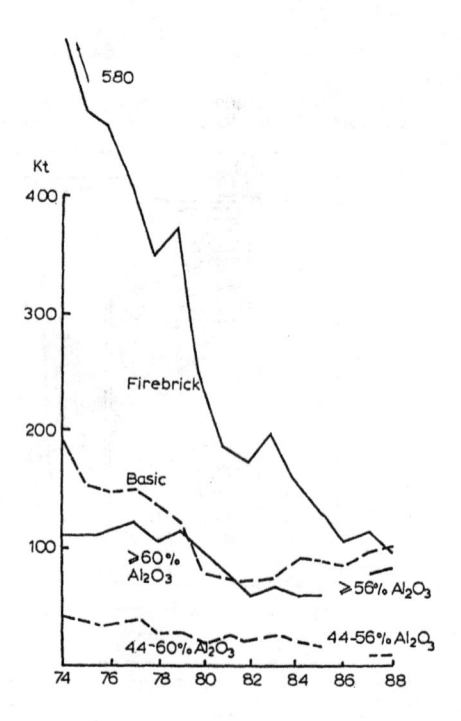

Fig. 7. Trend in refractory material sales by UK manufacturers since 1974

However, costs have not reduced to the same degree because of the move to higher grade refractories for reasons of both steel quality, long life components and continuity of operations (Figure 7). The requirement for still higher quality products has led to an increase in the proportion of steel requiring multistage refining. It is likely that there will be a levelling off in tonnage requirements but a further call for higher quality and special grades whilst the traditional process route is continued.

3. FUTURE DEVELOPMENTS

The blast furnace is likely to remain the main iron producer for many more years but, as the number of stacks have reduced, blast furnace campaign lives of 10-12 years are expected. For such lives it is necessary to have high grade refractories, highly efficient furnace cooling and operations which will reduce wall working and eliminate severe thermal and mechanical shock. A feature of such long campaign lives is intermediate blowdown to facilitate mid term spraying of the stack and bosh. Furnace operations are changing to increase the rate of coal injection and levels of 150 kg/tonne are being achieved, 200 kg/tonne are in sight and up to 400 kg/tonne of injected coal is the target. To achieve such levels high rates of oxygen injection or plasma heating of the blast are necessary. Refractory technologists will need to keep alongside operators as these developments to the traditional furnace take place.

In order to make further savings in capital and energy costs alternatives to the traditional blast furnace are being pioneered. One such process is the COREX process developed by Korf Engineering and Voest Alpine which comprises two main units, a fluidized bed melting gasifier and a counter current reduction shaft. Raw coal and directly reduced iron, from the reduction shaft, are fed into the refractory lined melting chamber which operates at pressures up to 5 bar. In the chamber the coal is dried, charred and combusted by oxygen injected through radial lances. A second such process is the Converted Blast Furnace (CBF) under development by British Steel and Hoogovens (Figure 8).

Partially reduced iron ore and coal are reacted with oxygen in one chamber of a two chamber melting unit. The molten metal flows into a second chamber for collection prior to tapping using conventional blast furnace technology. Gases from the melting chamber are cleaned and cooled to a temperature suitable for injection into the upper shaft where direct reduction of iron takes place. This configuration of equipment is so designed that it can directly substitute an existing blast furnace. Whilst it is not envisaged that new refractory developments will be required for such a process, the blend of oxidizing and reducing conditions, the flow between chambers and the maintained level in the melter gasifier may call for special selection.

Other process developers are seeking to introduce ferrous burden and raw coal into a single reactor, similar to an oxygen steelmaking vessel, to produce a liquid iron. The waste gases from the reaction of coal and oxygen are further combusted in the upper reaches of the vessel to preheat the incoming raw materials. Such a development requires good heat transfer from the secondary combustion reaction to the descending charge, without the associated refractory wear normally associated with such combustion reactions which have resulted in the demise of similar such processes in the past. Whilst refractory materials may not be able to cope with such combustion reactions a water-cooled reactor probably could. Thus for the success of such new developments, which aim to reduce the dependence of steelmaking on coke and coking coals in favour of raw coal feed, it may be necessary to use water cooling above the liquid metal line as is used in the blast furnace, electric arc furnace and newly developed energy optimizing furnace, EOF. It may also be necessary to consider recourse to such an alternative in the upper reaches of ladle arc furnaces if wear proves to be an insurmountable problem using conventional refractories.

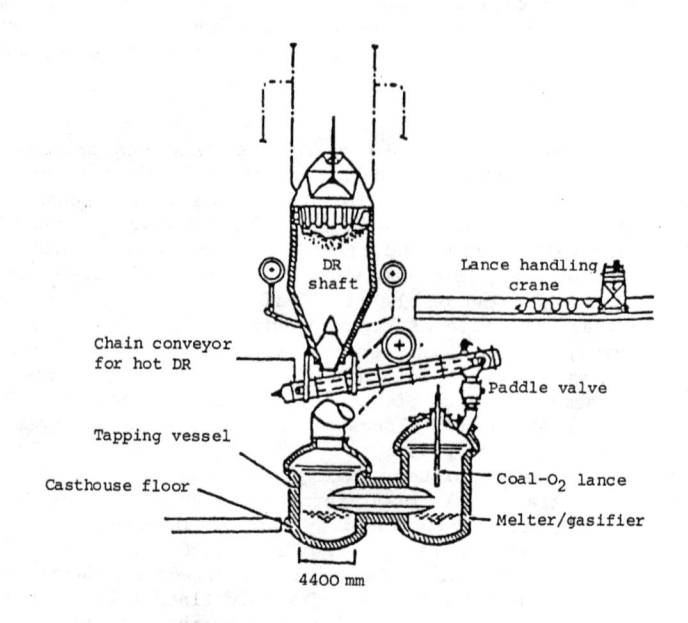

Fig. 8. Hoogovens No.3 blast furnace conversion to CBF

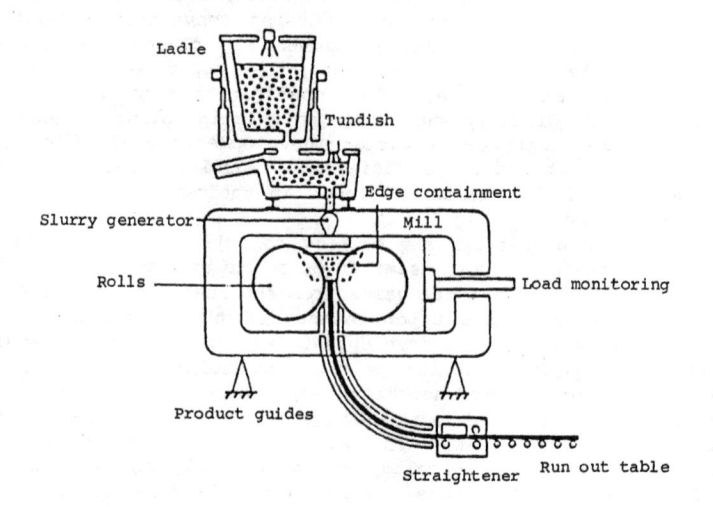

Fig. 9. Direct strip casting demonstration plant

Another objective of the steelmaker in order to reduce energy costs is that of direct linking of the caster to the rolling mill. Many steelmakers now operate a degree of hot charging, 500-800°C, where hot semis from the caster are transferred in insulated hoods or specially designed rail cars to the reheating furnace. For complete hot connection it will be necessary to introduce insulation on the caster itself, particularly at the slab edges where cooling is most severe for, until the skelp is severed from the continuous casting strand, its residence time is related to casting speed and it cannot be insulated under hoods. Requirements for successful hot connection were covered at the recent Conference on 'Achieving the Hot Link', which included design requirements on the casting machine (Ref.3). Several steelmakers are building casters in close proximity to the rolling mill to facilitate hot connection, whilst others have demonstrated that, with high speed, well insulated shuttle trains, 45 t steel slabs can be transferred between caster and mills at 1100°C such that a minimum degree of edge heating is required prior to rolling. Llanwern Works of British Steel have developed a novel stacker reheat furnace to meet their particular requirements of caster-mill linking.

Other developments aimed at eliminating the primary mill, improving cast structure and reducing energy costs are thin slab and thin strip casting. Low head vertical casters producing slab 50-60 mm thick, which is reduced to 30 mm thick during solidification on the strand, are close to application and casting of thin strip 2-5 mm thick is being researched intensively particularly for low tonnage products which are difficult to roll. Such qualities include silicon steels and some grades of stainless steels (Figure 9). The steelmaker will need to satisfy quality requirements related to segregation, central soundness, through thickness and surface qualities of steels produced by these new routes. The refractory technologies will have new challenges particularly in development of metal feeding systems, close linked tundishes, on-strand insulation or new reheating furnace configurations.

In this paper an attempt has been made to address current and future problems of the steelmaker in ensuring the success of his industry in meeting the ever more stringent quality and cost requirements of customers in the face of competition from other materials. This has resulted in lower quantities of liquid steel for each product tonne, and each tonne being considered a 'special' steel. No attempt has been made to discuss refractory materials themselves as this is the remit of speakers in later sessions. The steelmaking and refractories industries have been inextricably linked from the outset and similar reductions in tonnages but increased quality requirements apply to refractories.

REFERENCES

1. EMI, T. and IIDA, Y., Scaninjet III, 'Refining of Steel by Powder Injection', Lulea, Sweden, June 1983.
2. BAKER, R., 'Secondary Steelmaking for Product Improvement', Metals Society, London, October 1984.
3. CARR, R.A., WALTERS, J.H. and HEWITT, E.C., 'Achieving the Hot Link', Institute of Metals, London, May 1989.

EVOLUTION IN THE USE OF REFRACTORIES IN THE FRENCH IRON AND STEEL INDUSTRY

MICHEL BEUROTTE
Sollac Dunkerque, France

Summary

The French iron and steel industry has set itself the major goal of improving the quality of its products while reducing production costs. Refractories as a whole have made a contribution to reducing costs while at the same time adapting to changes in traditional steel production routes and the development of new processes.

During the current decade the consumption of refractory products in the French iron and steel industry has dropped significantly. This improvement is partly due to closure of the most obsolescent installations but also, and more particularly, to the advances made in various development fields, on which this paper will concentrate.

1. Development of research structures in the iron and steel industry.
2. Development of partner relationship with suppliers.
3. Good cooperation between operator and refractory maker and research into long-lives.
4. Installation of refractory materials.

We must continue along these lines if we wish to remain efficient and competitive.

1. INTRODUCTION

Like most industries, the French iron and steel industry has to compete on a world scale. In order to consolidate its markets and break into new ones it is developing products which are more efficient, more consistent, more advanced, concentrating on developing steels which:

- possess high mechanical characteristics;
- have improved surface properties, and
- allow for more demanding forming and assembly methods.

Metallurgical processes are constantly evolving, and the various technologies interact in a variety of ways. Advances need to be made simultaneously in all fields therefore.

Of course, the refractory maker is directly affected by this process. He has to satisfy the needs of his customer - the operator who produces the hot metal, the steel, the semi-finished product, i.e. the person who finally sells a product or a service.

The refractory maker needs to manage the 'refractory process' in close conjunction with the 'metallurgical process', which induces him to use:

Another objective of the steelmaker in order to reduce energy costs is that of direct linking of the caster to the rolling mill. Many steelmakers now operate a degree of hot charging, 500-800°C, where hot semis from the caster are transferred in insulated hoods or specially designed rail cars to the reheating furnace. For complete hot connection it will be necessary to introduce insulation on the caster itself, particularly at the slab edges where cooling is most severe for, until the skelp is severed from the continuous casting strand, its residence time is related to casting speed and it cannot be insulated under hoods. Requirements for successful hot connection were covered at the recent Conference on 'Achieving the Hot Link', which included design requirements on the casting machine (Ref.3). Several steelmakers are building casters in close proximity to the rolling mill to facilitate hot connection, whilst others have demonstrated that, with high speed, well insulated shuttle trains, 45 t steel slabs can be transferred between caster and mills at 1100°C such that a minimum degree of edge heating is required prior to rolling. Llanwern Works of British Steel have developed a novel stacker reheat furnace to meet their particular requirements of caster-mill linking.

Other developments aimed at eliminating the primary mill, improving cast structure and reducing energy costs are thin slab and thin strip casting. Low head vertical casters producing slab 50-60 mm thick, which is reduced to 30 mm thick during solidification on the strand, are close to application and casting of thin strip 2-5 mm thick is being researched intensively particularly for low tonnage products which are difficult to roll. Such qualities include silicon steels and some grades of stainless steels (Figure 9). The steelmaker will need to satisfy quality requirements related to segregation, central soundness, through thickness and surface qualities of steels produced by these new routes. The refractory technologies will have new challenges particularly in development of metal feeding systems, close linked tundishes, on-strand insulation or new reheating furnace configurations.

In this paper an attempt has been made to address current and future problems of the steelmaker in ensuring the success of his industry in meeting the ever more stringent quality and cost requirements of customers in the face of competition from other materials. This has resulted in lower quantities of liquid steel for each product tonne, and each tonne being considered a 'special' steel. No attempt has been made to discuss refractory materials themselves as this is the remit of speakers in later sessions. The steelmaking and refractories industries have been inextricably linked from the outset and similar reductions in tonnages but increased quality requirements apply to refractories.

REFERENCES

1. EMI, T. and IIDA, Y., Scaninjet III, 'Refining of Steel by Powder Injection', Lulea, Sweden, June 1983.
2. BAKER, R., 'Secondary Steelmaking for Product Improvement', Metals Society, London, October 1984.
3. CARR, R.A., WALTERS, J.H. and HEWITT, E.C., 'Achieving the Hot Link', Institute of Metals, London, May 1989.

EVOLUTION IN THE USE OF REFRACTORIES IN THE FRENCH IRON AND STEEL INDUSTRY

MICHEL BEUROTTE
Sollac Dunkerque, France

Summary

The French iron and steel industry has set itself the major goal of improving the quality of its products while reducing production costs. Refractories as a whole have made a contribution to reducing costs while at the same time adapting to changes in traditional steel production routes and the development of new processes.
During the current decade the consumption of refractory products in the French iron and steel industry has dropped significantly. This improvement is partly due to closure of the most obsolescent installations but also, and more particularly, to the advances made in various development fields, on which this paper will concentrate.

1. Development of research structures in the iron and steel industry.
2. Development of partner relationship with suppliers.
3. Good cooperation between operator and refractory maker and research into long-lives.
4. Installation of refractory materials.

We must continue along these lines if we wish to remain efficient and competitive.

1. INTRODUCTION

Like most industries, the French iron and steel industry has to compete on a world scale. In order to consolidate its markets and break into new ones it is developing products which are more efficient, more consistent, more advanced, concentrating on developing steels which:

- possess high mechanical characteristics;
- have improved surface properties, and
- allow for more demanding forming and assembly methods.

Metallurgical processes are constantly evolving, and the various technologies interact in a variety of ways. Advances need to be made simultaneously in all fields therefore.
Of course, the refractory maker is directly affected by this process. He has to satisfy the needs of his customer - the operator who produces the hot metal, the steel, the semi-finished product, i.e. the person who finally sells a product or a service.
The refractory maker needs to manage the 'refractory process' in close conjunction with the 'metallurgical process', which induces him to use:

- materials which are more stable, in order to satisfy the need for pure and clean steel;
- more reliable refractories, making it possible to plan precisely the duration of campaigns and to schedule relinings to avoid downtime and ensure continuity of production;
- products with enhanced properties to increase the productivity of installations, which greatly depends on the longevity of linings (converters, ladles, electric furnaces, tundishes) or the heavy-duty refractories (stoppers, sealing plates, submerged nozzles);
- and finally, to optimise costs - a vital factor in staying competitive.

All these advances have contributed to reducing the specific consumption of refractories (Figure 1).

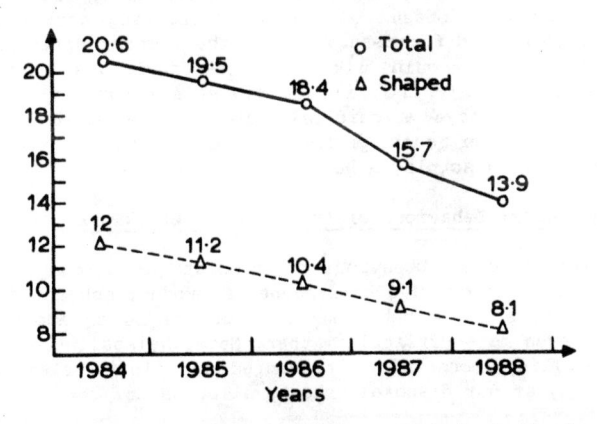

Fig. 1. Consumption of refractories kg/t steel

In order to be more efficient and more responsive to developments we have to reorganize ourselves, our working methods and our approach to problems, based on the following four elements.

2. REFRACTORY RESEARCH IN AN IRON AND STEEL COMPANY

Given that refractories are central to the steelmaking process, installation reliability and metal quality, it is important to have a consistent research policy aimed at understanding the mechanisms of wear and how they affect metal quality.

This research policy must dovetail with that which suppliers apply to their products and should help develop technical cooperation between supplier and user. In our case, our research efforts cover four main areas.

2.1 Modelling the Behaviour of Refractory Products in Service

This means finding out the main factors affecting product behaviour in a given environment. It concerns the essential properties of a product, and requires that we have as much information as possible on a product and in the form of a 'technical properties data sheet'. It also involves monitoring operating conditions, and therefore we need to make plant operators realize how they can influence product behaviour. This realization will be determined largely by how we can demonstrate the importance of certain parameters, support our case with comprehensive studies and explain cause-and-effect relationships.

One such study involving Sollac-Dunkerque's converters, and carried out by a team of plant operators, refractory specialists and suppliers, has made it possible to optimize operating conditions and product choice. Studies of this nature should be carried out in other fields of the steel industry.

2.2 Understanding the Mechanisms of Corrosion in Refractories

Corrosion, although sometimes associated with thermomechanical phenomena, is the main cause in the destruction of refractory materials. Determining the main factors governing corrosion dynamics is one of the main aims we have set ourselves, because, apart from reactions taking place under equilibrium conditions, which are easier to understand, full control of a piece of equipment more often comes from controlling corrosion dynamics. There is therefore a need for research into the factors affecting corrosion dynamics, and here I would like to cite just one example vividly illustrating what I mean: the size of magnesia crystallites of the same chemical composition plays a cardinal role in the speed at which wear occurs. And many other such parameters not found in the traditional technical data sheets also play a key role.

2.3 Thermomechanical Behaviour of Products and Brickwork

The thermomechanical behaviour of products is a complicated matter because a refractory is often a heterogeneous product subjected to stresses difficult to evaluate. The behaviour of products in an empirically built structure is an even more difficult matter. Nevertheless, studies of damaged materials, modelling constraints presented by finite elements and the computing capacity at our disposal, should allow us to:

- understand the different types of behaviour, and
- apply this knowledge to creating products (information for suppliers) and brickwork.

2.4 Potential of New Ceramic Materials

Given that the metallurgical factor is becoming increasingly sophisticated and metal must be kept free from contamination once produced, the direct cost of refractories can be negligible when compared with their indirect impact, and sometimes it pays to employ certain very high-performance materials in strategic locations. Therefore, we need to know the exact potential of certain materials, or combinations of materials, for resolving specific problems (e.g. clogging in continuous casting). Materials such as BN, AlN, ZrO_2, TiB_2, TiN, Si_3N_4 and combinations of these with Al_2O_3, C, MgO and certain metal additives need to be evaluated, and for this we require theoretical research and laboratory simulations. In certain cases, full-scale trials might be called for. However, in such cases - as happens in other less complex ones - we must guard against being carried away and going for too high a level of quality.

2.5 Technical Infrastructure Established

To do this work the French iron and steel industry has set up two teams of refractory specialists and provided the necessary resources. One team works at the IRSID in Maizières-lès-Metz and the other at the Dunkirk CRDM on the Sollac-Dunkerque works site.

These two teams are working on complementary and harmonized programmes, and are looking for support from laboratories at universities, the CNRS or major engineering colleges.

It is estimated that the amount spent on research by the French iron and steel industry is equivalent to about 1% of the direct cost of the refractories used.

To develop this research further the most advanced equipment (including computers) is either available or in the course of being obtained, making the following possible:

- thermomechanical models characterization up to 1 600°C in a neutral atmosphere;
- simulated corrosion up to 1 700°C in a controlled atmosphere;
- micro-analysis, etc.

3. DEVELOPING A PARTNER RELATIONSHIP WITH SUPPLIERS

Contrary to the position of most major European steel producers, the French iron and steel industry has no financial links with producers of refractory materials. We do not have any natural partners, and thus our suppliers are very diversified.

In the light of the rapid changes in the metallurgical process over the past few years, we felt it necessary to enter into closer relationships with experienced suppliers in order to speed progress and respond more effectively to our customers' needs.

We have therefore progressively developed a partner relationship with various French or foreign producers, and the progress achieved varies from case to case.

3.1 Partner Relationship for Quality

This is the first stage, and can be summed up as follows: each partner is responsible for his own area: the producer is responsible for the materials he delivers and the user is responsible for the conditions under which he uses them.

This might seem to be a simple idea nowadays, but it required a major change in customer-supplier relations. It was not all that long ago that we were working in a highly empirical fashion: it was very difficult to discover the cause of bad performance, nor was it clear who was responsible. Rapid progress is not very easy under such conditions.

The first step was to draw up product specifications; this involved discussions between customers and suppliers about product property data sheets with a view to:

- determining the main properties of a product at a given quality level and for a given use (these properties then being guaranteed);
- defining the limits of any authorized fluctuations from these reference values.

As a result it was possible to draw up specifications incorporating an undertaking by the supplier to submit proof of the quality of the products delivered, both in absolute value terms and as regards reproducibility.

This now means that the customer no longer has to engage in time-consuming quality control tests, since the supplier communicates the relevant data to him directly. All he needs now is to arrange a less stringent procedure to ensure that the guarantee obligations are being adhered to.

Parallel to this we developed a non-destructive ultrasonic testing method, which is simple and less time-consuming. A larger number of measurements can be carried out using this method and a sound idea of the homogeneity of a delivery obtained, which is vital for achieving the desired result. Of course, such non-destructive tests are no substitute for the other checks carried out during manufacture, neither do they give any indication of the properties of the products manufactured, but they do make it possible to check product homogeneity.

It took many months of discussion before this system was accepted by the suppliers. Some suppliers are now installing the requisite equipment and we are starting to compare results. This should allow the supplier to find out more quickly what can go wrong during the manufacturing process. Soon such ultrasonic testing will be carried out by the manufacturer, and the results will be passed on to the user in the same way as the findings of chemical analyses of products and the associated data on physical and mechanical properties.

The second stage in the partnership-for-quality scheme is to introduce Quality Assurance. This stage has only just got off the ground, and our intention is to develop it initially for heavy duty pieces of equipment and then apply it rapidly to all items.

No major difficulties should be encountered in this stage, since it is simply an extension of the previous stage and follows on logically from what we said earlier: each partner is responsible for his own sector.

3.2 Partners in Product Selection

Not too long ago it was the manufacturer of refractory products who proposed what materials should be used to solve the problems facing iron and steelmakers. Such refractory materials may have already been used at other sites or they may have just been developed, but we have not received much information on the actual service conditions.

Yet products should be chosen to suit the conditions under which they are to be used, and it is the user who has most information on this, both as regards the present and the future situation. Once presented with the facts, experts can work out the main product constraints (drawing on their store of knowledge), and determine the product properties from, and/or base their choice on, a detailed description of the product provided by the supplier.

Thus, the definition of a product evolves from cooperation between three people, the user, the expert and the manufacturer, who - on the basis of the primary materials at his disposal and his production resources - is in a position to know whether he can guarantee the required product.

This is increasingly leading to manufacture of 'special' products suited to a given use, especially in the case of heavy duty equipment. It is therefore increasingly the user who advises the manufacturer on the type of materials to be produced, which could have a major impact on supplies of raw materials in particular. It is therefore essential for relationships between the two partners to be as open as possible to enable them to adapt as quickly as possible to changes in the metallurgical process.

3.3 Partner Relationship in Research

This third stage in the partner relationship can really only be undertaken if a high level of mutual confidence exists. Exchange of results is nothing new and no divergences have occurred. Significant progress has been made on the basis of our joint accomplishments. The two partners are now convinced that this method of working will produce the quickest advances.

The research and development programme is drawn up at an annual meeting of some ten or so people from the two companies during which we discuss changes in performance, which tools are not performing as well as they should, forthcoming developments in the metallurgical process, and what improvements are required on the supplier's side, etc. Such meetings lead to very fruitful exchanges, and sometimes we are hard pressed to deal with all the items on the agenda. The research programme and the aims to be achieved are drawn up jointly. Research is currently under way, for example, on low-carbon dolomite bricks to limit recarburization, new materials for lining converters and a high-performance porous plug.

Thanks to our partnership with refractory product companies we can now cooperate in the same way as financially integrated companies. This ensures that French iron and steelmakers remain very open towards the outside. We live in a rapidly changing world where working with the same people day in, day out, could lead to sclerosis. The French iron and steel industry cooperates with various partners, and as a result the risks of it shutting itself off are minimal.

4. GOOD COOPERATION BETWEEN PLANT OPERATOR AND REFRACTORY MAKER, RESEARCH INTO LONG-LIFE PRODUCTS

4.1 Plant Operator - Contributor to Progress

It is vital for plant operators to be aware of the important contribution they can make to end performance. We require suppliers to deliver products which are as consistent and as sophisticated as possible, and we attach great importance to how they are packaged. But even materials conforming to the specifications can lead to failure if the plant operator subjects them to conditions differing from those initially intended.

Take the example of what happened at Sollac-Dunkerque's No.2 steel plant. Two identical converter linings gave 1 620 and 2 323 heats respectively in one year. What had happened in the intervening period was that the steel plant's production rate had been changed, the number of interruptions reduced and the iron content of the slag cut significantly.

The following table gives the main figures:

Number of Heats		1 620	2 323
Heats/shift		6.51	8.54
Proportion of shift with low-rate throughput		63	25
Tapping temperature	(°C)	1 659	1 662
Carbon content	(%)	0.042	0.046
Iron in slag	(%)	20.5	17.1
CaO	(kg/t)	43	39.8
Dolomite)	(kg/t)	10.1	10.4

Using the NSC statistical model to correlate converter behaviour with operating conditions gives a first-order explanation for the difference of 700 heats between the two campaigns, with slag oxidation appearing as the main factor.

OPERATING CONDITIONS	WEIGHTING IN NSC CORRELATION (in heats)	DIFFERENCE BETWEEN CAMPAIGNS	THEORETICAL SUPPLEMENTARY POTENTIAL
Heats per day (h/d)	+ 1 → + 29.8	+ 6.09	+ 181
Tapping T° (°C)	+ 1 → − 28	+ 3	− 84
Carbon content (%)	+ 0.001 → + 27	+ 0.004	+ 108
Iron in slag (%)	+ 1 → − 151	− 3.4	+ 513
CaO (kg/t)	+ 1 → + 12	− 3.2	− 38
Dolomite (kg/t)	+ 1 → + 72	+ 0.3	+ 22

What is more, use of this statistical correlation model to analyse the technical and economic performance of a converter lining makes it easier to pinpoint what is due to operating conditions and what is due to the refractory materials.

Since the plant operator is probably the person able to contribute most to the production outcome, it is vital for him to know the properties of the refractory materials he is using.

This good cooperation between plant operator and refractory maker was clearly a decisive element in improving converter lining performance at Sollac-Dunkerque and Fos (Figure 2).

Sollac-Dunkerque
CONVERTER BEHAVIOUR

Sollac Fos/Mer
CONVERTER BEHAVIOUR

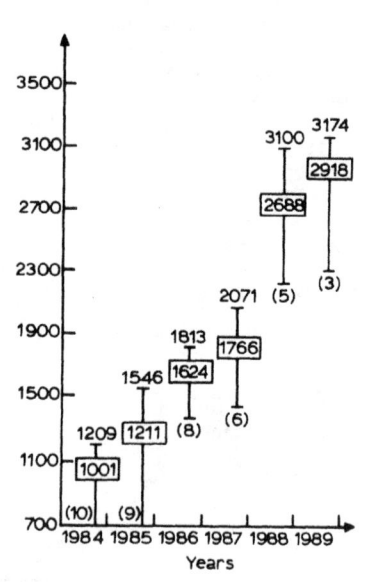

Fig. 2. Converter behaviour

The progress made has increased plant availability, thus allowing the two steel works to raise production at a time of economic boom.

4.2 Refractory Maker - Project Partner

The refractory maker must be involved from the outset in any project concerning the use of refractory materials. The equipment must be designed so as to limit the thermomechanical stresses on the refractory materials used, and increasing attention must be paid to meeting the need for pure and clean steel.

If refractory problems are taken into account when designing equipment, better solutions can be chosen and better results obtained.

5. INSTALLATION OF REFRACTORY MATERIALS

So far I have tried to show that all the different players have an important contribution to make to refractory performance, thus leading to improved equipment availability and also reduced costs sometimes, and favourably influencing metal quality and purity.

All this has been achieved by making products with improved performance characteristics and using different refractory qualities in one and the same installation to ensure that wear is as uniform as possible.

All these changes have made the task of installation teams more complicated - they need to adhere meticulously to increasingly stricter procedures to ensure that the desired performance is obtained. At the same time the French iron and steel industry is making increasing use of outside firms for installation work.

A few years ago we tried to develop 'progress contracts' with one or two firms, with the aim of fostering relations based on advantageous long-term confidence and joint interests. We wish to work in a consistent and continuous fashion with firms which, apart from having the requisite competence and experience, are motivated by a desire to modernize and improve their productivity and standard of service.

After three years of practical involvement I can only say that the results are inadequate and patchy. Each side is aware that joint interests are involved, but the technological changes made have not gone far enough.

Installation of refractory materials is a trade still involving a lot of manual work and handling. Skill and professional pride are still essential features required of bricklayers. Training was given in an attempt to make progress, to show that when modern materials are installed instructions must be followed to the letter and that any shortcomings can be very costly.

However, such training has not been sufficient to boost the bricklayer's image. This trade does not attract many school leavers, and it must therefore be modernized in order to bridge the gap separating the bricklayer of today from the bricklayer of tomorrow.

Developments in using materials which do not have to be manually handled are based not only on the advances made in making concrete and the properties obtained, but also on mechanized installation techniques, which make the job less physically demanding and make it possible to follow the required procedures to the letter. In many cases the possibility of using a material which required no manual handling for a given piece of equipment arose because the manufacturer introduced equipment making for easy and reproducible installation.

It is vital that firms installing refractory materials modernize, because their survival depends on their ability to advance in the fields of quality and productivity. And this they can only do by making the work less physically arduous and more interesting in order to attract young people, who are often a source of fresh ideas and thus of progress.

There is very likely a need to develop the same kind of partnership arrangement with such firms in order to help them restructure and progressively mechanize the work. There is no doubt that refractory makers serving the iron and steel industry should now be addressing this problem.

CONCLUSION

Refractory materials are very complex. It has become increasingly clear that they determine to a large extent the productivity of iron- and steelmaking equipment and the quality of the steel.

The advances made by French iron and steelmakers come from teamwork, with each partner contributing his know-how in pursuance of the common goal. Such teamwork must continue if we are to face up to the challenges of tomorrow.

The production results obtained at the various sites are converging. Where capacity is identical, direct costs are becoming comparable. We must now get to grips with the indirect costs of refractories, because this is probably what accounts for the difference.

Session IV

Chairman: E. MARINO (*Centro Sviluppo Materiali, Rome, Italy*)

ADVANCES IN RAW MATERIALS AND MANUFACTURING TECHNOLOGY
IN THE PRODUCTION OF REFRACTORY LININGS
FOR PRIMARY AND SECONDARY STEELMAKING OPERATIONS

P. Williams, D. Taylor and J. S. Soady

Steetley Refractories Limited
Steetley Works
Worksop S80 3EA
Nottinghamshire
England

Summary

The desirable properties required from shaped refractory products are reviewed with respect to the environment associated in primary melting and secondary ladle steelmaking processes. The most compatible raw materials for these processes are based on magnesia-carbon and doloma-carbon products. With the introduction of magnesia-carbon refractories containing natural flake graphite into Europe in the late 1970's to early 1980's the relatively low density, small crystal size magnesias traditionally used in pitch bonded and fired direct bonded bricks required further improvements related to increasing the magnesia grain bulk density, the periclase crystal size and modifying the second phase distribution and chemical composition. Current efforts are related to increasing the periclase crystal size of synthetic magnesias to above 150 μm. Natural flake graphite has been utilised in a finer flake form due to improved manufacturing techniques. Binder systems based on pitches and phenol formaldehyde resin systems continue to evolve. Modified resins have allowed the successful development of carbon bonded doloma products which have given improved thermal shock and abrasion resistant properties over fired ceramically bonded doloma brick. Health and safety legislation currently being introduced will dictate development in binder systems through the 1990's.

The actual manufacturing route for shaped basic refractory products has been advanced to include stringent quality management systems, quality training and use of statistical process control techniques as well as increased capital investment in highly automated equipment. These moves have enabled the user to receive highly consistent refractory brick shapes. Better use from refractories will result from improved supplier-user relationships.

1.0 INTRODUCTION

In developing and manufacturing shaped refractories for bulk primary and secondary steelmaking operations an appreciation of the steelmaking environment must be considered. This environment can be thought of as a combination of thermal, mechanical and chemical stresses superimposed on the common base of time. In the processing of steel these stress factors are inherent and beyond the control of the refractories manufacturer. However, the refractory manufacturer has had to continue to improve the quality, suitability and consistency of refractory products to withstand this arduous steelmaking environment and keep pace with process developments in a commercially acceptable manner.

This paper outlines an approach from which shaped basic refractories are considered from a raw material selection and manufacturing control viewpoint and some of the recent advances that have occurred, which include quality assurance management systems, statistical process control techniques and the necessary capital investment.

2.0 THE STEELMAKING ENVIRONMENT AND DESIRED REFRACTORY PROPERTIES

For converters and electric arc furnaces the thermal environment encompasses the high temperatures associated with the use of electrical energy for scrap melting and exothermic (heat generating) reactions inherent in the basic oxygen steelmaking process, the thermal cycling of vessels and resultant stresses set up within constrained circular structures.

The mechanical environment is associated with shell distortion or movement generated during vessel manoeuvres which cause tensional and compressional stress on the lining structure. The lining also has to withstand severe impact (abrasion) due to charging of scrap and erosional forces created during hot metal charging and during the blowing operation within converters.

The chemical environment generates corrosive oxidation reactions within the lining through slag development and its infiltration into the refractory hot face.

Not all sections of a refractory lining structure will be exposed to the same level of combined thermal, mechanical or chemical environments and many refractory designs allow for zoning by quality and thickness.

Table 1 summarises the desirable properties required from refractory materials associated with primary processing units and Table 2 illustrates the environment associated with the various types of secondary steelmaking activity. Ladle product refractory properties are similar to those described for the primary melting process but in addition may have the ability for the refractory lining to contribute to steel cleanliness.

Refractory Property Requirements - Primary Melting Units
Ability to withstand highly fluid oxidised slags and fume.
Able to conduct heat - improved thermal conductivity.
Resistant to slag ingress - low apparent porosity, small pore size distribution.
Abrasion resistant/high hot strength.
Resistant to oxidation reactions.
Resist thermal shock damage.
Compatible with process slags.
Low cost energy route for manufacture.

Table 1: Desired refractory properties for use in primary steelmaking melting units.

Effects of degassing, injection and stirring	Supplementary heating and vacuum
Increased residence time.	Extended residence time.
Greater turbulence and abrasion.	Turbulence and abrasion (high velocity metal/slag flow regimes).
High vessel tap temperatures to allow for heat loss during treatment.	Localised high temperatures.
	High vacuum conditions.
Reactions with basic or synthetic slags.	Reaction with synthetic slags.

Table 2: The process environment associated with ladle steelmaking operations.

Many of the desired properties outlined in Tables 1 and 2 can be achieved by refractory products based upon magnesia, doloma and magnesia-chrome refractories. The magnesia based systems are generally combined with carbon based components including natural flake graphites.

Over the past decade refractory advances have been concentrated on the development of magnesia-carbon, doloma-carbon, magnesia-chrome and high alumina products. This has resulted in a marked decline in working lining refractory consumptions and to the volume of refractories used. Table 3 illustrates the refractory consumptions typically associated with the working linings of primary steel processes and ladle systems within the United Kingdom. The rapid advances in ladle steelmaking technology in recent years has necessitated a rapid advance in ladle refractory technology where, within the overall context of steelmaking, refractory consumptions are at their highest. This is an obvious area where refractory producer and user co-operation pay dividends.

Process	Refractory Consumption kg/ts
BOS converter	1.7
EAF (sidewall/slagline/roof)	2.1
Ladles - standard ladles	3-4
- supplementary heating/vacuum	6-8

Table 3: Refractory consumption of working lining refractories related to process.

Refractory advances in the bulk steelmaking sector have continued to occur through incremental product improvements but "resources" such as raw materials, capital equipment and people have perhaps shown the greater magnitude of advance over the last decade within the constraints imposed by quality management systems, statistical process control and process capability/availability which ensure the most effective use of these resources.

3.0 RAW MATERIAL DEVELOPMENTS

   3.1 Magnesia

   A prime objective of the refractory manufacturer has been to identify and aid the improvement of cost effective raw materials. One area of raw material development that has been subjected to intense research has been that related to magnesia production.

   Magnesia raw material developments have progressed both through a natural and synthetic route and are summarised in Figure 1.

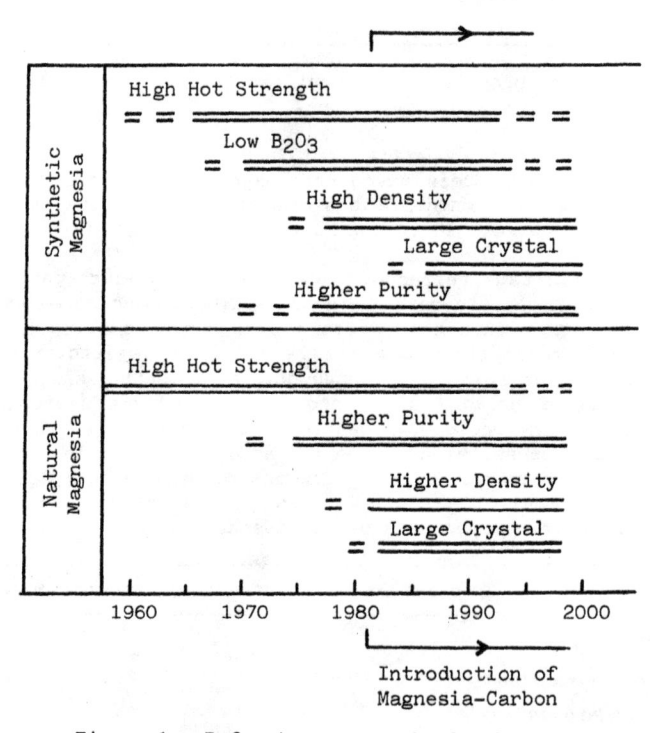

Figure 1: Refractory magnesia developments.

   Prior to 1975 magnesia raw material development efforts were concentrated largely on the removal of boric oxide from seawater route synthetic magnesias and improving the purity (MgO content) of both natural and synthetic magnesias. These developments allowed direct bonded magnesia bricks to be made with good hot strength properties for a variety of converter and furnace applications.

   With the subsequent introduction and well documented developments of magnesia-carbon refractories by the Japanese in the mid 1970's, and their introduction into Europe during the 1980's, it became apparent that the relatively low density, small crystallite size magnesias traditionally used in pitch bonded and fired direct bonded bricks required further improvements.

From a performance viewpoint in magnesia-carbon refractories several parameters of magnesia quality were identified as being critical:-

(i) Bulk density

(ii) Periclase crystallite size

(iii) Second phase distribution and chemical composition

Bulk density is considered to be significant in terms of its influence on magnesia grain porosity. The grain bulk density of magnesia rarely approaches the theoretical value of 3.56 g/cm^3 unless it is a high purity fused material. Magnesias produced commercially by a shaft kiln or rotary kiln firing route (sintered) have bulk densities ranging from 3.41 to 3.47 g/cm^3. These bulk density differences from the theoretical value are due to open (interconnecting) and closed porosity and lower density second phase components. The open porosity is about half of the total grain porosity. Slag attack occurs more rapidly along crystal boundaries and it is the intergranular porosity located at the periclase crystal/grain boundaries which is detrimental to slag resistance.

As the magnesia grain density increases, the intergranular porosity disappears to be replaced by a lower level of intragranular porosity. Density values above 3.40 g/cm^3 are normally required to achieve this effect.

As the size of the periclase crystals within the magnesia grains also increases, there is a corresponding decrease in crystal surface area and open porosity which makes the magnesia less reactive to infiltrating iron oxide rich slag, particularly in magnesia-carbon refractories operating above 1650°C. Large crystal size magnesias are normally those associated with a mean 3-d periclase crystal size >100 μm.

Often there is a proportion of closed porosity that remains trapped within the periclase crystals. This appears to arise as a consequence of the process of crystal growth where crystal boundaries are annihilated by the migration of lattice defects which create vacancies in their wake to form closed pore systems within the new crystal. This action appears to influence grain density development. A magnesia encompassing large crystal size and high density is unlikely to be formed by conventional sintering (deadburning) techniques.

Figure 2 illustrates some of the features related to open and closed porosity and crystal size development.

Figure 2(a) shows a typical standard crystal magnesia (65 μm) with clearly defined intergranular porosity. In Figure 2(b) the resultant crystal growth has trapped many of the pores within the periclase crystals (closed porosity). Figure 2(c) shows even larger crystal growth developed within a natural magnesia. Recent advances in firing technology should allow synthetic magnesias to approach this level of crystal size. However, this will be gained at the expense of a reduction in density. For completion, Figure 2(d) illustrates a high quality fused magnesia product with only part of a grain boundary showing, the periclase crystals being >800 μm in this particular example.

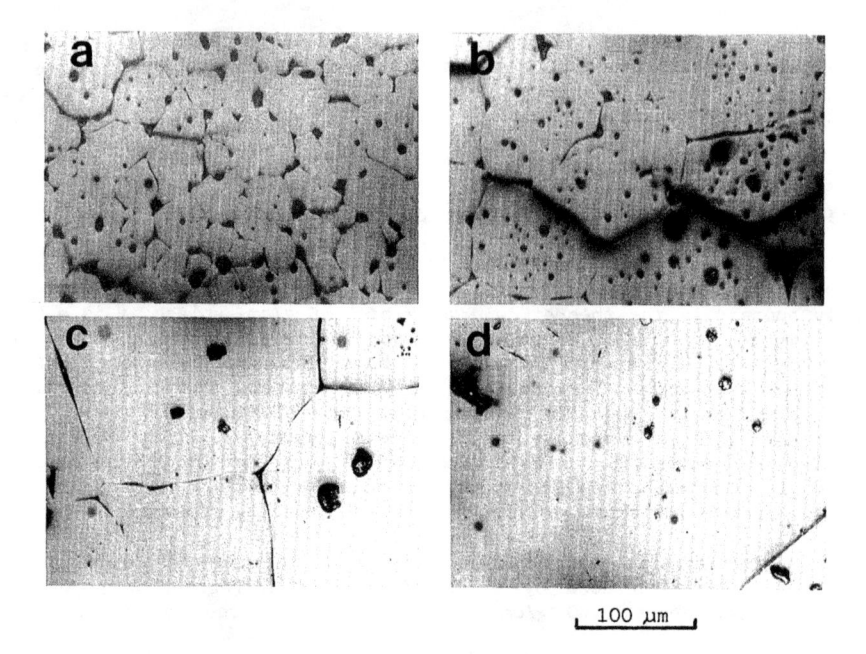

Figure 2: (a) "Standard" synthetic magnesia
(b) Large crystal synthetic magnesia
(c) Natural large crystal magnesia
(d) Fused high purity magnesia

Control of the second phase chemistry and distribution is a further development area that magnesia producers have concentrated on. By raising the lime (CaO) to silica (SiO$_2$) ratio to above 3 to 1 or producing an almost stoichiometric 2 to 1 dicalcium silicate phase, the second phase chemistry has been modified to resist slag reactions. Developments of high purity (>99% MgO) magnesia have also helped minimise the quantity of second phase material formed within the magnesia grain.

By selecting magnesias similar to those illustrated in Figure 2 it is possible to design refractories to be both cost and performance effective in primary and secondary steelmaking operations.

Efforts to produce larger crystal magnesias by conventional or shaft kiln firing technologies is likely to dominate magnesia development into the 1990's. Figure 3 illustrates the effect of increasing periclase crystal size on wear rates for magnesias of identical chemistry.

Variations in grain bulk density did not show an equivalent trend.

Magnesia raw material development in the 1990's will play a more dominant role in controlling and reducing refractory wear in full magnesia-carbon lined ladles involved in increased secondary steelmaking activities.

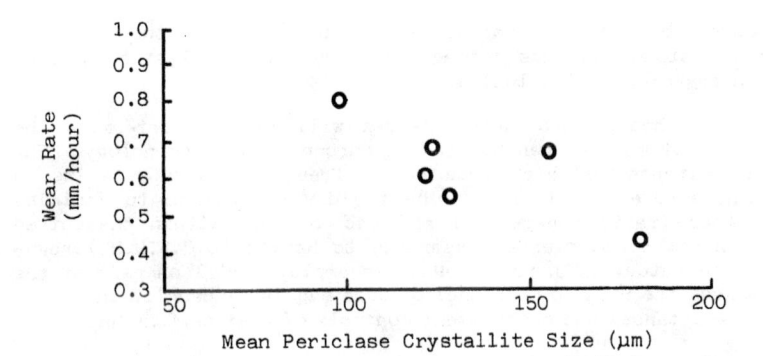

Figure 3: The relationship between mean periclase crystal size and wear rate in magnesias of the same composition based on rotary slag test studies.

3.2 Flake Graphite

Without magnesia-carbon refractory development over the past decade it is unlikely that steelplants would have been able to achieve the current rate of high steel throughput. The carbon sources used in magnesia-carbon or similar carbon containing refractory products is mainly natural flake graphite. The world's largest exporter being China. The quality of flake graphite has tended to range between 85% to 96% retained carbon, so that the influence of the aluminosilicate ash products is minimised at service temperatures. Advanced manufacturing techniques have enabled finer grades of flake graphite to be utilised instead of the more expensive coarser flake types.

Unlike Japanese producers, who are using 20% retained carbon products, many European refractory manufacturers are using only 10% to 15% carbon content products in steelmaking vessels, with 20% retained carbon products used largely in electric arc furnace slaglines and hotspots. Ladle systems using magnesia-carbon refractories rarely exceed 13% retained carbon levels in the slaglines and 5% to 8% retained carbon in sidewalls and bottoms.

Recent difficulties in supply, very high price increases coupled with greater internal consumption by the Chinese could have a retarding effect on magnesia-carbon developments. With demand outstripping supply, the early 1990's will prove to be an interesting period for flake graphite product development.

3.3 Binder Systems

Medium to high carbon containing refractories are bonded with either pitches or phenol formaldehyde based resin systems. Developments have been directed towards oxidation resistant and high strength binders. The development of new binders has allowed the advance of doloma into ladle and steelmaking refractory systems, and has increased the flexibility of basic refractory usage. Compared with the fired ceramically bonded doloma product the carbon bonded brick has proved to have excellent thermal shock and abrasion resistance and has become widely used in steel stream impact areas of secondary steelmaking ladles.

Increasingly, health and safety issues will tend to determine the types of binder systems that can be used in carbon bonded technology. The utilisation of pitches and high free phenol/free formaldehyde resins in refractory manufacture are likely to be rigidly controlled to minimise health risks to refractory personnel and end users. Within the United Kingdom the Control of Substances Hazardous to Health (C.O.S.H.H.) regulations become mandatory in January 1990. An employer will as part of the regulations assess the risk to personnel of each substance handled and keep records of the substances and measurement controls of dust or fume hazards.

4.0 MANUFACTURING DEVELOPMENTS

The manufacture of shaped refractory products has been subjected to a series of measures which have led ultimately to refractory performance improvements. These measure can be broadly defined as:-

(i) Quality Management Systems

(ii) Quality Training and Statistical Process Control

(iii) Increased Capital Investment

The most expensive raw materials and the best manufacturing equipment does not necessarily guarantee that superior refractory products will automatically be produced. One of the greatest assets a company has are its people and the quality of training and direction they receive. As part of a quality orientated direction a quality assurance management system has been introduced which conforms to the ISO 9002 international quality standard. A quality assurance management system may be integrated into shaped refractory production in a manner similar to that illustrated in Figure 4.

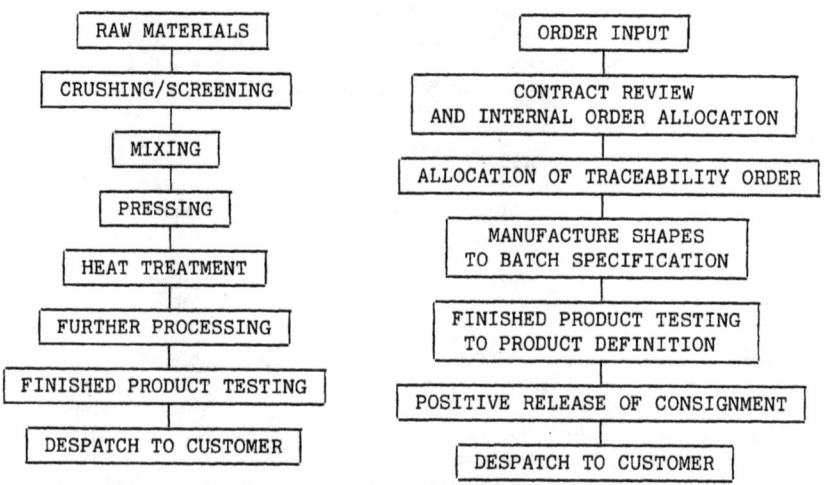

Figure 4: Shaped refractory production and the key elements to the quality assurance system.

Quality assurance is a system rather than a product/process orientated function which ensures that all areas which can affect product quality are structured to assure that the final product is always fit for its intended purpose. One major benefit to refractory users has been an improved consistency of products they receive.

One element of the quality assurance system that has involved extensive operator training is in areas relating to statistical process control (SPC) techniques. Such techniques can be applied in many parts of the production process. For example, an important acceptance test parameter which is used to positively release carbon containing products is that of "coked" apparent porosity. Customers receive carbon bonded products with apparent (open) porosities of <3.0%. After installation and preheat the apparent porosity of the carbon bonded refractory product increases due to loss of binder volatiles. In service the product may develop apparent porosity values between 10 and 15% depending upon the type of product. In order to simulate this condition on a reproducible test basis, brick samples collected under an acceptance sampling scheme are heat treated under non-oxidising conditions at a temperature of 1000°C. A typical SPC control chart is illustrated in Figure 5 for coked apparent porosity values for a batch of magnesia-carbon ladle bricks. SPC techniques can be applied to monitor the consistency of a particular product type from the period of manufacture to its performance in service.

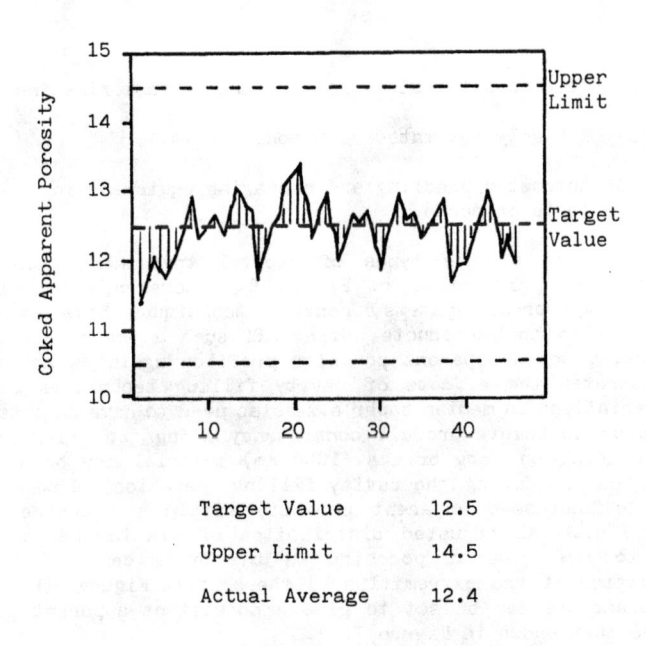

Target Value   12.5
Upper Limit   14.5
Actual Average   12.4

Figure 5: An SPC control chart showing batch coked apparent porosity for magnesia-carbon ladle bricks.

Even with the buoyant steel production at the end of the 1980's and the high demand for refractory products there has still been a need to control manning levels and improve product consistency. This action has been achieved by increased capital investment in automated processes.

(a)   (b)

Figure 6:   Example of capital investment in the refractories industry.

    (a) A highly automated 3200 tonne press.

    (b) Automatic handling and packaging equipment for ladle products.

Examples of some of the types of capital investment made in the refractories industry are shown in Figure 6. However, even with the installation of new press systems, control techniques have had to be further developed so that products coming off such a press have optimum properties. Using coked apparent porosity profiles again as an example, Figure 7 illustrates the effects of cavity filling techniques on brick properties. Variations in median taper size also need to have an optimised filling technique to ensure product consistency along the brick profile. To eliminate bending on long bricks (1000 mm) material may be withdrawn from the brick centre during the cavity filling operation. However, this action leads to increased apparent porosity within the centre of the brick, Figure 7(a). An adjusted distribution of mix in the cavity may however lead to low apparent porosity in the brick centre but higher apparent porosities at the extremities of the profile, Figure 7(b). Press operational parameters can be set to give a consistent apparent porosity profile such as that shown in Figure 7(c).

```
                    12
                    10                              (a)

Coked Apparent Porosity %
                    12
                    10                              (b)

                    12
                    10                              (c)

                    Porosity Profile
```

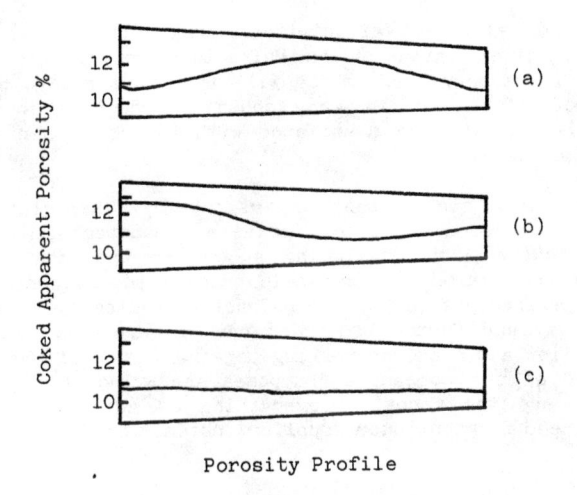

Figure 7: Development of fill techniques for a modern press to give consistent "coked" apparent porosity values along brick sections up to 1000 mm in length.

In this way, the refractory user can obtain a greater consistency of product. Better trained operators are able to carry out routine SPC measurements on single and multi-cavity presses. This feature may be seen with reference to Figure 6(a) where dimensional and weight tolerances are being checked before being plotted onto a shift control chart. Any corrective actions necessary are made by the operator to bring products back into the control limits.

5.0 CONCLUSIONS

The quest for refractory raw material improvement/advancement has been occuring continually in common with metallurgical process developments. If "raw materials" are taken to include not only those associated within the refractory product but also better trained personnel, stringent quality management systems and continued capital investments, then the advances in shaped refractories technology have been quite dramatic over the last decade.

Within the context of shaped basic refractory production, advances in carbon bonded products will progress further through developments in large crystal sinter magnesias and more user friendly binder systems for both primary and secondary steelmaking operations. With the more tightly controlled refractory manufacturing processes a much greater consistency of product has already resulted. Further efforts will be needed to reduce the relatively high consumption of refractories used in ladle steelmaking systems.

The types and relative availability of raw materials will continue to have some practical influence on refractory costs. The continued improvements in refractory performance will therefore always be tempered with cost effectiveness but in many cases a reduction in refractory consumption will generally be associated with higher rather than lower cost refractory products.

It has also been stated that the refractory manufacturer is only able to exert influence on the raw material and manufacturing control of his product. However, by developing a greater understanding of the customers process environment he can help educate the refractory user to obtain the best performance out of the refractory. Clearly this education process is two way and further advances can be made only with both the refractory supplier and customer working together. One of the statements that is included on our company's "Purposes and Principles" document is that "Customers are the focus of everything we do" which again lays emphasis on the need for continued supplier/customer commitment.

HOW DOLOMITE REFRACTORIES CONTRIBUTE TO CLEAN STEELMAKING

R.D. SCHMIDT-WHITLEY

Director of Development
DIDIER S.I.P.C.
102, rue des Poissonniers,
75018 Paris, France

Summary
    Dolomite refractories are increasingly used in the steel industry in order to fulfill demands for better steel quality. The advantages of an inert practically SiO_2-free refractory such as dolomite are described for ladle refining operations, i.e. deoxidation, desulphurization, decarburization, alloying and inclusion reduction. Environmental problems are also discussed. It is shown how clean steel and a clean environment can be obtained with newly developed dolomite refractories at an acceptable cost.

1. INTRODUCTION

Refractories based on natural dolomite have been used for over a century in Europe. For a long time, the main reason for the success of dolomite as a refractory material was its low cost compared with other available products. Over the last decade however, the metallurgical qualities of dolomite are increasingly coming to the forefront and the use of dolomite refractories is being more and more dictated by the necessities of clean steelmaking. More recently still there has been a rising concern over the environmental aspects of the use of unfired pitch bonded basic refractories, and therefore also of dolomite based products.
    Thus the demand for clean steel and a clean environment has stimulated research and lead to a series of new developments in the field of dolomite bricks and of dolomite monolithics.

2. DOLOMITE REFRACTORY PRODUCTS

There are three main groups of dolomite refractory products used for vessel linings: fired dolomite bricks, unfired pitch or resin bonded bricks, and unshaped ramming and filling materials.
    The raw material for refractory dolomite products is sintered dolomite, which is obtained from natural raw dolomite by burning. The dolomite stone is quarried, crushed, partially washed and then introduced as lumps into a shaft kiln for burning at 1900-1950°C. The resulting sintered dolomite has a total impurity content of about 2,4 % and is a mixture of minute periclase and lime crystals, figure 1.

2.1. Fired Dolomite Bricks

In the plant making fired dolomite bricks, the sintered dolomite lumps are crushed, milled and then screened into grain fractions which are stored in silos. According to the grade which is to be produced, various fractions are weighed and introduced into a mixer together with a non-aqueous temporary binder phase. After shaping and firing, bricks are checked for aspect and dimensions. Quality control and quality assurance measures are an integral part of the manufacturing procedure. The production process has been described in detail elsewhere (1).

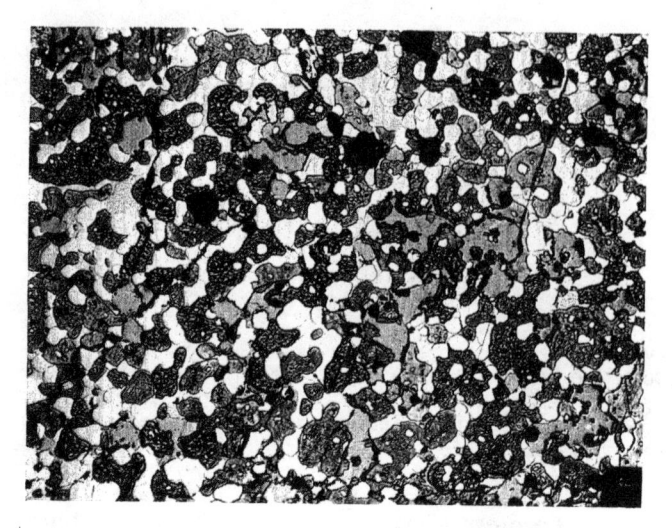

Figure 1: Natural dolomite sinter

Fired dolomite products with varying composition are available in order to give the best service performance under manifold service conditions, see table 1.

	Units	DOVAL N	DOVAL D	DOVAL NZ	DOMAG 55
Description		Standard	Dense	Dense zirconia toughened	Dense magnesia enriched
Chemical analysis					
MgO	%	39,5	39,5	38,5	55
CaO	%	58	58	56	42,5
ZrO$_2$	%	–	–	3	–
Physical properties					
Bulk density	g/cm^3	2,79	2,83	2,85	2,87
Apparent porosity	%	17	16	16	16
CCS	N/mm^2	50	80	50	45
HMOR (1400°C)	N/mm^2	12	16	14	12
Main applications		Ladle	Ladle AOD	Ladle SL AOD	Ladle SL AOD

Table 1: Properties of fired dolomite bricks.

The standard grade is used in steel ladles. For applications such as the AOD converter, where an increased corrosion resistance at very high temperatures is necessary, a more dense grade is available. Compared with the standard product, porosity and permeability are lower, the hot modulus of rupture is higher, but thermal shock resistance is reduced.

A high resistance against attack by slags rich in iron oxide or fluorspar is obtained with a magnesia enriched grade containing 55 % MgO. This grade is mainly used in ladle slag lines and high wear areas of AOD converters.

An increased thermal shock resistance of standard dolomite and of magnesia enriched dolomite is achieved by zirconia structural toughening combined with a precise control of pore size and grain size distribution. Corrosion resistance is increased due to a reduction in open porosity. Thermal shock resistance is increased due to microstructural modifications.

Fired dolomite bricks are protected against hydration by a dense sintered skin. This is why they are much more hydration resistant than pitch or resin bonded dolomite. In adverse climatic conditions it is however necessary to protect even fired dolomite bricks against hydration. Dipping of bricks in tar is effective in reducing the speed of hydration, but tar fumes during processing and heating up are environmental problems. An alternative method of protection is by dipping in wax (2). This type of treatment has the advantage of being free from side effects such as fume emission and steel contamination by carbon.

2.2. Unfired dolomite bricks

Unfired dolomite bricks are obtained by mixing dolomite grain fractions with a pitch or resin binder. After shaping, the bricks are heat treated in a tempering kiln where they obtain their special properties, table 2.

	Units	DOVAL T	DOVAL R	DOVAL R-X35	DOVAL R-X36
Description		Pitch bond	Resin bond	Hybrid resin bond	Hybrid resin bond low C
Chemical analysis					
MgO	%	39,5	39,5	39,5	42
CaO	%	58	57,5	57,5	55
C	%	2,5	3,5	2	1
Physical properties					
Bulk density	g/cm^3	2,89	2,85	2,92	2,92
Apparent porosity	%	5	6	6	6
CCS	N/mm^2	60	100	90	90
Main applications		Ladle Converter	Ladle	Ladle	Ladle

Table 2: Properties of unfired dolomite bricks.

Unfired pitch bonded dolomite bricks are the most economic dolomite product. But they contain tar volatiles and carbon, and this can sometimes create problems with environment and steel quality for ELC grades. Bricks with varying levels of carbon or magnesia additions are available for special uses.

Unfired resin bonded dolomite products are a recent addition to the range of dolomite bricks. For tempered products there are no tar volatiles and practically no fume formation during heating up and use. In many steelplant this type of brick has replaced pitch bonded dolomite due to lower fume emission and also increased life. Grades with hybrid resin bond and extra low carbon content are also used in certain cases.

2.3. Unshaped dolomite materials

A complete lining is invariably composed of shaped and unshaped products. Ramming and filling materials based on dolomite are generally obtained by mixing dolomite grain with additives and a temporary binder such as tar, resin or other organic liquids. In some cases it is necessary to avoid any temporary or carbon containing binder so dry products are also available, see table 3.

	Units	DOLSET T5	DOLSET HC5	DOLSET HP5	DOLSET H3	DOLSET S1
Description		Ramming tar bond	Ramming organic bond	Plastic organic bond	Filling organic bond	Filling dry
MgO	%	39,5	39,5	39,5	39,5	39,5
CaO	%	58	58	58	58	58
C	%	1	1	2,5	-	-
Grain size	mm	0-5	0-5	0-5	0-3	0-0,2

Table 3: properties of unshaped dolomite materials

3. CLEAN STEEL

Over the last few years, new fields of application and increased quality demands have made it necessary to improve steel properties. The most beneficial effects on steel quality are obtained by:
- precise control of deoxidation
- more efficient desulphurization
- accurate levels of alloying elements (Si, Al, C...)
- the removal of inclusions

3.1. Deoxidation

In ladle refining the precise control of the oxygen content of the steel is one of the keys to steel quality. Deoxidation requires the presence of slags with very low levels of FeO, Fe_2O_3, MnO, P_2O_5 and SiO_2. Such slags are also necessary to ensure efficient alloy additions and enable extensive desulphurization. Any reaction between the refractory lining and the steel which could lead to a rise in the steel oxygen content is likewise undesirable.

Refractory materials containing oxides of iron, chromium and silicon have a strong tendency to feed oxygen into deoxidized steel. This tendency is barely influenced by the possible existence of mixed oxides, since the instable oxide dissolves preferentially (3).

The ladle lining is a source of oxygen especially in the case of acid refractories such as silica sand or fireclay. In steel deoxidized with aluminium there is a reaction between the dissolved aluminium and the lining with reduction of silica and formation of alumina:

$$4\,[Al] + 3\,SiO_2 \rightleftharpoons 2\,Al_2O_3 + 3\,[Si].$$

But also neutral or basic refractories may supply large amounts of oxygen to the steel bath. Figure 2 illustrates the loss of Al in a 50 kg induction furnace under argon for various refractory linings (4). The Al loss is small in the case of lime or dolomite bricks, but is a great deal larger for bricks containing the less stable components Cr_2O_3 (magnesite-chrome) and SiO_2 (zircon and high alumina).

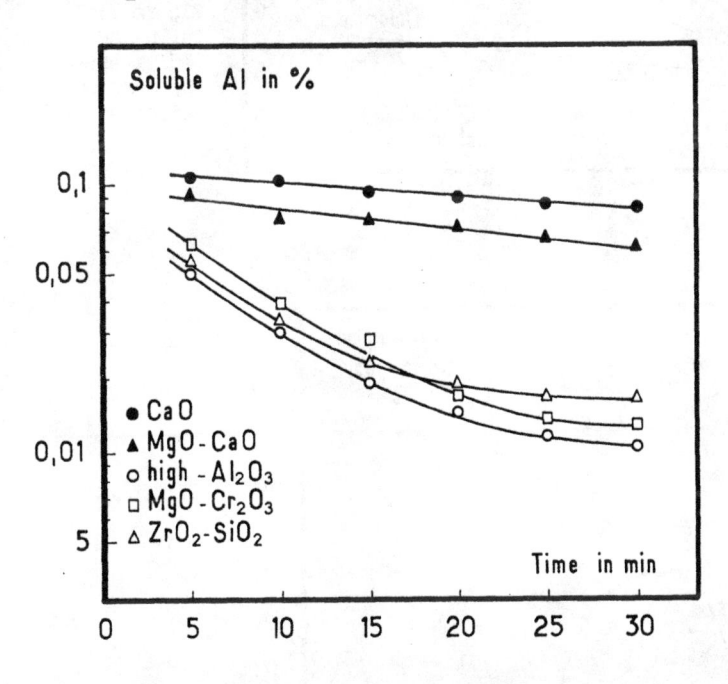

Figure 2: Influence of refractory materials on the deoxidation of steel, measured by the change of soluble aluminium with time (1600°C, 50 kg induction furnace) (4)

3.2. Desulphurization

During the desulphurization of steel, sulphur must be transferred from the melt to the refining slag. This is achieved with low viscosity slags which have a high sulphur capacity and a low oxygen potential. Lime-saturated molten calcium aluminate with some added fluorspar is a very efficient reagent for the desulphurization of Al-deoxidized steel, according to the reaction

$$3\,CaO + 3\,[S] + 2\,[Al] \rightleftharpoons 3CaS + Al_2O_3.$$

Desulphurization is generally carried out in dolomite lined ladles because desulphurization efficiency is higher than in high alumina ladles. Also considerably lower end sulphur values can be obtained (5), figure 3.

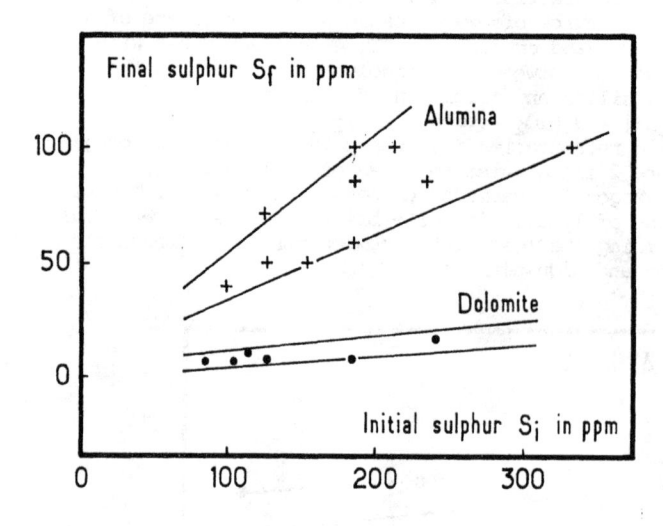

Figure 3:
Influence of the ladle refractory on the desulphurization efficiency (5)

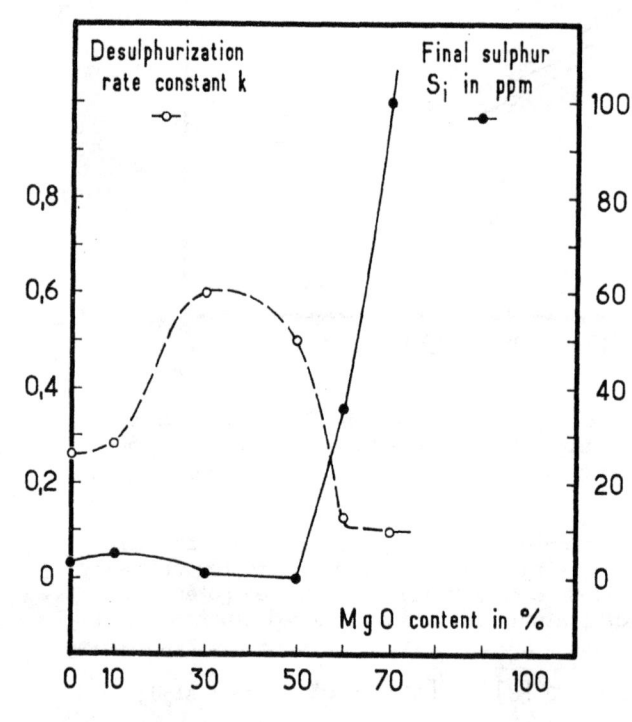

Figure 4:
Influence of the MgO content of lime-periclase crucibles on the desulphurization rate constant and on final sulphur content (1600°C, molten iron under Ar) (6)

A series of laboratory experiments with synthetic lime-periclase refractories have been recently conducted by Degawa et al.(6) in order to investigate the influence of the MgO : CaO ratio on desulphurization efficiency. In the range of periclase contents from 0 to 70 %, it was shown that desulphurization rates are highest for compositions between 30 and 50 % MgO, figure 4. This means that the naturally occuring dolomite has an ideal composition from the point of view of desulphurization. Tests with fired natural dolomite brick samples have given metallurgical results equivalent to those with synthetic products (7). The only limitation is that impurities such as Fe_2O_3, SiO_2, Al_2O_3 and Cr_2O_3 must be kept to a low concentration.

3.3. Ultra low carbon

Requirements for steel with extremely low levels of carbon are steadily growing. When producing very low carbon content steels (i.e. C < 50 ppm) it is not sufficient to remove the carbon, the carbon level must also be maintained low during subsequent refining. The sources of carbon are numerous, and include electrodes, alloys, fluxes and also carbon containing refractories.

Carbon contamination by refractories is due to the dissolution of binder carbon or added carbon in pitch bonded, resin bonded or impregnated products. Carbon pick up depends on the carbon level in the brick and on the rate of wear.

In a French oxygen steelplant using teeming ladles lined with pitch bonded dolomite, carbon pick up values of 20 to 30 ppm have been measured during refining of steel with a target value < 50 ppm. In a Japanese steelplant, carbon pick up in a ladle furnace with a refractory containing 6 % C has been determined to be 0,2 ppm/minute (8). Decarburization rates were also measured with refractories of higher carbon content. With magnesia carbon brick containing 17 % C the decarburization rate was only half that obtained with a magnesia dolomite brick containing 6 % C. Final carbon levels were therefore significantly higher for steel refined in ladles lined with the 17 % C refractory, figure 5.

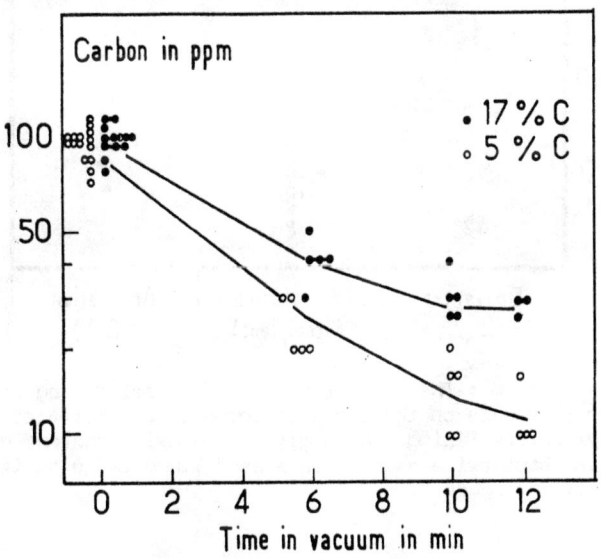

Figure 5: Decarburization rate and final carbon level for ladle slag line refractories containing 17 % and 5 % carbon (8)

Fired refractories, and particularly fired dolomite, are absolutely carbon free and are therefore used in ladles and in AOD converters when carbon pick up is to be completely avoided. Unfired dolomite brick with extremely low carbon content is now also available and can be used in less critical applications. With sufficiently low brick carbon levels it is possible to decarburize and sinter the brick hot face during high temperature treatment without significant loss in brick properties. Carbon contamination is then limited to the first few heats of a new campaign.

3.4. Other Alloying elements

Close analysis tolerances can only be obtained at a reasonable cost when refractories are used that are thermodynamically stable in contact with steel. In this respect lime, dolomite and magnesia based refractories are far superior to silica or chrome containing products. Refractories containing silica are potential sources for oxygen and silicon, whereas magnesite-chrome refractories will liberate large amounts of oxygen and chrome (9).

Silicon : When accurate and low levels of Si are needed, it is common practice to use dolomite ladles. With sand casting ladles in an LD plant, an increase of Si content from 0,005 % in the steel at the end of blowing to an average of 0,020 % in the finished steel is observed. Heats from ladles with dolomite linings show a much lower level, with an increase from 0,005 % to only 0,008 % Si (10), figure 6.

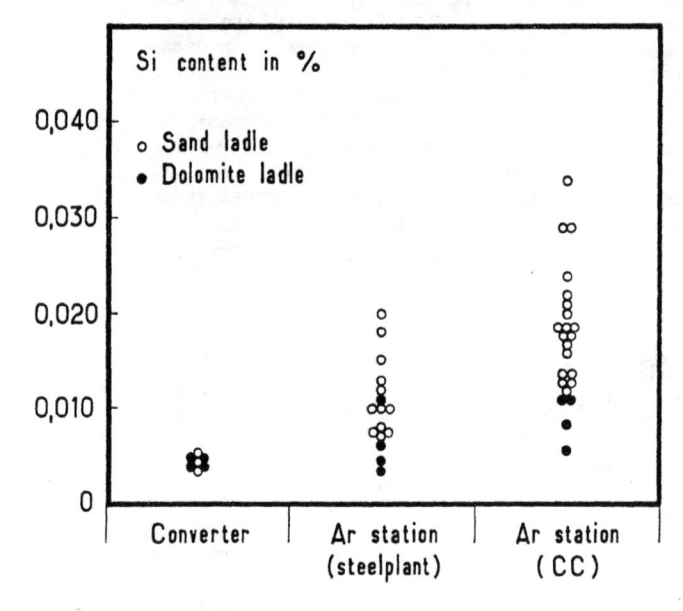

Figure 6: Influence of the ladle refractory on the silicon concentration of steel at various process steps (10)

Aluminium : The Al melting loss of steel during inert gas treatment not only depends on the top slag composition, but also on the nature of the refractory lining. For a given iron oxide content of the slag, the Al loss is about twice as high in a sand lined ladle as it is in a dolomite ladle (10), figure 7.

Figure 7:
Influence of the ladle refractory on the aluminium loss during Ar-stirring (10)

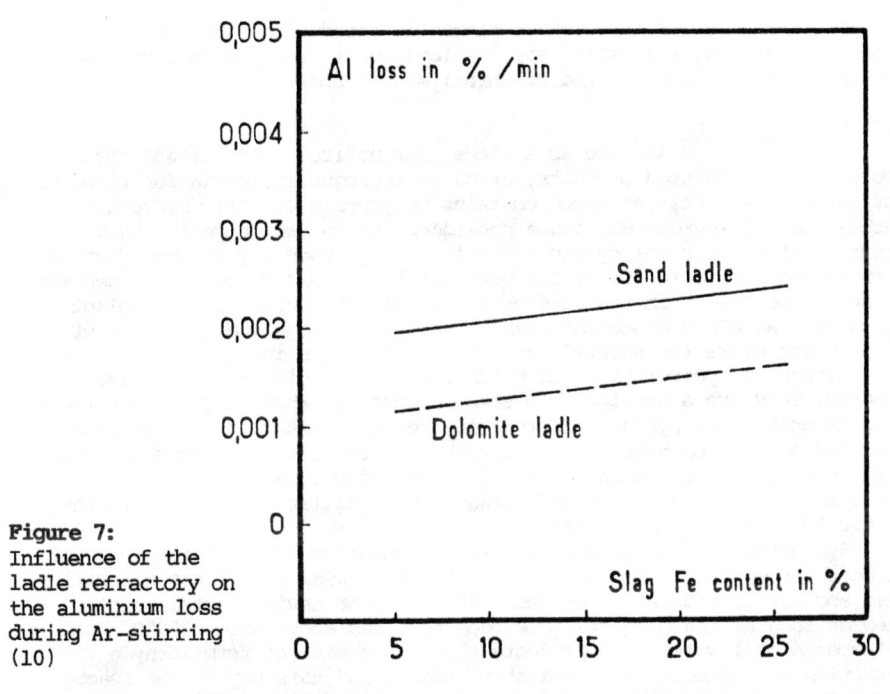

Chromium : In the same way as carbon-free materials must be used for ultra low carbon steel qualities, it is necessary to avoid the use of chromium containing refractories such as magnesite-chrome bricks if Cr pick up is to be avoided.

3.5. Inclusions

Non-metallic inclusions are formed when an element dissolved in the liquid steel is able to reduce a refractory oxide component of the lining. Silica and chromic oxide are especially reactive with most of the elements found in steel, and in particular with manganese and aluminium. Alumina, magnesia and lime refractories are more stable and should therefore be preferred.

According to thermodynamic data, the calcium introduced into steel for inclusion shape control reacts with all the main refractory constituents such as SiO_2, Cr_2O_3, but also MgO and Al_2O_3. In reality the calcium will generally combine with the inclusions in suspension in the liquid steel before coming into contact with the lining. So reactions with the refractories are found to be very limited (11). Dolomite linings will however give a higher calcium yield and will be less attacked at the slag line than high alumina linings. Dolomite therefore constitutes the best choice.

4. CLEAN ENVIRONMENT

European legislation defines a large number of dangerous chemical substances which have been proved or are assumed to create a hazard to environment or health when certain concentrations are exceeded. In such a case the products which contain such substances must be clearly marked with an appropriate hazard warning. Spent refractories can also be the cause of potential hazards consecutive to waste disposal.

The refractories industry is presently faced with two important environmental challenges in steel applications due to the problems created by the use of coal tar pitch and of chromium compounds.

4.1. Coal tar pitch

Coal tar pitch is used as a binder for unfired dolomite and magnesite products, for unshaped products, or as an impregnating media for fired or unfired bricks. Coal tar pitch contains polycyclic aromatic hydrocarbons such as benzo(a)pyrene which are considered to be carcinogenic. Many hundreds of similar polycyclic aromatic hydrocarbons may be formed on the burn in and during the first few heats of ladles and converters lined with these refractories. In the absence of adequate burners and anti-pollution equipment, an added discomfort can be created through the formation of thick black smoke for several hours during the burn in.

Bitumen or petroleum based pitches are available as alternative binders. They are a considerable step forward, as they only contain trace amounts of benzo(a)pyrene. But they do create at least as much smoke as coal tar pitches on heating up. Petroleum based oils are also sometimes used instead of pitches as binders for unfired bricks or unshaped products. One of the main disadvantages limiting large scale use is the rather low resulting mechanical strength.

Synthetic heat setting resins are a recent addition to the range of binders for unfired dolomite bricks. Phenolic based resins are generally used and are considered to be acceptable from an environmental point of view as long as the products have been tempered above about 200°C. With untempered or low tempered products there is a risk of formaldehyde formation during heat up in the steel plant, and this may be the reason for unfavourable comment on the use of phenolic resins (12).

This subject is still open to debate, but special care should be taken with some of the less expensive resins which are obtained as by-products of other chemical processes. These resins of complex and ill-defined composition are often used in so-called chemically bonded dolomite bricks.

When a basic refractory is indicated, fired natural dolomite products are of particular interest. They do not contain any carbon or hydrocarbons, nor indeed any other toxic or volatile substances. Preheating is therefore completely fume and trouble free, and without any smell. Wax is also fume free on preheating, and wax immunized fired dolomite has the added advantage of an increased hydration resistance.

4.2. Chromium compounds

The trivalent chromium compounds such as chrome ore and chromic oxide have gained wide acceptance as additions or principal ingredients in refractories such as magnesite-chrome and alumina-chrome. Trivalent chromium is considered to be nontoxic and therefore these refractories do not constitute a health hazard. However, studies have indicated that the hazardous hexavalent chrome can be formed in the $MgO-Cr_2O_3$ system at high temperatures in contact with a lime source. Hexavalent chrome can be leached out of waste refractories by the action of rain and ultimately find its way into drinking water. Spent magnesite-chrome refractories should therefore be considered as a potentially hazardous solid waste, and steps should be taken to dispose of this material in an appropriate landfill.

5. OVERALL POSITION OF DOLOMITE

The advantages and disadvantages of the most currently used ladle refractories are resumed in table 5, taking into account the main properties such as inertness towards steel, environmental pollution, hydration resistance, thermal shock resistance, resistance to slags, and price. In this comparison, products based on natural dolomite are preferable because of their inertness to steel and because of their price.

Property Refractory	Inert to steel	Atmosphere pollution	Hydration resistance	Thermal shock resistance	Basic slag resistance	Price
Silica/fireclay	O	◐	●	●	O	●
Zircon	O	◐	●	●	O	●
High alumina	◐	●	●	●	O	●
Magnesite (unfired)	●	◐	●	O	●	O
Magnesite carbon	◐	◐	●	●	●	O
Magnesite-chrome	◐	O	●	◐	O	O
Synthetic dolomite	●	●	O	◐	●	O
Natural dolomite	●	●	O	◐	●	●

O bad   ◐ medium   ● good

Table 5: Advantages and disadvantages of various steelplant refractories.

A more detailed picture of the various types of natural dolomite refractories can be gained from table 6. Price considerations speak for pitch or resin bonded unfired dolomite. Inertness to steel and environmental considerations favour low carbon resin bonded unfired dolomite and carbon free fired natural dolomite products.

Property Refractory	Inert to steel (C)	Atmosphere pollution	Hydration resistance	Price
Unfired pitch bonded	O	O	O	●
Unfired resin bonded	O	◐	◐	◐
Fired	●	●	●	O
Fired immunized	●	●	●	O

O bad   ◐ medium   ● good

Table 6: Advantages and disadvantages of natural dolomite bricks

6. CONCLUSION

The increasing demands for cleaner steel cannot be fufilled in ladles lined with conventional silica containing refractories such as sand, fireclay, zircon or high alumina. It is therefore necessary to use basic linings. Amongst the basic products available, it has been shown that dolomite refractories are the most quality and cost efficient.

Due to a growing concern for the environment. the use of refractories is also being investigated from the point of view of pollution and health. Pitch bonded dolomite is not expensive but constitutes a hazard to health if the binder is coal tar based, and at least a source of atmospheric pollution if the binder is petroleum pitch based. Resin bonded dolomite is a reasonable alternative from both the health and pollution aspects of use. The safest choice is of course fired dolomite which is not hazardous. not polluting, and less reactive with steel than any other known industrial refractory.

7. REFERENCES

(1) J.H. CHATILLON, R.D. SCHMIDT-WHITLEY: Clean steel and a clean environment with fired dolomite. Proceeding International Symposium on Refractories, Chinese Society of Metals, Hangzhou (1988) 433/451.
(2) K. NAZIRI, R.D. SCHMIDT-WHITLEY, K. WIELAND: The operating behaviour of dolomite linings in sintering zones of cement rotary kilns. Taikabutsu 40,2 (1988) 36/38.
(3) J. HÄRKKI, R. RYTILÄ: Reoxidation caused by the refractory materials. Proceedings 5th International Conference on Ladle Metallurgy, Mefos, Luleå (1989) II, 251/279.
(4) T. KISHIDA, S. KITAGAWA, S. SUGIURA: Proc. 7th Japan-Germany Seminar, Verein Deutscher Eisenhüttenleute, Düsseldorf (1987) 167/180.
(5) G. CARLSSON et al.: Fachausschussbericht 2.023, Verein Deutscher Eisenhüttenleute, Düsseldorf (1985) 45/48.
(6) T. DEGAWA, S. UCHIDA, T. OTOTANI: Development of CaO-MgO refractories and their effects on refining mechanism of extremely clean steel. Proceedings 2nd International Conference on Refractories, Tokyo (1987) 842/856.
(7) T. OTOTANI, T. DEGAWA, S. UCHIDA: Calcia and calcia magnesia refractory refining for metal, alloy and compound. Proceedings 5th International Conference on Ladle Metallurgy, Mefos, Luleå (1989) II, 181/223.
(8) T. KISHIDA, T. KURISU, H. USHIYAMA, T. YAJIMA: Secondary metallurgy to meet demands from both continuous casters and customers for the optimization of plant economy. Proceedings 5th International Conference on Ladle Metallurgy, Mefos, Luleå (1989) I, 577/596.
(9) F. HAUCK, J. PÖTSCHKE: Arch. Eisenhüttenwesen 50 (1979) 145/150.
(10) E. HÖFFKEN, H.D. PFLIPSEN, W. FLORIN: Measure taken to avoid the entrainment of converter slag during tapping and their effect on rephosphorization, Al melting loss, and the degree of purity of the steel. Proceedings International Conference on Secondary Metallurgy, Verein Deutscher Eisenhüttenleute, Aachen (1987) 124/136.
(11) G. GUENARD, M. NADIF: Behaviour of treatment ladle refractories and their influence on steel quality. Rev. Métallurgie-CIT 85,3 (1988) 219/229.
(12) J. STRADTMANN, G. MLAKER, R.C. THOMAS: Basic linings in casting and treatment ladles. World Steel and Metalworking 8 (1986/87) 80/89.

USE OF NONOXIDE CERAMIC MATERIALS IN METALLURGY

T. BENECKE

Elektroschmelzwerk Kempten GmbH
Herzog-Wilhelm-Straße 16
D-8000 München 2

Summary

The growing interest in nonxoide ceramic materials for applications in metallurgy, the state of the art and future prospects for CaB_6, ZrB_2, TiB_2, SiC, B_4C, Si_3N_4, AlN and BN in ferrous and nonferrous metal production are described with special regard to steel making and steel making refractories.

Nonoxide ceramic materials are characterized by the fact that they contain one of the following elements: boron, carbon, silicon or nitrogen. Thus, it is the matter of borides, carbides, silicides and nitrides which are distinguished into metal compounds (example: titanium diboride) and compounds of the elements boron, carbon, silicon and nitrogen (example: boron carbide).

The history of the nonoxide ceramic materials started with silicon carbide and its hardness. In the meantime, the use of these materials extends far beyond purely abrasion or wear resistant applications. Also in metallurgy, nonoxide ceramic materials become of increasing importance. Oxide materials alone cannot meet the growing demands for ever increasing lining lives. The impact of modern process technology requires more sophisticated properties. In general, it is not one characteristic only that makes a nonoxide material interesting for metallurgy. It is the property combinations that play a decisive role [1]: High melting or decomposition temperatures (exceeding by far those of metals) with excellent thermal conductivities, low thermal expansion coefficients and, principally, excellent thermal shock resistance. In addition, titanium diboride and zirconium diboride offer high electrical conductivity, which in contrast, is especially low in aluminium nitride and boron nitride. Nonoxide materials offer almost always excellent stability in oxidizing gases. Silicon nitride is highly resistant to molten nonferrous metals whereas boron nitride resists molten steel. On the other hand, there is silicon carbide. Although it is the most important nonoxide refractory material in the nonferrous and iron metallurgy, it is soluble in liquid iron. "Reaction kinetics" is the key to this apparent contradiction. It is thanks to this characteristic that SiC is used as charge material in the secondary metallurgy of steel and, above all as alloying, deoxidation and pre-inoculation agent in the cast iron production, but, it is also an indispensable constituent (as so-called antioxidant) in Al_2O_3/SiC/C monolithics and bricks in the blast furnace area.

	CaB$_6$	ZrB$_2$	TiB$_2$	SiC	B$_4$C	Si$_3$N$_4$	AlN	BN
Crystal structure	cubic	hexagonal	hexagonal	α,hexagonal (β=cubic)	rhombohedr.	α,hexagonal (β=cubic)	hexagonal	α,hexagonal (β=cubic)
Melting point °C	2235	3245	3225	Decomp.2760	2450	Decomp.1850	Decomp.1850	2730 (Subl.)
Thermal expans.coeff. 10^{-6} K^{-1} (∅ 25-1000 °C)	6,5	6,8	7,7	4,7	4,5	3,1	4,8	7,5 (∥) 0,8 (⊥)
Thermal conductivity Wm^{-1} K^{-1} (25/1000 °C)	70/	130/90	120/80	100/50	35/20	30-40/15-20 4-13/2-8(p)	200/80	45/25 (∥) 50/30 (⊥)
Electrical resistivity Ω cm (25/1000 °C)	10^{-4}/	10^{-5}/	10^{-5}/10^{-4}	10-100/2-30	0,1-10/0,1-4	10^{12}/10^7 10^{15}/10^8(p)	10^{15}/10^9	10^{12}/10^7
Hardness Mohs	8 - 9	9	9,5	9,5	9,6	9	8	1 - 2 (β=10)
Max.service temp. °C -oxydizing atmosphere - inert	700 2200	1100-1400 3200	1100 3200	1650 2200	800 2250	1300-(p)1500 1400-(p)1600 (only in N$_2$)	700 1900	800-1000 2500-3000
Resistance to molten - nonferrous metals - iron	– –	++	+ trials	+ –	– –	+ –	+ trials	++
Refractory component for iron and steel melts		×	trials	×	×	×	trials	×

(∥) and (⊥) = measured parallel and perpendicular to the BN laminations
p = porous

Table 1. Properties of CaB$_6$, ZrB$_2$, TiB$_2$, SiC, B$_4$C, Si$_3$N$_4$, AlN, BN (indicative values)

The subjects of the following report are the versatile properties and property combinations of a total of 8 borides, carbides and nitrides and their actual and potential applications in metallurgy (table 1).

BORIDES
Calcium hexaboride, CaB_6

Calcium hexaboride is a true hard material with a melting point well over 2.000 °C. Despite this fact, it is used as an addition to molten copper alloys only, especially for high conductivity copper castings. Since CaB_6 is also soluble in liquid iron, there are attempts to use the particulary light boron-calcium compound for boron micro alloying of steel melts employing the cored wire technique.

Calcium hexaboride shows excellent behaviour in plasma spraying [2]; is this a first indication for a possible refractory application?

Zirconium diboride, ZrB_2

being also very hard and high melting is corrosion resistant. Developments in connection with liquid nonferrous and ferrous metals are reported from Japan [3]. These include, for continuous casting of steel, trials with zirconium diboride as anti-clogging material; fast responding ZrB_2 thermocouple protection tubes for temperature measurements in the tundish; and for the tundish, tests with a ZrB_2 reheating electrode. Both these last applications illustrate the high thermal and electrical conductivity of zirconium diboride. Similar metallic conductivities are typical of many diborides, also of

Titanium diboride, TiB_2

wich is very similar to zirconium diboride with regard to this and other properties. TiB_2 is mainly used for the production of so-called evaporation boats for the metallization of films and papers, a process where aluminium is evaporated under high vacuum by passing electric current directly through these boats. Evaporation boats are a ceramic composite: Aluminium nitride is added for the control of the electrical conductivity and boron nitride is added to improve machinability.

Another speciality of diborides is the combination of high electrical conductivity and excellent wettability by liquid metals and, at the same time, excellent corrosion resistance to metal melts. This is the reason for their use, not only in metallization technology, but also for developments aiming at a cathode material based on TiB_2 for aluminium melting electrolysis [1].

In the meantime, also in steel making, first tests with titanium diboride are taking place.

CARBIDES
Silicon carbide, SiC

Silicon carbide, of all the nonoxides, is a real mass-produced material. It is also the oldest and most versatile nonoxide. Therefore, it is the best known nonoxide ceramic material also in metallurgy [1]. In molten iron, SiC is soluble and plays an important role as alloying agent with special metallurgical effects because of its dissolution kinetics, whereas it is insoluble in nonferrous metal melts. Its corrosion resistance, in connection with high thermal conductivity and thermal shock resistance has made silicon carbide very important for the metallurgy (extraction, refining, remelting, casting) of copper, zinc and aluminium and their alloys. A great numer of specially bonded silicon carbide refractory materials such as bricks, crucibles, nozzles and other parts are being used. The latest development is particulate reinforced SiC/Al composite materials produced by adding silicon carbide microgrits to liquid aluminium.

The high hardness and abrasion resistance, excellent thermal conductivity, thermal shock resistance and almost universal chemical stability (alkalis) have led to silicon carbide's most successful refractory use in the steel industry, namely in lining of modern high performance blast furnaces, where applications range from the middle shaft to the tuyeres. The solubility of silicon carbide in liquid iron is compensated by its excellent thermal conductivity resulting in a protective layer of solidified slag on the surface of the SiC bricks.

Furthermore, due to its excellent thermal conductivity and thermal shock resistance, silicon carbide is used for foam ceramic filters that are already in worldwide use in the production of nodular cast iron. In this case, the solubility of SiC is advantageous for melting the returns together with the SiC filters used.

As mentioned in the beginning, another refractory application of SiC is as a so-called antioxidant or oxidation inhibitor. As such it is used for the reduction of carbon loss in $Al_2O_3/SiC/C$ blast furnace runner mixes, tap hole clays or ASC bricks for torpedo ladles. Here, the contribution of the silicon carbide is its reactivity. The reaction mechanism with the oxidic gas phase has well been examined. According to

$$\begin{array}{rcl} SiC + CO & \longrightarrow & SiO\,(g) + 2\,C \text{ and} \\ SiO\,(g) + CO & \longrightarrow & SiO_2 + C \\ \hline SiC + 2\,CO & \longrightarrow & SiO_2 + 3\,C \end{array}$$

there is both the formation of SiO_2 and carbon resulting in a protection of the refractory by closing the pores and compensating for the carbon loss.

For some years now another carbide has undertaken to conquer the market as oxidation inhibitor,
Boron carbide, B_4C
Highly appreciated as absorber material for thermal neutrons and for its high hardness (it takes the third place on Mohs' hardness scale), boron carbide is being used mainly in nuclear technology and for lapping and polishing operations. Although boron carbide is used in large quantities for lapping, it is not used for making grinding or cut-off wheels. What prevents its use for this application is now seen in a different light. What we refer to is the oxidation sensitivity of boron carbide which starts reacting with oxygen at 800 °C. It is this property that led to its discovery as antioxidant in special, high quality carbon bonded refractory bodies [1].

The exact reaction mechanism is under examination at the moment. It is quite possible that there are analogies between B_4C and the allied SiC, between the gaseous oxidation products silicon monoxide (SiO) and boron monoxide (BO) as well as between SiO_2 and the low melting B_2O_3. The addition of boron carbide enhances, above all, the erosion resistance of the refractory bodies. Current, principal fields of application are
- the continuous casting area (Al_2O_3/C based refractories) where monobloc stopper, shroud and submerged nozzle are involved,
- the converter (MgO/C bricks) with the scrap impact area in particular.

Further applications for boron carbide as an antioxidant in iron making, primary and secondary steel making are under development.

NITRIDES
Silicon nitride, Si_3N_4

The best known refractory applications of silicon nitride in the steel industry are the Si_3N_4 bonded silicon carbide bricks used in blast furnaces, produced by nitriding the metallic silicon containing SiC compacts. Also it is known, that silicon nitride powder is directly added in the production of shaped and unshaped refractories. Less known are pure silicon nitride components. They are used to advantage where high strength at elevated temperatures and/or high wear resistance are required under corrosive and abrasive conditions. The extreme hardness, the excellent thermal shock resistance and its fracture toughness, which is extraordinarily high for a ceramic material (impact and wear resistant ball and roller bearings made of Si_3N_4!) have opened up a great number of applications in the forming of metals. Typical examples are guide wheels and plates in wire manufacture and caps for ultrasonic measuring heads for hot steel sheets and pipes. A hard metal or oxide ceramic protection cap withstood the attack of 500 pipes whereas a silicon nitride cap could handle 20,000.

To use as induction hardening fixtures for so-called slide shoes, materials must exhibit excellent electrical insulation properties, high mechanical strength and thermal shock resistance as well as high resistance to abrasive wear. Silicon nitride has almost completely replaced other materials for this application.

Like SiC, Si_3N_4 is soluble in iron but not in nonferrous metal melts. This fact, combined with silicon nitride's excellent high temperature properties, is being exploited for numerous components where silicon nitride is in contact with molten aluminium, zinc or silicon. In connection with aluminium, examples for Si_3N_4 applications are: Die casting, welding nozzles especially for TIG welding, thermocouple protection tubes, plates for slide gates [1].

Aluminium nitride, AlN

Aluminium nitride is characterized, like boron nitride, by a bad electrical but a good thermal conductivity. One aluminium nitride application has already been mentioned: Its use as high-ohmic constituent for the control of the electrical resistivity in TiB$_2$ evaporation boats for high vacuum metallization of aluminium. In addition, aluminium nitride can be found in the melting metallurgy of aluminium in the form of crucibles, tubes and other shapes. AlN crucibles are used for molten salts and bases, and for titanium alloys.

Research is done on aluminium nitride for refractory use in steel making. Depending on the surrounding conditions, it is possible that AlN will exhibit good resistances. The presence even of small concentrations of oxygen and the formation of layers of Al$_2$O$_3$ or alons could play an interesting role.

Boron nitride, BN

There are two modifications of boron nitride, the cubic modification which is very hard and therefore used in abrasive applications; and the generally available hexagonal modification which, as an exception among the nonoxide ceramic materials, is particularly soft. Hexagonal BN has the same layer lattice as graphite. Like graphite, boron nitride can be easily machined; and BN has excellent lubricating properties that are exploited in applications such as high temperature slide bearings, forging of hard special alloys or mould release agents in casting and pressing of glass or nonferrous metals.

One of the properties boron nitride is best known for in metallurgy, which it also has in common with graphite, is its bad wettability by nonferrous metal but also steel melts. Thanks to their excellent corrosion resistance in contact with nonferrous metals, boron nitride sintered shapes are used in the form of crucibles, thermocouple protection tubes and insulating sleeves, tubes and pump components for liquid metals.

One distinct difference to graphits is boron nitride's corrosion resistance in contact with molten steel. This advantage over carbon which, as we know, is soluble in liquid iron, has paved the way for boron nitride's use in steel works and especially in the continuous casting area [1].

BN break rings for horizontal continuous casting of steel

The boron nitride break ring is placed where the molten steel leaves the tundish and enters the horizontal water-cooled copper mould. Due to BN's non-wettability, the strand shell starting to solidify is easily released from the ring when the strand is drawn off. Also other properties of boron nitride play an important role is this: The high thermal shock resistance of BN when the molten metal comes into contact with the comparatively cold ring at the beginning of casting. In addition, the excellent machinability, which allows one to fit the ring exactly into the given shape, and also the fact that BN can cope with the high demands on wear resistance satisfied today by a ceramic composite based on boron nitride with oxide and carbide additions.

BN powder as constituent in (vertical) continuous casting refractories

The application concerned is the black submerged nozzle based on alumina/graphite. Two zones are problematic:
- the liquid steel level in the mould subjected to corrosive attack and
- the inner wall of the nozzle with its susceptibility to

clogging with Al_2O_3.

The wear of the Al_2O_3/C submerged nozzle, in the liquid steel level, is due to the mould oscillation. It causes dissolution of the carbon by the molten steel laying bare the Al_2O_3 skeleton; which then causes dissolution of this skeleton by the casting powder slag.

At the beginning of the 1980's, first publications by Piret (cit. in [1]; see also [4]) and by Jeschke, Lührsen, Oberbach and Stallmann (cit. in [1]) reported on the considerably reduced wear of alumina/graphite submerged nozzles in the liquid steel level if this nozzle area is reinforced or enriched with boron nitride. However, during the years following this, zirconia was generally used as wear protection. Nevertheless, developments continued and in 1988, Höffken, Lax and Pietzko came up once more with an Al_2O_3/C submerged nozzle enriched with boron nitride in the liquid steel level employed in the biggest German slab caster [5]: The lower wear resistance of boron nitride, as compared to zirconia, is more than compensated by the slabs' reduced susceptibility to develop longitudinal cracks. This has been explained by the better thermal conductivity that boron nitride has over zirconia which improved the heat flow to the mould wall, thus creating more uniform solidification conditions.

Boron nitride is also important for fighting nozzle clogging. In this case, it is used in the form of boron nitride inserts. There are a number of alternatives, e.g. inserts of ZrO_2/CaO, Sialon or flushing with gas; these are all possibilities that have advantages but also considerable disadvantages. This is the reason why intensive work is done to find anti-clogging solutions; the chances for boron nitride seem to be good due to its non-wettability which here too plays an important role. The subject becomes all the more important the smaller the diameters of the submerged nozzles and the thinner the wall thicknesses. This also extends into the future. What we are speaking of is

BN for near net shape casting developments

With thin slab casting the future of near net shape casting has begun. Casting of thin slabs requires small moulds and, consequently, small and thin walled submerged nozzles. For the first time, submerged nozzles of this type were exhibited at the fairs "METEC" and "THERMPROCESS" in Dusseldorf, in 1989.

According to Parbel et al in 1988 [6] not only the properties and effects of the various constituents of a submerged nozzle should be considered but also their overall benefits. This is why they evaluated various mixes with zirconia or boron nitride in view of corrosion resistance in contact with steel and steel/casting powder slag, anti-clogging behaviour, price and thermal shock resistance. On the whole, the combination $Al_2O_3/C/BN$ got the highest marks. Therefore, it is not astonishing that submerged nozzles presented in recent publications [7, 8] also contained boron nitride, while thin walled submerged nozzles shown at the fairs in Dusseldorf even contained BN over the entire part.

It is the goal of many pilot projects to find a way to skip the intermediate step of producing thin slabs and instead produce strip directly. Here too, boron nitride figures among the material alternatives, either as refractory constituent or as boron nitride ceramic.

REFERENCES
(1) T. Benecke: Nichtoxidische sonderkeramische Stoffe in der Eisen- und Nichteisenmetallurgie. VDEh-Seminar "Feuerfeste Werkstoffe in der Eisen- und Stahlindustrie", Bad Neuenahr (1987)
(2) Pers. inform. S. Janes, Institut der Feuerfestindustrie, Bonn (1989)
(3) K. Kuwabara: Properties and applications of ZrB_2 ceramics. Taikabutsu 39 (1987), p. 697-698; cf. K. Sakai, Ceramics Japan 24 (1989), p. 526-532
(4) J. Piret: La corrosion des busettes immergues de coulee continue. Journees sur la corrosion des materiaux refractaires et ceramiques par les metaux et les phases condensees liquides. Université de Nancy (1988)
(5) E. Höffken, H. Lax and G. Pietzko: Development of improved immersion nozzles for continuous slab casting. Proceedings 44th Int. Conf. Cont.Casting, Brussels (1988)
(6) W. Parbel, F. Schruff, B. Bergmann and M. Weiler: High quality refractory materials - the key to modern continuous casting. Proceedings 44th Int. Conf. Cont. Casting, Brussels (1988)
(7) R. Oberbach, H.G. Stallmann and W. Wlach: Hochwertige Feuerfest-Werkstoffe und -Systeme - ein Schlüssel zur modernen Stranggußtechnik. Steel and Metals Magazine 27 (1989), p. 278-283
(8) H.-J. Ehrenberg, L.Perschat, F.-P. Pleschintschnigg, W. Rahmfeld and C. Praßer: Casting and cast rolling of thin slabs at the Mannesmannröhrenwerke AG. Metallurgical Plant and Technology 3 (1989), p. 52-69

EVOLUTION OF THE TECHNOLOGY AND USE OF REFRACTORIES IN ITALY AND IN OTHER EUROPEAN COUNTRIES

P.L. Ghirotti

NUOVA SANAC S.p.A. - Via M.Piaggio 13-GENOVA
ITALY

Summary

The object of this report is to offer a survey on the present development situation in the sector of refractory products and related applicat= ions in Europe and Italy. We must expect the decline in the volume of the global demand for refractories and the trend towards types of higher techn ology refractories to continue in the next years.

In the last ten years ('77 - '87) in Italy we've gone from the speci= fic consumption of steel industry refractories of 20.2 Kg/t to 13.3 Kg/t. As to the other user-industries, the consumption values are steadier.

Remarkable importance will be given - in relation to the recent chang es being made in steel making and aiming at cost reduction, higher quality, production elasticity - to refractories for pig iron pre-treatment vessels (torpedo car and pig iron ladle), to converter refractories with combined blowing techniques (top and bottom blowing) and above all to refractories for post-treatment and teeming ladles and to continuous casting refracto= ries.

Attention shall be paid to the development of the future metallurgic processes,, many of which aiming not only at the constant object of obtain ing purer and cleaner steels at a low cost, but also solving the problem of raw material sources and energy saving (scrap iron utilization in con= tinuous process, "ore smelting technology", direct combination of contin= uous casting with hot-rolling: CC-DR).

As to refractory industry, the technology of computer aided manufact= uring systems (CAM) and automation may lead to a significant advance in product quality and in productivity.

Wherever bricks and monolithic materials are used, the principles of technical ceramics and high technology shall be more and more applied. The further development will turn more and more to particular and specific ap plications, to the realization of products allowing time and energy saving in production and installation, to the safeguard of the ecological aspects of production and installation, to the technique and growth of mechanized and automated installation, to furnace repair techniques.

All this will even more require an unceasing team-work between engin= eers and technicians of steel industry, furnace building and refractory manufacture.

1. GENERAL TRENDS

This article stresses the latest developments as well as the future trends in the development and application of refractory products in Euro= pe and in Italy, in connection with the technological changes occurred in the consumer-industries.

Steel industry is known to be the main consumer of shaped and unshaped refractories. The remaining consumer-industries, namely lime, cement, glass, ceramics, non ferrous metal, chemical and petrochemical industries, as well as power plants, all together reach about 40% of the total refractories consumption.

The share of the unshaped refractory products employed by the latter consumers practically amounts to 15-20% only; namely the 80-85% of all unshaped materials are employed in steel industry. In all the highly industrialized countries the demand for refractory products has sharply declined, while the decrease in the shaped product consumption has been even more remarkable.

The reason why unshaped products have mantained a relatively steady market share is mainly to be ascribed to a wider employment of such products in ladles, in foundry wear and safety linings, in blast furnace runners (see fig. 1).

The typologies and qualities of the refractory products that make up the present demand show a clear trend towards higher quality products, which is emphasized by the significant increase of the average price paid for one ton. product.

The price of the different typologies goes from about 0,5 cents of ECU/Kg for the simplest unshaped product up to 10 ECU/Kg for bricks or special high quality shaped products. The types of refractories required to meet the different application conditions have multiplied since 1960. In steel industry, which is the main refractories consumer, the consumption both of shaped and unshaped products is declining.

The percentage share of monolithics, however, is increasing, thanks to the favour they have met in numerous applications. The total decline is due to the technological changes in steel making. Fig.2 reports the trend of the specific refractories consumption (Kg/t of steel) in steel industries in Italy (until 1987) and in Japan (until 1985).

Towards the end of 1980s Europe has therefore reached the exceptional goal of 15 Kg of refractories consumption per 1 t of steel produced (9 Kg/t for bricks, 6 Kg/t for monolithics).

The European refractories industry is small, compared with other industries. In 1988 the sales of refractories industry in Western Europe amounted to as much as 2,804 million ECU (+ 7.3% VZ 1987) and 294 million ECU in Italy (+ 2.4% VZ 1987).

The number of people employed in this industrial sector amounts to 35,000 units in Europe and to 3,170 in Italy (9% of PRE countries). The plants generally belong to small and medium firms.

In Italy operate about 60 companies. In Italy two companies cover about 60% of the Italian market turnover. At present, in Europe, 15 companies out of the approximate 300 reach 70% of the total production. The reorganizing processes that the European refractories industries are undergoing, due to the continuous decrease in demand, have brought to the grouping of plants and companies in order to survive and lower costs and to bear the costs of research and development as well as of the plant engine=

TREND IN SALES (ton and %) OF THE DIFFERENT CLASSES OF REFRACTORY PRODUCTS IN EUROPE (PRE) AND IN ITALY (1980-1988)

Country	Year	Fireclay t×10³	%	High alumina t×10³	%	Silica t×10³	%	Basic t×10³	%	Special t×10³	%	Insulating t×10³	%	Monolitics t×10³	%	Total Production t×10³
EUROPE	1980	1894	30.3	564	9.0	85	1.4	1204	19.3	263	4.2	116	1.9	2116	33.9	6242
	1988	849	17.2	498	10.1	60	1.3	1133	22.9	spec.+ ins.:		571-11.6%		1819	36.9	4930
ITALY	1980	283	35.9	114	14.4	14	1.9	99	12.5	87	11.0	29	3.7	163	20.6	789
	1988	107	21.3	89	17.7	14	3.0	81	16.2	28*	5.6	38	7.6	144	28.7	501

* Classification in 1988 some materials are not considered as refractory products.

— Fig. 1 —

DEVELOPMENT OF SPECIFIC REFRACTORY CONSUMPTION IN THE STEEL INDUSTRY IN ITALY AND IN JAPAN

(Kg. of refractory for ton of steel: KG/TON)

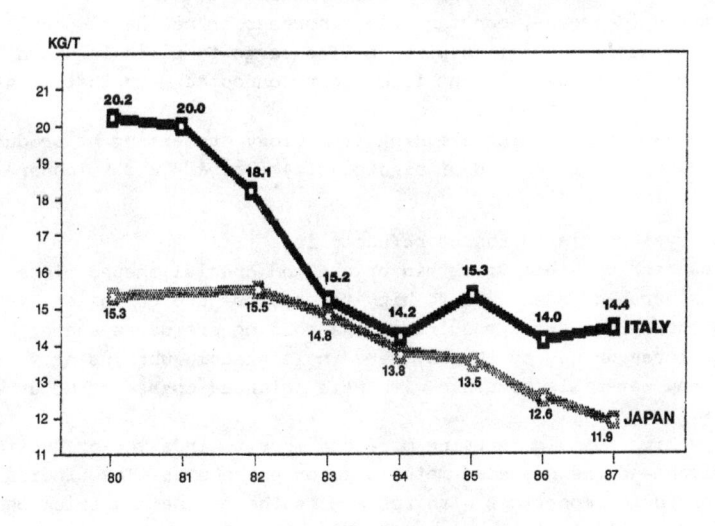

— Fig. 2 —

ering investments that the technological requirements of new products are increasingly demanding.

2. DEVELOPMENT IN RAW MATERIAL EMPLOYMENT

The various types of refractory products are manufactured with a wide range of raw materials. Some of these are acquiring greater and greater importance, namely those with low flux content, synthetic materials, such as various types of corundum, alumina, zirconia, chrome oxide, ceramic fibres, different carbonaceous materials, silicon carbide.

The employment of synthetic raw materials is certainly going to furtherly increase, especially if the production of some of them can be effected with more economical processes. Moreover, alongside with traditional oxide-based raw materials, there will be raw materials based on carbon, nitrogen, boron, as additives to be used in special composition (see fig. 3).

In short, as to the employment of raw materials, the following tendencies are emerging:
- increase in the use of purer and purer synthetic magnesite;
- increase in the use of spinel $MgO-Al_2O_3$;
- increase in the use of semi-stabilized zirconia;
- increase in the use of graphite;
- use of super-fine refractory powders based on oxides and non oxides;
- use of metallic powders as additives;
- increase in the use of binders and binding systems that allow the manufacturer of bricks to be used without firing processes and/or that allow to obtain a corrosion-resistant binding phase (the resin binder is an example).

3. DEVELOPMENT IN THE REFRACTORY PRODUCT SECTOR

The trend (decrease, constant use, increase or new development) of refractory product consumption divided into large families is shown in fig. 4 - Fig. 5. The development is closely connected with that of steel industry, the mayor consumer.

The latest trends in steelmaking technology are aiming at producing steels of different types and of high quality ("purer" and "cleaner") at increasingly lower costs.

3.1. Main developments in shaped refractories

It has been possible to obtain bricks and special shaped pieces with a most considerable resistance to high temperature and corrosion by means of manufacturing processes requiring high molding pressures and high firing temperatures which, by the way, are indispensable when using synthetic and purer raw materials combined with well balanced coarse structures of the matrix.

Energy high cost is carrying more and more towards the production of unburnt pitch and resin-bonded materials, or phosphates. The binding phase is still of basic importance with respect to the refractory brick characteristics and particularly to slag attack.

BASIC RAW MATERIALS FOR REFRACTORIES

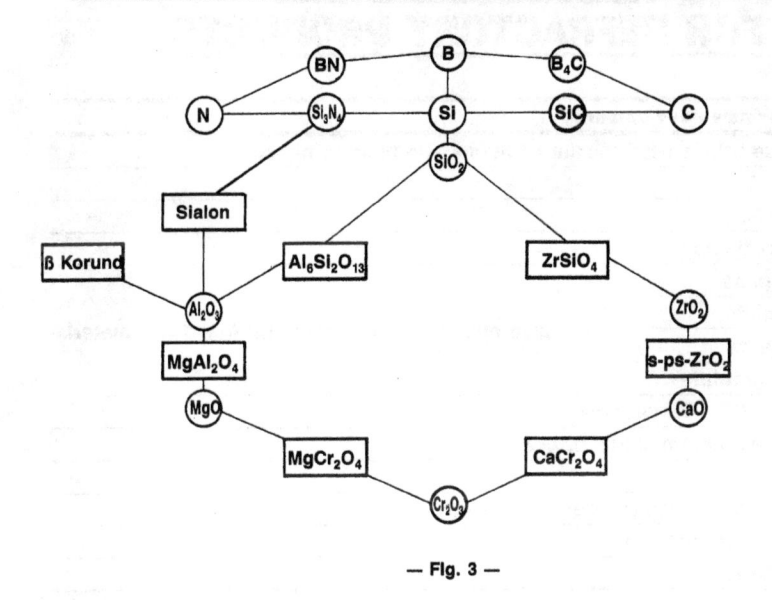

— Fig. 3 —

TREND OF DEMAND
FOR REFRACTORY PRODUCTS

Decreasing	Constant
Fireclay bricks	Silica bricks
Burned Magnesia bricks	High alumina bricks (45 ÷ 90% Al_2O_3) Andalusite bricks
Burned magnesia Chrome bricks	Dolomite bricks
Bauxite bricks	SiC Bricks
Siliceous unshaped material	Insulating fire-bricks
Ramming mixes	Castable and repairing mixes
	Electrofused bricks and shapes
	Zircon and zircon-mullite materials

— Fig. 4 —

TREND OF DEMAND FOR REFRACTORY PRODUCTS

Increasing or new developments
Alumina-base bricks with special additions and binding phase
$Al_2O_3 - Cr_2O_3$
$Al_2O_3 - Si_3N_4$ (Sialon)
Unfired brick types
Al_2O_3-graphite
MgO-graphite   also mixtures of sintered and fused raw material
ZrO_2-graphite
Dolomite-graphite
$Al_2O_3 - SiO_2 - SiC - C$ materials
Magnesium aluminium spinel bricks
$MgO - ZrO_2$ bricks
Zirconium oxide shapes (partially stabilized)
Ultra-low cement castables
Vibrating mixes
Ceramic fibers
Recrystallized SiC shapes

— Fig. 5 —

TYPICAL WORKING LINING FOR TORPEDO CAR

Plants	Capacity	Barrel	Cone	Trunnion	Nose	Average life ton. pig-iron/ campaign
ILVA	250	Al_2O_3-SiC-C	Al_2O_3-SiC-C	Al_2O_3-SiC-C	High Al_2O_3 castable	250.000
ILVA Bagnoli	200	High Al_2O_3/ xAl_2O_3-$ySiO_2$	xAl_2O_3-$ySiO_2$	xAl_2O_3-$ySiO_2$	High Al_2O_3 castable	200.000
ILVA Piombino	200	mAl_2O_3-$nSiO_2$	mAl_2O_3-$nSiO_2$	mAl_2O_3-$nSiO_2$	High Al_2O_3 castable	170.000

x ~ 46%   y ~ 50%   (chamotte)
m ~ 65%   n ~ 31%   (andalusite + bauxite)

— Fig. 6 —

Improvements have been made with various means, by employing, for instance, the phosphate bond in andalusite bricks for torpedo cars and in bauxite bricks for incinerators, the nitride and sialon binding phase produced in situ in SiC and corundum bricks employed in B.F. linings.

The fundamental materials in steel industry are now the products based on oxides and carbides with graphite content, mainly the Al_2O_3-C and MgO-C bricks and special shapes.

The increasing diffusion of pig iron pre-treatment system has made indispensable the development of materials suitable for torpedo car linings. The resin-bonded materials based on Al_2O_3 - SiO_2 - SiC - C, which offer a satisfactory resistance both to the slag combined effects due to dephosphorization and desulphurization, and to oxidation, are widely used in Italy.

The selection of the appropriate refractory product for a special application is most important. The drawback of these products lies in their thermal conductivity, which is over 3 times higher than that of the traditional fired products with Al_2O_3 content and based on bauxite.

The metal shell temperature may reach over 300°C. The ILVA steel plant in Taranto is the only works in Europe that, as happens in Japan too, has 100% of its torpedo cars (nr. 46) lined with Al_2O_3 - C - SiC material (see fig. 6).

Magnesite-graphite bricks with over 7% carbon content have become more and more important in the linings of converters, electric arc furnaces, ladle slag lines. Nowadays in Italy about 70% of converter linings is made with MgO-C material.

In Italy the consumption of magnesite-carbon for electric arc furnaces is also remarkable, due to the relevant amount of steel produced in E.A.F. in this country, as previously said. The caracteristics of this class of products have been improved thanks to detailed studies on the effects of different types and quantities of graphite and special additives, such as SiC and BN, metallic additives such as Si and Al, Mg and alloys with said elements.

Fig. 7 shows the characteristics of some MgO-C materials, which are more commonly used in Italy for converter and electric arc furnace linings.

As regards the typologies of the ladle linings adopted in Europe, we must distinguish basically between integrated steel-plants (with blast furnace, converter) and electric arc furnace steelplants. Generally integrated steelplants use high alumina/bauxite/andalusite content bricks for wall lining and bricks based on tar-bonded magnesite-carbon for slag line lining.

The entirely basic dolomite or magnesite lining offers potential advantages, such as endurance to the increasingly higher temperatures required and above all steel purity and "cleanliness", but the thermal operating conditions in the cycle that ladles undergo in integrated plants, often impede to fully exploit the relevant technical characteristics of such linings.

Electric steelplants, generally equipped with smaller-sized ladles and where more rapid cycles can be performed, make extensive use of entir

REFRACTORY PRODUCTS MgO AND MgO-C ADOPTED IN CONVERTER AND E.A.F. LINING IN ITALY (resin bonded)

Reference	(4)	(5)	(6)	(7)	(8)	(9)	(10)
Type of product	MgO 97-C6-R	MgO 97-C11-R	MgO 97-E3-C11-R	MgO 97-E5-C11-R	MgO 99-C14-R	MgO 99-E3-C14-R	MgO 99-E5-C14-R
Base Components	Magnesia Graphite	Magnesia Graphite	Magnesia Graphite	Magnesia Graphite	Magnesia Graphite	Magnesia Graphite	Magnesia Graphite
Type of Bonding	Resin	Resin	Resin	Resin	Resin	Resin	Resin
Chemical Analysis on raw material (Magnesite) MgO%	97,0	97,0	97,5	98,0	99	98,5	98,5
CaO%	<2,0	<2,0	1,8	1,7	0,6	<1,0	<1,0
Fixed carbon %	6,0	11,0	11,0	11	13,0	13,0	13
Bulk density g/cm^3	2,92	2,95	2,96	2,96	2,90	2,90	2,94
Apparent porosity %	5	5	5	5	4	4	4
Cold crushing strength kg./cm^2	>400	>350	>400	>400	>350	>350	>350
Modulus of rupture kg./cm^2 at 25°C	>90	>90	>90	>90	>70	>80	>70
kg/cm^2 at 1250°C	60	50	60	50	40	50	40

— Fig. 7 —

TYPE OF TEEMING LADLE REFRACTORY LINING ADOPTED IN ITALIAN STEEL-WORKS

Type of Steel plant	Electric mini steel plant	BOF Steel plant	Ilva-Taranto Steel plant
Ladle capacity	50-130 t	110-160 t	300-350 t
Wall	Tar bonded dolomite	Bauxite or andalusite	Rammed zircon
bottom	Tar bonded dolomite	Bauxite or andalusite	Andalusite
slag zone	Magnesia-Carbon	Magnesia-Carbon	Magnesia-Carbon

— Fig. 8 —

ely basic linings, mainly in tar-bonded dolomite; however, the use of special chemically bonded magnesite products is also spreading. The slag-line, also in the case of basic linings, is in magnesite or magnesite-carbon. The particular situation of ladle linings in Italy is shown in fig. 8.

In particular, the ILVA works in Taranto employs monolithic zirconium silicate containing materials for ladle walls (rammings and thixotropic castables).

As to secondary refining vessels, the use of basic material is widespread. The selection of a lining mostly depends on the type of metallurgic process carried out. An estimate of the chemical and physical influence factors is required.

Fig. 8 shows the characteristics of the most used basic materials employed in ladle linings, and fig. 9 gives an example of the typical linings used in the various secondary refining systems.

In the field of the sliding gates, the latest technological development, as regards composition, is based on the carbon bond. Plates are manufactured with high alumina raw materials bond with a carbon and silicon carbide combination.

Various types of resins (phenolic, novolac) and pitch are used as binders. Zirconium oxide-containing materials will have a further development in the future; at present they are employed for special uses, such as tundish nozzles, plates for ladle and tundish sliding gates used for highly aggressive steel casting (e.g. extra low carbon, etc.).

Special manufacturing techniques are required to produce refractories for gas insufflation into metallurgic vessels and continuous casting. The largest market share in Europe goes to permeable refractory elements with diffused porosity. They mainly contain alumina or fired magnesite with 25% porosity.

Apart from steel industry, other consumer industries have been introduced to the use of special refractories. Spinel-containing magnesite bricks in alternative to magnesio-chromite materials, have developed for cement rotary kiln linings.

These new products are costly (1.8+2 times the cost of standard magnesio-chromite). In any case, it is necessary to estimate the technical-economical balance resulting from their employment. The basic bricks are generally composed of about 80% magnesite and about 20% MA spinel. The same type of bricks is also used in the intermediate layer of glass furnace heat exchangers.

Molten silica or vitreus silica is used in manufacturing the rollers of furnaces for the heat treatment of transformer plates and silicon plates; now it is also employed for non-ferrous material casting accessories.

The recristallized silicon carbide, fired at ultra-high temperature, is employed in the lining of the car charging floors of ceramic industry tunnel kilns and in general in the lining of charging floors used in high temperature firing, due to its excellent hot mechanical behaviour.

Mullite and alumina rollers are used for car firing in the so-called "roller kilns". The new activities in the ceramic "foam" sector are obtaining a promising success. Their peculiarity lies in their tridimensional

WEAR LINING REFRACTORIES FOR BASIC LADLES OF SECONDARY METALLURGY

Equipments Zone of wear lining	Stirring station + powder/wire injection	Ladle furnace and VAD	VOD
Slag zone	Pitch or resin-bonded Magnesia-carbon (> 90 MgO)	Resin or pitch-bonded Magnesia-carbon (> 97% MgO)	Direct-bonded cosinter magnesia-chrome (60% MgO)
Wall	Pitch-bonded dolomite Specific chemically-bonded magnesia	Pitch-bonded dolomite Specific chemically-bonded magnesia	Direct-bonded magnesia-chrome (60% MgO)
Bottom	Pitch-bonded dolomite Specific chemically-bonded magnesia	Pitch-bonded dolomite Specific chemically-bonded magnesia	Direct-bonded magnesia-chrome (60% MgO)
Impact area	Pitch-bonded magnesia (vacuum pitch-impregnated > 97% MgO)	Pitch bonded magnesia (vacuum pitch-impregnated > 97% MgO)	Direct-bonded magnesia-chrome (60% MgO)

— Fig. 9 —

HIGH TECHNOLOGY MATERIALS

Materials	Applications
SiC and SiC composites	Blast furnace shaft brick, through materials, torpedo lining, ceramic heat exchanger
Si_3N_4 sialon ans Si_3N_4 composites	Blast furnace, through materials Break ring for horizontal C.C Crucibles, Dies
BN	Crucibles, Boats, Sheath for thermo-sensor
Carbon composites	Brick of refining furnace for steel Tube and nozzle for continuous casting
Oxides	Parts for burner, Nozzle, Inner pipe Regining furnace lining for high alloy steel

— Fig. 10 —

reticulated structure with interconnected pores; besides, they have excel=
lent thermal resistance and high corrosion resistance, due to the employ=
ment of ceramic powders of high purity and high sintering power.

The most important applications of ceramic foams are realized in steel
filtration, continuous casting, filtration of hot dust-containing gases,
nozzles for turbulence-free casting, materials for various furnaces. The
introduction of ceramic foams has been sped up by the necessity of saving
energy owing to their low heat retention and to their excellent thermal
shock resistance.

Naturally their applications are restricted to those uses where ero=
sion and corrosion attack is negligible or non-existing. The field of ther
mal insulation has been undergoing an important and steady evolution after
the introduction of refractory ceramic fibres.

More recently, fibres obtained by chemical process have been realized:
- 95% alumina fibres for uses up to 1600°C
- carbon fibres whose use is increasing in compound products
- zirconia fibres.

Technical ceramics or advanced ceramics or high technology ceramics is
the object of an enormous development effort all over the world, it is clos
er to fine ceramics than to refractory materials.

A lot of companies producing refractories in Japan are engaged in the
diversification of productions and since the early '80s they have been mak=
ing investments intended for the research and production in the field of
technical ceramics.

Some materials, which may be identified with technical ceramics, are
employed in the traditional application sector of refractories: SiC, Si_3N_4,
sialons, BN, AIN, AION.

Also the fibres of such materials are used separately or in the form
of graphite composites. Figures 10, 11, 12 show examples of high technolo=
gical ceramics materials in refractory applications, materials for structur
al ceramics and components of functional ceramics.

In 1986 the Japan market had already reached 656 milliard Yens, namely
about 2.5 times the refractory market (253.3 milliard Yens). In Italy, dif=
ferently from what happened in Japan, only few refractory industries have
directly or indirectly, through associated companies, invested in initiati=
ves connected with technical ceramics.

The progress in the knowledge of technical ceramic principles will give
impetus also to the future development of technologies and manufacturing
processes of the so-called traditional refractories.

3.2. Main developments in unshaped refractories

Recently, they have gained even more interest as their employment in
construction and repair allows to save time, energy, money. Their installa
tion is easy and rapid, which means less global cost. The significant deve
lopment in installing techniques: casting, vibration and gunning, in parti=
cular the possibility of automation, have contributed to increase their
importance.

STRUCTURAL CERAMICS

Application	Material
Mechanical	
Tool materials, diamond tools C (w) BN tools	Al_2O_3, ZrO_2, SiC, C(w) BN, Si_3N_4, Diamond
Thermal	
Heat resistant materials Handling jig for semi-conductor	Al_2O_3, Si_3N_4 - BN, Carbon, SiC, AlN
Chemical	
Corrosion resistant materials Catalist carriers	Al_2O_3, SiC

— Fig. 11 —

FUNCTIONAL CERAMICS

Electro-magnetic

IC package (Surdips, multilayer), Capacitator element
Substrates, Piezo electric, Quartz resonator
Thermistor, Baristor, Magnetic material

Chemical, medical:

Gas sensor element, Bio-ceramics

Optical

Charge coupled device, optical fiber

Others

Super conductivity

— Fig. 12 —

The binders used in unshaped materials are of great importance. At first ceramic binders were employed, later followed by chemical-inorganic binders and hydraulic alumina-based cements. Organic binders are also used, they develop a carbonaceous structure, when in operation, besides lending a temporary bond.

Practically there are no limits to all the possible combinations between binders and the raw materials forming the refractory aggregate.This extraordinary flexibility of unshaped materials, as regards the fluctuating situation of their demand, has brought some modern steel-plants to use unshaped materials up to 50% of their total refractory consumption.

In Japan some steel plants use even 60% of unshaped materials. In Italy the average percentage of unshaped material employment is 35%. In the latest years castables based on hydraulic calcium aluminate binders have evolved towards low cement products (from 2 to 5%) and, quite recently,towards ultra-low cement content ($< 1\%$).

The reduction of hydraulic binder quantity, the use of ultra-fine powders, such as colloidal silica, as well as chemical additives to help fluidization, have allowed to develop materials having exceptional mechanical strength also in the range of medium temperatures (500°C and 1000°C),which is known to be the weak zone of classical refractory castables.

The strong compactedness of these new castables, due to the scanty amount of water required for their application, means low porosity and permeability, which favour good corrosion resistance.

It must be said that, in parallel with the development of unshaped high density products, generally unshaped products have more and more often recourse to raw materials containing graphite, silicon carbide, zircon.Such new special unshaped products, having particular features, are called "high technology castables".

Their typical application in steel and non-ferrous industries, in foundries, vary from the zones subject to severe erosion-corrosion attacks to complete linings for B.F. runners (compositions $Al_2O_3 - SiO_2 - SiC - C$), steel ladles (zircon-containing mixings) and furnace fixtures.

4. LATEST AND FUTURE TRENDS IN STEEL MANUFACTURE TECHNOLOGY AND PROBLEMS TO SOLVE IN THE REFRACTORIES SECTOR

4.1. Trends for refractories connected with the present requirements of steel manufacture processes.

We have pointed out that the greatest attention, as regards the refractory technology, has been directed, in the siderurgic field, towards the following areas:
- pig iron pre-treatment
- converter combined blowing processes
- secondary refining and teeming ladles
- continuous casting.

Tables 1, 2, 3, 4 summarize, for the different single areas, the main subjects of study to be investigated in order to improve the refractories performances.

TECHNICAL PROBLEMS TO SOLVE WITH REFRACTORIES FOR PIG IRON PRE-TREATMENT

Applications	Present Situation			Condition changes	Problems to solve	Subjects of development
	Lining zone	Material	Life (hits)			
Torpedo car	Slag line	Al_2O_3 SiC-C	720	Splitting of pig iron pre-treatment (in particular de-Si treatment)	Increase of operation costs High thermal conductivity of SiC-C containing linings	Development of lower cost materials Measures against spalling of Al_2O_3 - SiC - C Materials Resistance to SiC and C oxidation
	Remaining lining	Al_2O_3-Cr_2O_3 Al_2O_3 Al_2O_3-SiC-C				
Pig iron ladle	Slag line Walls Bottom	44% Al_2O_3	830	Change in slag composition (by adding de-P and de-S treatment)	Life decrease	Use of materials based on Al_2O_3 - C - SiC Improvement of products with 44% Al_2O_3 Cost reduction
Lances		High alumina	30	Increase of pig iron T (injection of gaseous O_2)	Life decrease	Widening of compositions with high Al_2O_3 content Improvement of bond conditions Optimization of design (Iron structural work - refractories)

— Table 1 —

TECHNICAL PROBLEMS TO SOLVE WITH REFRACTORIES FOR THE PROCESS OF COMBINED INSUFFLATION CONVERTERS

1. Extension of the life of tuyere and bottom bricks; materials with high spalling resistance (use of fine ceramics technology).

2. Improvement of corrosion resistance:
   — use of pure raw materials (MgO, graphite)
   — increase of bulk density.

3. Lower cost of refractory linings:
   — use of advanced repair technique (e.g.: flame gunning)
   — use of natural raw materials (zoned lining)

4. Methods for preventing the distortion of iron structural work:
   — development of cooling technique.

— Table 2 —

BASIC TECHNICAL PROBLEMS TO SOLVE WITH LADLE REFRACTORIES

Future trends	Technical problems
Improvement of ladle reliability in general	— Use of shaped and unshaped basic materials. Development of new materials. Technical application definition for unshaped bricks (drying, temperature rise, etc). — Development of bottom (e.g. preshaped materials) and nozzle materials. — Development of insulating materials with high corrosion resistance. — Optimization of ladle lining thickness. — Definition of hot repair techniques (gunning repairs, etc.). — Techniques for monolithic material and brick evaluation

— Table 3 —

BASIC TECHNICAL PROBLEMS TO SOLVE WITH CONTINUOUS CASTING REFRACTORIES

Tundish Refractories	Submerged nozzle refractories
Measures for adding adjustment functions: - components - temperature	Development of highly corrosion-resistant material: - Development of material for nozzle used for low killed steel
Removal and prevention of mixing of inclusions in steel: — Removal: Development of effective refractories — Prevention of mixing: - Prevention of reaction with molten steel. - tight closing of tundish.	Improvement of equipment reliability: — Establishment of non-destructive inspection technology of nozzle — Investigation and action against thermal shock cracks on nozzle
Establishment of continuous use (hot cycle) technique: — hot cleaning — hot repairing	Establishment of repair techniques: — Gunning, etc.
Improvement of maintenance: — Mechanization of dismantling operation — Mechanization of work (board coating)	Establishment of technique for preventing nozzle clogging
Study of capacity and profile	Use of new ceramics
Low cost refractories	Effective refractories capable of blowing stable gas
	Low cost refractories

— Table 4 —

4.2. Trends for refractories connected with the future requirements of steel manufacture processes.

In Europe and in Italy as well steel production will be increasingly centered on diversified qualities. The objective of purer (very small C,P, S, N contents) and cleaner (inclusion of very small dimensions: $< 20\mu$) steels will be constantly pursued, the demand for high alloy steels will increase.

The possibility of developing some new siderurgic processes will largely depend on the possibility of developing adequate refractory products. Among the new siderurgic processes we shall mention:
- process of continuous steel production by using scrap iron as the main raw material (problem of resource and energy saving);
- process called "smelting reduction" (reduction in Mn and Cr minerals inside and outside - e.g. electric arc furnace - and reduction in iron mineral in B.F.);
- direct combination of continuous casting and hot-rolling (CC - DR);
a process which is going to have great development in the future, due to the relevant resource and energy saving, to the possibility of reducing slab stocks of increasing flexibility and delivery rapidity.

The future development of CC - DR is likely to be a direct casting into the product final shape, through the continuous casting process and the exclusion of hot rolling: a rivolutionary technology indeed;
- high speed rotating continuous casting;
- vertical continuous casting.

The two last mentioned processes have been developed, in the U.S. and Europe, on the basis of national projects of study.

For these future projects there still are many problems to be solved as regards refractories: development of compound materials for the diversification of steel production processes, use of basic materials for steel purity and cleanliness, problems of air insulation by gas injection and the so-called direct processes (e.g. CC - DR), the use of unshaped refractories for reducing costs.

REFRACTORIES TO MEET FUTURE STEELMAKING REQUIREMENTS

C W HARDY
British Steel - Technical, Teesside Laboratories
Grangetown, Teesside, UK

P G WHITELEY
GR-Stein Refractories Ltd., Central Research Laboratories
Worksop, Nottinghamshire, UK

Summary

Important refractories developments are needed for steel production in integrated iron and steelplants. Improved materials and design/construction techniques are highlighted for extending blast furnace campaign life together with the application of new intermediate repair techniques. Further improvements in refractory linings for hot metal transport/transfer ladles will be needed with the increasing requirement for hot metal pretreatment.
Operating conditions in basic oxygen steelmaking converters should be less severe but operators will require increased campaign life which could result in selective introduction of water-cooling to reinforce the refractory lining. Secondary processing of steel is assuming increasing importance and results in a demand for improved refractory linings which are able to combat an increased range of slag composition.
To meet these increased demands the refractory manufacturers and their customers are placing greater emphasis on independently assessed Quality Assurance and the application of Statistical Process Control techniques during manufacture to improve quality consistency and known continuity of supply.

1. INTRODUCTION

The blast furnace/oxygen converter process route will continue to be the major source of liquid steel worldwide in the foreseeable future. Increasing demands will be placed on the production units in terms of high productivity and improved cost/thermal efficiency, which in turn place greater requirements on the refractories and the way in which they are applied.
Greater emphasis is placed on refractories associated with ironmaking in this presentation as there are a number of others more specifically relating to steelmaking applications.

2. BLAST FURNACE

Annual pig iron production worldwide has ranged 450-550 million tonnes since 1975 with the western world contributing around 300 million tonnes per annum in the same period. Most of that tonnage is made in blast furnaces which, over the years, have drastically reduced in number, but have dramatically increased in size and productivity. They include large modern furnaces that are specifically designed to operate at 8000-10 000 tonnes hot metal per 24 hours period (tpd) and a large number of smaller furnaces which were initially working at around 1000-1500 tpd but are now upgraded and operating at 2500-3500 tpd.

To achieve the requisites of high productivity associated with improved cost/thermal efficiency, it is necessary to keep the furnaces in regular operation over long time periods and to keep downtime for relines and repairs to a minimum.

As a result, it is necessary for the ironmakers, the furnace designers and the refractory manufacturers to work in close liaison to produce a furnace with a water-cooled lining to achieve a campaign life of 8-12 years compared with the typical 5-7 years currently achieved.

2.1 Hearth Pad and Walls

The developments that have taken place on the Redcar blast furnace are of interest. The furnace is the largest in Western Europe and has already produced 15 million tonnes hot metal in its first campaign (10.1979-04.1986) on a liquid undercooled/peripheral stave cooled carbon hearth. The essentials of the initial hearth design and the new design of the current hearth are shown in Figure 1.

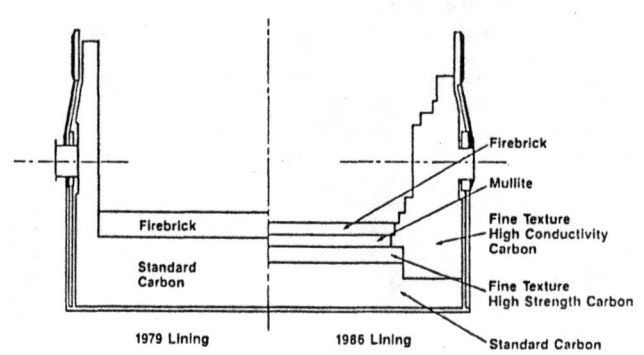

Fig. 1. Redcar blast furnace hearth

In general the hearth in the first campaign performed well and, as predicted, there was considerable residual thickness in the pad with the penetration having only occurred into the upper layer of carbon. The firebrick ceramic layer on top of the pad was known to have remained intact for a considerable period. However, a soft zone was found behind the working face in the hearth sidewalls.

A target was set for 10 years trouble free life for the second campaign. The design of the centre part of the hearth pad has remained virtually unchanged but there have been changes to materials used in likely iron contact areas and the overall depth of the pad has been reduced. This has been achieved by replacing the lower half of the firebrick protection lining by volume stable high alumina brick of high mullite content and reinforcing the top layer of anthracite base carbon blocks in the pad with a more iron erosion resistant grade micropore carbon. The hearth wall construction has been redesigned with increased lining thickness in the area of the tap-holes and lower part of the walls. Again the newer micropore carbons have been installed but in this area, the blocks have higher graphite content to increase the thermal conductivity (10 W/mK compared with

4 W/mK) and to make a product which is considered to be more durable and more capable of withstanding thermal variations. As a final measure the carbon is protected by high alumina and firebrick in the lower hearth wall.

The newer micropore grades have modified textural properties resulting in very fine pore structured materials of low permeability and high strength as shown in Table 1.

TABLE 1. Typical properties of carbon used in blast furnace hearth

Physical Property		Standard CARBON	Fine Textured ANTHRACITE	CARBON ANTH./GRAPH.
Apparent porosity	%	14	14	15
Pore diameter <5µm	%	28	90	95
Bulk density	g/cm³	1.50	1.54	1.61
Cold compressive strength	N/mm²	41	52	48
Permanent linear change - 2hr/1500°C (red)	%	+0.1	+0.1	-0.1
M.M. abrasive index		130	130	150
Thermal conductivity at mean temp.	W/mK 200°C 400°C 600°C 800°C	5.2 6.0 6.4 6.8	5.2 6.0 6.4 6.7	13.5 13.2 13.0 12.4

Typically the proportion of pores <5 microns in diameter is increased to over 90% compared with 25% in the equivalent standard products as shown in Figure 2.

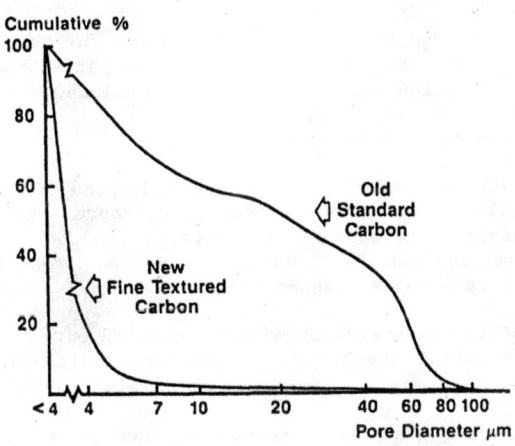

Fig. 2. Pore size distribution

Resistance to iron penetration is determined by loading iron into drilled blocks which are heated to 1500°C and pressurizing the iron to force it into the carbon. This technique is adopted to reproduce an equivalent ferrostatic head of 1.46 m iron together with a blast pressure of 2 kg/cm².

Comparisons of the amount of penetration are made after cooling. The benefits that can be achieved in both resistance to iron penetration and molten alkali attack by the introduction of micropore carbons are shown in Figure 3.

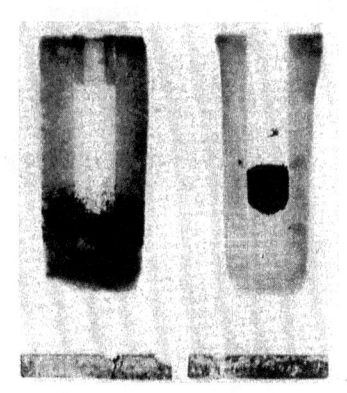

Standard carbon   Fine textured carbon

Fig. 3. Resistance to iron penetration and molten alkali

Whilst adopting this approach, due consideration must be given to the deformation characteristics of these new products, especially when carbon blocks up to 3000 mm in length are to be installed to a 0.5 mm joint tolerance. If they are too brittle and the stress/strain generated in service is too high, localised wear could be exaggerated by micro-cracking mechanisms.

2.2 Tuyere Belt

Increased use of tuyere injectants will place further demands on the refractory construction, and so extra care and attention is required regarding material selection and installation techniques - especially with regard to expansion allowances.

Various balances have to be achieved:

i   tight well installed structure with ease and speed of installation;
ii improved alkali and thermal shock resistance against excellent oxidation resistance to steam atmospheres;
iii minimum joints against high quality physical properties in large blocks as well as machine pressed bricks.

Preassembly of tuyere surrounds will be carried out at manufacturers' plants to a greater extent whether large precast alumina-chrome blocks or smaller nitride/sialon bonded silicon carbide or high alumina brick shapes are used.

Extra care is required with refractory design to ensure minimal difficulty in replacing tuyere coolers during the campaign.

2.3 Bosh, Belly and Lower Stack

This area of the furnace is frequently referred to as the high heat zone and is considered to be the critical part of the refractory lining, i.e. under normal circumstances the campaign life will depend on the refractory performance in this area. The main problems affecting the installed refractory lining in service are:

alkali and chemical attack,
high heat load,
thermal fluctuations,
oxidation - more likely at tuyere level and the lower part of the bosh,
abrasion from descending burden and ascending gases - more likely in the lower stack.

TABLE 2. Comparison of properties of blast furnace refractories

	SILICON CARBIDE	SEMI-GRAPHITE	CARBON	90-95% Al_2O_3	DENSE 45% Al_2O_3
Thermal shock resistance	V. Good	V. Good	Good	Fair	Poor
Cooling efficiency	Good	V. Good	Good	Fair	Poor
Alkali resistance	V. Good	Fair	Fair	Good	Poor
Resistance to iron dissolution	Fair	Poor	Poor	Good	Fair
Corrosion FeO	Good	Good	Fair	Fair	Fair
Resistance CaO	Good	V. Good	V. Good	Good	Fair
Abrasion resistance	V. Good	Poor	Fair	V. Good	Good
Oxidation resistance	Good	Poor	Poor	V. Good	V. Good

A comparison of some of the commonly used refractories is shown in Table 2 in terms of their relative ability to resist various parameters likely to be encountered in service. Of these silicon carbide products are being increasingly used in the high heat zone. These are fired bricks which are based on fused α-SiC grain and one of these matrix-binding systems:

i. β-SiC
ii Si_3N_4
iii SIALON

Both nitride bonded and β-SiC bonded products are well established in this application and Sialon bonded silicon carbide is being evaluated.

As shown in Table 3, all three types of product are typified by low porosity, high mechanical strength (resulting in tight texture and excellent resistance to abrasion) and high thermal conductivity - the latter in combination with the high strength, low thermal expansion and modulus of elasticity, ensure that these refractories have superior thermal shock resistance over most other blast furnace refractories. Total oxide contents are low which results in improved slag resistance. All three types have good resistance to attack by alkali vapour but some are more prone to molten alkali attack than others, which has led to the introduction of the Sialon bonded products.

TABLE 3. Silicon carbide refractories for blast furnace high heat zones

Bond System		β-SiC	Si_3N_4	Sialon/Si_3N_4
Bulk density	g/cm³	2.65	2.62	2.68
Apparent porosity	%	15.0	14.5	14.5
Modulus of rupture at 20°C in air	kg/cm²	435	434	470
Modulus of rupture at 1350°C in air	kg/cm²	425	427	476
Thermal expansion coefficient x 10^{-6}	°C	4.9	4.7	4.7
Thermal conductivity				
Mean temp.	400°C	34.5	24.5	29.6
W/mk	600°C	30.7	21.5	23.5
	1000°C	23.3	19.2	18.9
	1200°C	20.3	17.5	17.0
Molten alkali test (Bethlehem Steel Corp.)				
Weight change	%	+0.4	-9.0	+2.6
Residual MOR	kg/cm²	206	33	520
Change in MOR	%	-48.7	-92.0	-1.9
M.M. abrasion index		20	20	20
Young's Modulus	kg/cm²	1.4×10^6	1.7×10^6	1.9×10^6
Pore sign distribution				
80% pores have radius		<0.4	<2.0	<2.0
50% pores have radius (in microns)		<0.15	<0.75	<0.75

No known refractory will achieve the required campaign life in this area of the blast furnace unless it is designed to operate with an effective water-cooling system. This latter may take the form of heavy stave coolers (as at Redcar and Taranto), or sophisticated close-space copper internal plate cooling systems as used elsewhere in British Steel, Hoogovens etc. In both systems great care has to be taken to ensure:

i   maintenance of good structural stability and thermal contact between the refractory lining and the cooling system;
ii   prevention of high temperature gases flowing behind the refractories/cooling system;
iii adequate compressibility to minimize setting up excessive stresses/cracks in the steel shell.

Further work is needed to develop new and better materials/installation techniques in these critical applications.

An alternative method of extending the campaign life of the furnace is to carry out a number of mid-campaign repairs to the high heat zone refractory lining. The concept is not new but emphasis must be placed on speed and safety aspects of the repair, and the following were, until recently, the limited choice of alternatives:

i   Reline the furnace - costly and time consuming.

ii   Grout injection using the burden as the shutter - extremely quick but of limited effectiveness.
iii   Manual refractory spraying - empty and cool the furnace, scaffold and gun which involves too much downtime
- partially empty and seal a live furnace and gun manually from a platform (safety problems).

Blowing down of the furnace in very short periods (around 24 hours) has made quick refractory spray repairs feasible. The introduction of remote control gunning/spraying from outside the furnace has removed the problems of safety aspects, and up to 350/400 tonnes of spray material has been effectively applied in furnaces at Scunthorpe. Improvements in material quality and make-up are required to make this technique more suitable for repairs in the high heat zone of the furnace.

Benefits claimed for this practice are:

i   lower initial capital investment
ii   lower operating fuel rates
- cooling members increasingly protected by refractory coatings
- restoration of the furnace profile resulting in improved burden distribution.

3. HOT METAL TRANSPORT AND TRANSFER

For many years the working linings of torpedo ladles used to transport hot metal from the blast furnace to the basic oxygen steelmaking plants were made up of either fired high alumina (53% to 86% Al_2O_3) or tempered/fired doloma. In recent years more emphasis is placed on hot metal quality in terms of low and controlled metalloid content (Si, P and S) which has resulted in increasing amounts of pretreatment being carried out in the blast furnace runner system, the transport and/or the transfer ladle.

The increased use of metal pretreatment has shown up the limitations of the conventionally used refractories which were prone to excessive joint attack and reaction slabbing, reduced campaign life and hence increased refractory costs.

N.B. De-siliconization
- high temperature
- low C/S ratio 0.8-1.2
- mill scale or other fine iron oxide additives.

De-phosphorization
- lower temperature
- high C/S ratio 2.5-4.5
- fluorspar and reactive lime additives.

De-sulphurization
- high temperature
- high C/S ratio >2.5
- can be achieved by a wide variety of additives, e.g. reactive lime
    calcium wire
    calcium cyanamide
    coated magnesium metal
    soda ash, etc.

Where possible, it is preferred to desulphurize the hot metal in the transfer ladle within the steelmaking plant to reduce emission problems and take advantage of existing fume extraction equipment.

In this respect, it is important to be able to test refractories quickly and cost effectively within the laboratory to establish the most suitable refractory working lining configuration for a given ladle practice.

Methods normally used involve expensive graphite crucible induction furnaces or gas fired rotary slag tests - the latter have been found to be very time consuming and, due to the limited attack, a poor test to differentiate between refractory qualities. More recently we have developed a modified crucible test in which both granulated iron and powdered slag are charged. The brick crucible is covered with a lid and fired at top temperature for three hours. The lid is removed, a further charge of slag added and the firing cycle is repeated. After cooling, the 'crucibles' are cut in two and examined.

Figure 4 shows the extent of slag reaction and penetration in a series of fired high alumina products including a chromic oxide doped synthetic alumina brick (Bricks 1 to 5). In comparison the improved performance of heat treated resin bonded alumina - silicon carbide - carbon composite bricks is shown in Figure 5 where:

Brick 6 - standard quality
Brick 7 - alternative standard quality
Brick 8 - alumina enriched for improved slag resistance
Brick 9 - further alumina enrichment.

Fig. 4. Slag test on fired high alumina

Fig. 5. Slag test on Al_2O_3-SiC-C bricks

Test firings were carried out under oxidizing conditions and comparison of Bricks 6 and 7 show the importance of using low grain porosity refractories, especially if the bricks are to be installed in the upper parts of a lining in a torpedo ladle that is known to be only used intermittently and is allowed to stand empty for significant periods.

Benefits of selecting the correct resin binder and use of suitable additives are seen in Figure 6 where:

Brick 6 - standard quality
Brick 6A - as Brick 6 with alternate resin system
Brick 6B - as Brick 6A with metal additives.

Fig. 6. Slag test - effect of binder

Further work is being carried out to investigate the air/slag/refractory and the metal/slag/refractory interfaces over a range of slag compositions.

Although it was initially considered that alumina-silicon carbide-carbon bricks would be required when extensive pretreatment of hot metal is carried out, we are already experiencing increased use of these refractories for conventional hot metal transport because of:

i   lack of open joints,
ii   limited penetration of slag into the brick (easier deslagging operations),
iii less tendency to crack behind and parallel to the hot face,
iv reduced wear rate resulting in a very much reduced need to carry out gunning maintenance,
v   improved performance of the impact pads,
vi improved ladle availability and as a result lower number of ladles required to be in operation.

4.   <u>BASIC OXYGEN STEELMAKING</u>

Future developments in fully integrated iron and steelmaking are very much aligned to a consistent supply of known quality hot metal of controlled low silicon, phosphorus and sulphur content which will be effectively decarburized in the basic oxygen converter and then refined into the required steel quality in the ladle, using the benefits of vacuum degassing and other secondary treatments as required.

4.1 Extended Campaign Life

On the basis of the above, it can be considered that the conditions under which the refractory linings operate should become easier and the demands on the refractory less difficult. However, the steelmaker will be looking for increased campaign life to ensure that the operation is more cost effective. Productivity rates are likely to increase and, in the longer term, the steelmakers will want to reduce the amount of gunning maintenance that is currently being used to increase campaign life.

Benefits of combined top and bottom blowing over top blowing are now well established for a number of steelmaking operations. Improvements will have to be made to increase the performance of the bottom agitation elements either by the introduction of replaceable elements or preferably by the development of improved more thermal shock resistant magnesia-graphite products with special additives to improve further oxidation resistance and mechanical strength.

At the same time as making developments to improve overall bottom life, consideration must be given to the 'knuckle' area which is unavoidable with dome-bottom converters. Stadium bottom constructions have given some improvement but there are still weak areas in the design. There may have to be radical changes to the complete design of the 'knuckle' area to enable a solid block working lining to be installed rather than the current practice of installing the bottom lining and wall linings as separate entities and installing a ramming make-up.

Wear characteristics of the barrel/trunnion working linings may change as the slag bulk will probably be reduced and hence lessen the chance of the working lining picking up a protective slag cover. Refractories with higher mechanical strength and improved oxidation resistance but at the same time having the resilience of existing magnesia-graphite bricks will be required.

Existing long campaign lives (1500-2500 heats) are already showing up problems of stability in the cone linings. Consideration has already been given to interlocking bricks and even tying-back the refractories to the shell, but we expect water-cooling will be introduced to extend campaign life. This step could easily affect the types of refractory that have to be used in conjunction with the water cooled members.

4.2 Clean Steel Through the Taphole

Operators will face further demands to produce 'clean steel' to meet even more stringent demands on quality. As a result, converter slag should be kept out of the ladle.

Increasingly steelmakers are looking towards improved, easy to use and dependable systems to prevent slag leaving the converter during:

i   initial tapping
ii   end of tapping cycle.

Typical systems in current use are shown in Figure 7 and comprise:

i   Intumescent plugs - to prevent slag flow on initial tilt.
ii   Slag dart - to prevent slag flow towards the end of tapping.
iii Induction coils - to detect slag in the metal flow.

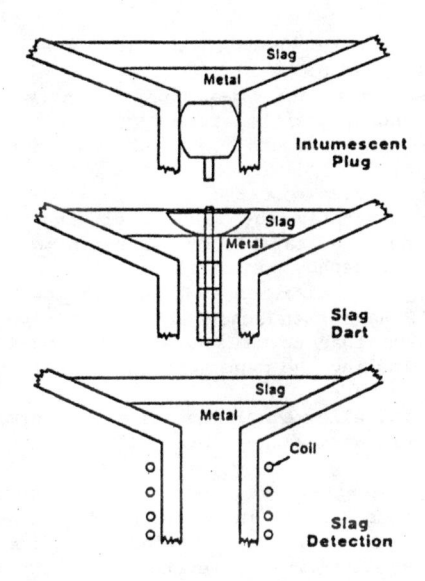

Fig. 7. Clean steel through the tap-hole

The intumescent plug is a semi-refractory plug which swells on application of heat to fill completely the irregular cross-section of the tap-hole. Because of its refractory nature, it has advantage in that it is placed well into the tap-hole in a position close to the hot face of the working lining.

During the blowing period the plug is capable of swelling in excess of 100% of its diameter to temporarily block the tap-hole. On initial tilting, the slag does not penetrate through the plug, but on further tilting the extra heat of the steel causes the plug to collapse which allows normal steel tapping to proceed.

The dart is another semi-refractory component that can float in the slag. Towards completion of tapping the vortex action pulls the dart into the tap-hole and effectively stops slag flowing into the ladle.

Inductor coils can be installed around the tap-hole turret to monitor the tapping operation to ensure minimal slag is tapped into the ladle.

Benefits likely to be accrued over extended time periods are:

i   cleaner steel - less slag bulk,
ii lower phosphorus reversion from slag,
iii improved alloy yields - especially savings in aluminium and ferro-alloy additives,
iv less wear in tap-hole and improved steel flow-stream characteristics,
v   improved ladle refractory performance,
vi more efficient use of synthetic slag additions to the ladle.

5. LADLE TREATMENT

Refractory consumption in this part of the steelmaking process exceeds that of any other area in the integrated iron and steelmaking process. The situation is becoming increasingly complex with the oxygen converter essentially being used as a decarburizer to produce consistent quality metal of known composition which can then be made into the required quality steel in the ladle utilizing one or more of the following:

i   inert gas bubbling to homogenize the steel analysis and the temperature in the ladle,
ii circulation degassers for gas (particularly hydrogen) removal, decarburization and composition trimming,
iii modified circulation degassers, e.g. RH-OB, for production of ultra low carbon steels,
iv lance injection systems (powder or wire),
v   ladle furnace - currently the fastest growing secondary steelmaking facility worldwide, and capable of all refining operations other than decarburization and dephosphorization.
    NB. Adequate dephosphorization to levels as low as 0.001%P from steel tapped at 0.07%P have been achieved through tapping from the oxygen converter at lower than normal temperatures on to a basic oxidizing slag and then readling the dephosphorized steel for final composition trim in a ladle furnace,
vi VAD/VOD for special alloy steel requirements (normally associated with electric arc steelplants rather than fully integrated plants).

In addition the percentage of steel that is continuously cast rather than poured into ingots is running in excess of 80% in Western Europe, which places further demands on refractories in ladle linings, e.g. cleanliness and increased time of steel in the ladle etc.

Ladle linings based on durable low cost bloating firebricks and slung sand are no longer acceptable. Currently high alumina, zircon/silica or doloma are used for the normal lining reinforced with either high quality alumina, magnesia-graphite or high quality magnesia-chrome products. In the future, there is the likelihood of increased use of magnesia base refractories or improved qualities of doloma and high alumina products in the working linings. Special quality high purity direct bonded magnesia-chrome refractories are particularly required for degassing vessels and associated processes.

Generally the improved refractories (with the exception of magnesia-graphite) are more rigid throughout the temperature range and they all have increased thermal expansion, higher heat content and thermal conductivities than bloating firebricks. All these characteristics are interrelated in that the operator is looking for:

i   consistent high life
ii low heat flow and low shell temperature
iii minimal skulling in the ladle
iv balanced low cost refractory performance
v   low stressing on the ladle shell.

As refining of steel in the ladle increases, there will be a demand for improved high density, higher thermal conductivity refractories to be installed in critical areas of the safety lining as well as the working lining. This will necessitate the use of insulation materials as a back-up to reduce the heat flow through the refractory lining and reduce the steel shell temperature as shown in Figure 8.

Introduction of backing insulation does result in the working lining operating with a reduced thermal gradient and hence at higher overall temperature. In addition the backing lining will be operating at significantly higher overall temperatures - especially when the working lining is badly worn in localized areas, e.g. slag line. This, together with the higher thermal expansion characteristics of the refractories, and the more effectively cooled steel shell can give rise to circumferential stressing problems either in the refractory lining or the steel shell. To minimize this, greater care will be needed in drying out/heating up newly lined ladles, and improved control during operation to ensure that ladles are kept hot and clean.

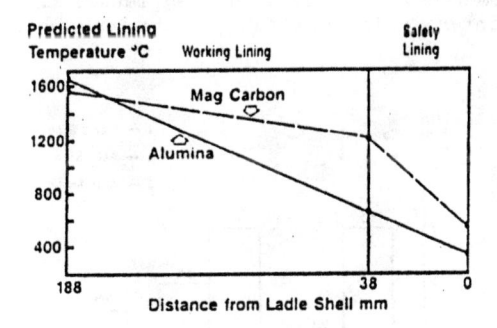

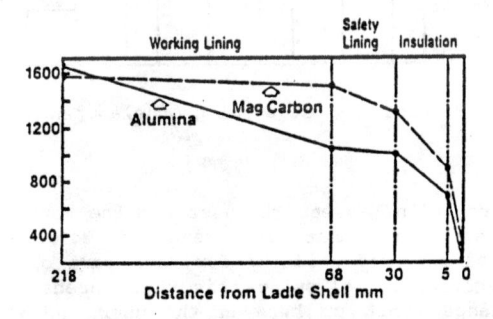

Fig. 8. Temperature distribution through ladle linings

A further problem that occurs with increased application of ladle steelmaking is the use of different synthetic slags to make specific steel qualities and the effects of different sequences. Synthetic slags of composition shown in Table 4 were reacted at high temperature with a chemically bonded chromic oxide doped high purity alumina brick.

TABLE 4. Synthetic slag compositions used in slag sequence testing (Fig.9)

		Basic	Slag Analysis Acid	Aluminous
Total Fe	%	5.8	2.2	9.4
SiO_2	%	16.1	36.0	7.1
MnO	%	4.3	4.4	5.1
MgO	%	4.5	4.1	5.8
CaO	%	58.5	39.3	42.7
Al_2O_3	%	10.8	14.0	29.9
C/S Ratio		3.9/1	1.2/1	6.4/1
C+M/S+A		3.0/1	1.1/1	2.2/1

Various sequencing programmes were carried out as shown in Figure 9 which indicate the cumulative attack by different slag compositions and the difference in slag penetration into the brick when, for example, there is a basic slag treatment followed by acid slag rather than the reverse being seen. Also of interest is the high penetration recorded with aluminous slags.

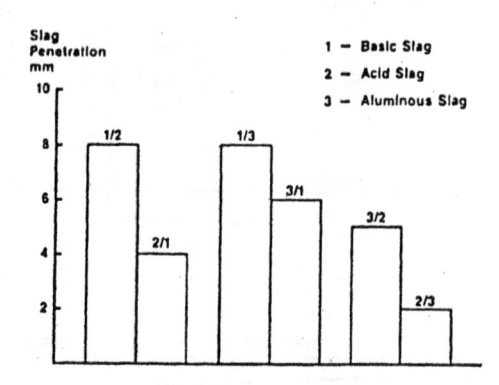

Fig. 9. Effect of sequencing on slag penetration

6. QUALITY IMPROVEMENT

Initial cooperation between the user and the manufacturer establishes the types of refractory that are required to meet the particular requirements of the refractory lining for each part of the process and at each specific steelplant. Following this there needs to be an increasing degree of confidence built up between the user and the manufacturer to ensure that consistent quality of the refractories is made and supplied to be installed and used in iron and steelplants.

In the United Kingdom, considerable effort and money has been expended in developing the whole Quality Assurance programme. This was initially carried out by the steel industry and interested manufacturers. It has now culminated with fully independent third party accreditation to EN 29000 and/or EN 29002.

There is a general awareness of the need for such systems on a fully international base. It is important to remember that such schemes audit the general business running of the suppliers, and it is equally important for the supplier and the user to agree 'product definitions' or specifications for the individual products that are to be covered by this scheme. In 1989, approaching 90% of British Steel's refractory requirements are supplied under EN 29000 series which results in more confidence in the quality and consistency of the product.

Currently manufacturers are being encouraged to implement and extend Statistical Process Control (SPC) techniques further to ensure that their products are made consistently to meet the agreed characteristics.

In summary, it could be said that in the United Kingdom, the steel and refractory industries are working to:

ZERO DEFECT PHILOSOPHY

Stage 1 - 2nd party user approval.
Stage 2 - Independent 3rd party accreditation.
Stage 3 - As per Stage 2 - further enhanced by Statistical Process Control.

ADVANCED MATERIALS FROM MICROCRYSTALLINE MAGNESITE
FOR THE MODERN STEEL MAKING INDUSTRY.

Z.E.FOROGLOU
FIMISCO
18-20 Sikelias Str., GR-11741 ATHENS

SUMMARY:
This paper outlines recent developments in the area of magnesia carbon refractories in particular of their lining performance in steel making as related to the crystal structure of the MgO component, i.e. MAGFLOT and other low-iron sinters of a natural origin.

INTRODUCTION:
All new or revised steel making techniques have set requirements for refractory products with improved resistance to wear and/or economic performance. The availability of such products is of importance both in establishing new technologies and in designing advanced refractory systems. The combination of a basic oxide such as MgO, a carbon carrier and a bonding agent based on pitch or a resin, proved to be the most suitable system to meet these requirements.

Magnesia carbon composite bodies show an improved resistance to thermal shock. In addition, the non-wetting of carbon components inhibits slag penetration and premature distruption of their structure, while the absence of slag penetrated zone diminishes hot face spalling during thermal fluctuations.

Magnesia carbon bricks were initially used for the lining of EAF. For many years they have been established as components in converter linings and related applications and had a significant impact on wear profiles.

The failure of MgO carbon linings occurs by corrosion and erosion while oxidation of the carbonaceous part of the brick impairs the strength of the refractory body. This reaction has been interpreted in terms of a decarburization rate and is also considered to relate to products formed between magnesia and carbon. It is known that the reaction rate is greatly affected by the structure and the impurities level of the raw materials involved. One way to increase the performance of the magnesia-carbon system is to select magnesia clinkers with built-in resistance to structural changes. The products presented in this report have been designed for the purpose, that is to say, to test selected components of magnesia carbon linings and evaluate their wear performance.[1] The results have shown a lengthening of the refractory lining resulting in an overall economy in the steel production.[2][3] The use of an all magnesia lining has been proved an economic solution offering a lining concept capable of meeting both the requirements of the new technology and those of a balanced wear.

2. RAW MATERIALS FOR BRICK MAKING:
The magnesia components of the linings presented in this paper consist of natural origin sinters, the conventional typ "K" and/or the type "F", derived from microcrystalline Grecian magnesite from deposits situated mainly on the Island of Euboea. The conventional sinter "K" is produced from magnesite concentrates beneficiated by heavy media, magnetic and photoelectric separations and subsequent dead-burning in a rotaty kiln. The sinter "F", a second generation product, is derived

from froth flotation concentrates of microcrystalline magnesite, caustic calcination of the up-graded ore, briquetting of the calcination product and finally dead-burning in a high temperature shaft kiln.

2.1. CHARACTERIZATION OF THE MAGNESIA's CLINKER:

The typical composition of the two grades of clinker "K" are given in Table 1

	E 21 A SPECIAL	E 21 A STANDARD
SiO_2 :	1.10%	1.35%
CaO :	2.10%	2.20%
Fe_2O_3 :	0.35%	0.50%
Al_2O_3 :	0.05%	0.05%
B_2O_3 :	< 0.01%	< 0.01%
MgO :	96.30%	95.80%
B.D.	3.45	3.45

Table 1: Composition of clinker "K".

The clinker type "F" is characterized by lower silica content and improved homogeneity while maintaining all desireable characteristics of the clinker "K".

The composition of the different grades of clinker "F" are given in Table 2:

	MAGFLOT SPECIAL	MAGFLOT-SA	MAGFLOT-A1
SiO_2 :	0.40%	0.50%	0.60%
CaO :	1.40%	2.10%	2.80%
Fe_2O_3 :	0.50%	0.50%	0.50%
Al_2O_3 :	0.05%	0.05%	0.05%
B_2O_3 :	< 0.01%	< 0.01%	< 0.01%
MgO :	97.50%	96.80%	96.00%
B.D.	3.43	3.43	3.42

Table 2: Composition of clinker "F".

The bonding phase in the clinker "F" consists mainly of dicalcium silicate distributed in thin partings or triangular interstices between the MgO crystallites. The high lime/silica ratio and the considerable amount of CaO present in the form of solid solution in the MgO matrix, (4) both contribute to the stability of the refractory silicate phase at high temperatures. The low alumina content and the absence of boron is a further safeguard in preventing the formation of low melting compounds. The size of the MgO crystallites of 150 to 200 microns and its shapes as expressed by the microstructure of the clinker is another determining parameter of the strength and the stability of the latter, depending critically on the raw materials used and the thermal treatment they were subjected to.

2.2. EVALUATION OF MAGNESIA CLINKER:

In assessing the potential of magnesia clinkers as components of MgO-carbon bricks, the clinkers were used for making test specimens along with other established grains (high purity synthetic magnesias and fused grains).

The results of this investigation have been presented in the First International Tokyo Refractories Conference. In the meantime, additional testing and field applications of composite bodies containing primarily MAGFLOT have shown that the microstructure of the clinker as expressed by the MgO size and shape of the crystallites its low porosity and the composition and distribution of the minor phases greatly decreases the rate of the brick wear. This type of microstructure with built-in resistance to structural changes is characteristic of clinkers derived from microcrystalline magnesite. The clinker "F" has shown a slag resistance comparable only to that of a fused grain.

3. BRICK QUALITIES AND SELECTION OF SAME

Different qualities and grades of magnesia carbon bricks have been used in recent years for lining converters as well as other steel making vessels. The following account on the brick lining of some representative vessels shows that the methodology presented in Tokyo for evaluating the magnesia clinkers and/or the composite bodies designed for the lining of the modern steel making vessels is a reliable one.

In most of the cases, the clinker "F" was the predominant magnesia component of the bricks used for the linings presented in this paper. Brick qualities and grades are designated as follows:

MC - pitch bonded magnesia carbon bricks (amorph carbon carrier)
MG - pitch bonded magnesia carbon bricks (crystalline graphite carrier)
RG - resin bonded magnesia carbon bricks
A,B,C,D - indicate residual carbon level as decreasing percentage from 15 down to 5%
1 - means antioxidant agent
L - indicates presence of fused grain as a minor component
T - indicates impregnation with pitch

In designing the converter linings consideration was given to the corrosive environment in the upper cone area, to abrasion in the charge pad area by the liquid metal and scrap, to erosion resulting from the movement of the metal in the bath and to the reactions of the slag with the refractory. Consideration was also paid to specific problems relating to vessels in various steel plants.

Modern steel plants favour technological innovations which, among other things, can influence the consumption of refractories. It was, therefore, necessary for the whole lining concept to develop a close cooperation between producer and user. Although the evaluation in discussion covers a short period, we feel justified for its presentation since it is acknowledged that these types of natural sinters have served with success the modern steel industry since the spread of the LD-process (early 60's).

In all the following cases, the linings have been designed with a zone concept, consisting of different grades and qualities for areas of differing stress.

3.1. STEEL PLANT NO:1, 3 X 230 TONS LD-LBE CONVERTERS

Since the adoption of an all magnesia lining in 1984, the converter lining life increased from 1350 to 2323 heats in April 1987- a European lining life record at that time. The composition of this lining has been presented in the Second International Conference on Refractories in Tokyo. In the same plant, in January 1989, a new European lining life record of 3434 heats was achieved, which brought about a further improvement in the overall cost per ton of steel. The composition of this lining is shown in Figure No. 1.

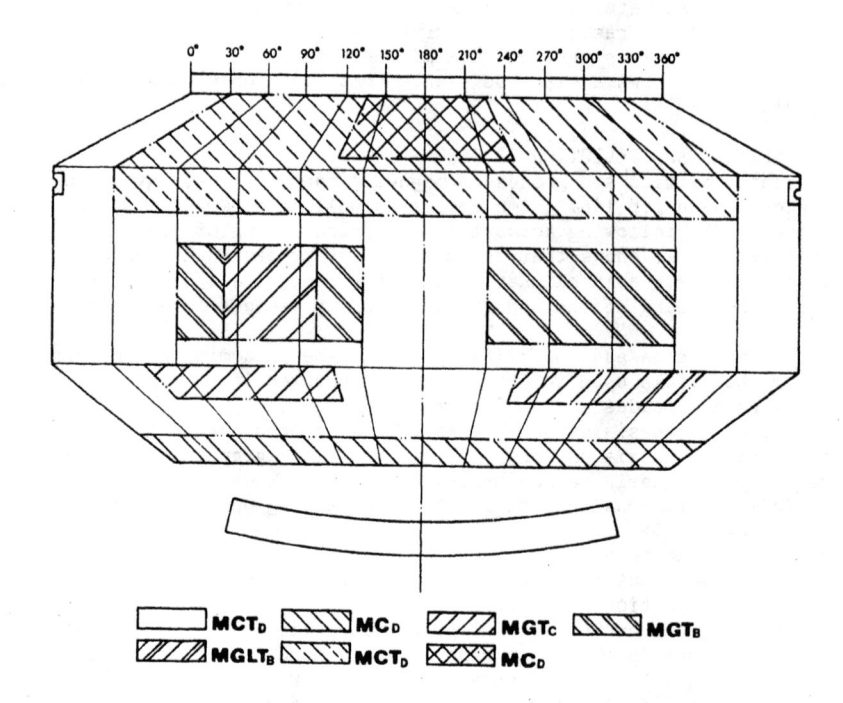

Figure No: 1

The predominant MgO component of the bricks used in this lining was sinter "F", and 30% of the vessel in the upper and lower cone areas was lined with qualities based on conventional sinters.

The residual carbon content, part of which was due to the pitch used as bonding and for impregnation, was on an average 7.0% out of which only 33% originated from graphite. Post impregnated bricks counted for 91% of the whole lining. These post impregnated bricks were characterized by a low open porosity, lower than 1%.

The chemical and physical properties of the bricks used are shown in Table 3.

	R.C./C.V.	B.D./σ	OPEN POR/σ	C.C.S./C.V.	MgO
MCT_D	6.10/6.14	3.13/0.023	1.10/0.439	553/11.58	96.70%
MCT_D	6.24/5.92	3.13/0.016	1.20/0.491	528/11.52	96.00%
MC_D	4.69/4.21	3.13/0.010	2.50/0.404	000/10.81	96.10%
MGT_C	9.72/7.32	3.07/0.018	1.50/0.397	338/11.60	96.10%
MGT_B	11.34/6.48	3.03/0.010	1.30/0.492	347/11.96	95.80%
$MGLT_B$	11.56/5.98	3.07/0.014	1.00/0.327	340/9.80	96.60%
MC_D	4.76/5.31	3.10/0.023	3.40/0.434	419/9.75	95.70%

Table No.3: Chemical and physical characteristics of lining No 1.

The MgO content ranges between 95,7 and 96,7%. Standard deviation and coefficients of variation, as measures of the spread of the individual values of the characteristics confirm reproducible processing and homogeneity of the products.

3.2. STEEL PLANT No:2, 2 X 300 LD-LBE CONVERTERS

This lining achieved, in April 1989, 3174 heats, which is a plant record lining life. The remaining thickness was enough for 100-300 additional heats. Its performance brought about a significant improvement of the cost factors. The development of the lining life in this plant was as follows:
Average lining life in 1986: 1630 heats
Average lining life in 1987: 1766 heats
Average lining life in 1988: 2688 heats (3100 was the highest performance)
April 1989 (FIMISCO LINING): 3174 heats

Figure 2 shows the composition of the lining.

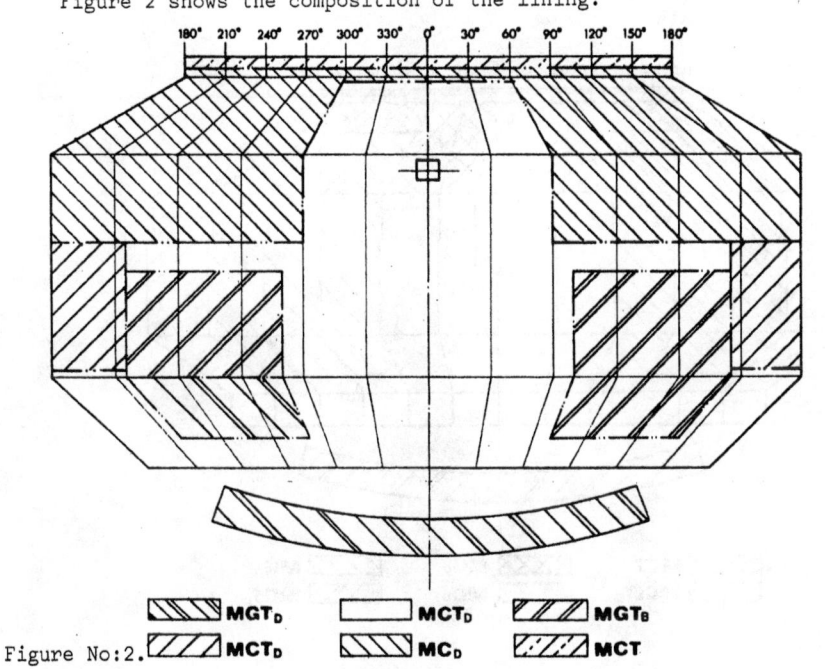

Figure No:2.

Bricks based on "F" sinter accounted for 25% of the lining and 77% of the whole lining were pitch post impregnated. The average residual carbon level was 6,8% and the MgO - graphite bricks represented only 38% of the whole lining.

The non impregnated bricks used had a low open porosity level, lower than 4%.

Table 4 gives the average values for the quality characteristics of the bricks used and the measures of the spread for the individual values.

	R.C./C.V.	B.D./σ	OPEN POR/σ	C.C.S./C.V.	MgO
MCT_D	5.67/5.43	3.13/0.017	0.9/0.376	562/10.60	95.90%
MC_D	4.96/4.38	3.10/0.022	3.7/0.726	401/7.43	95.70%
MCT_D	5.71/4.88	3.14/0.023	0.6/0.327	670/9.94	96.80%
MGT_D	7.12/8.50	3.08/0.019	1.0/0.238	370/11.36	96.10%
MGT_B	12.41/3.10	3.05/0.022	1.1/0.430	375/11.78	96.00%
MCT	1.98/3.42	3.11/0.011	1.6/0.307	1040/5.395	96.20%

The MgO content of the magnesia component ranges between 95,7 and 96,8%.

3.3. STEEL PLANT No:3, 1 X 170 TONS LD-LBE CONVERTER

The lining presented here reached in May 1988, 1870 heats, signifying a further improvement of the previous plant record performance of 1678 heats in 1987. This performance resulted in a new reduction of the cost. The composition of the lining is shown in Figure 3.

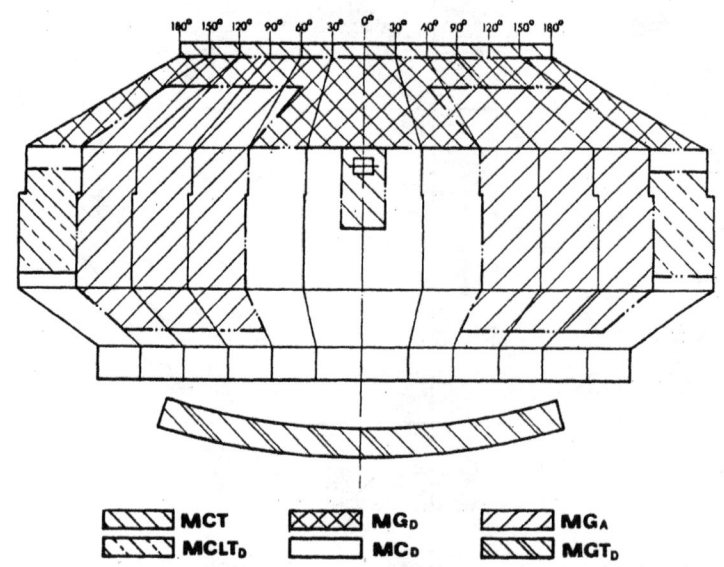

Figure No:3.

All the bricks were based on a "F" sinter. Post impregnated bricks counted for 20% of the whole lining. The residual carbon level on an average was 7,5% out of which 73% was from MgO - graphite bricks.

The average values of the qualities characteristics and the spread of individual measurements are given in Table 5.

	R.C./C.V.	B.D./σ	OPEN POR/σ	C.C.S./C.V.	MgO
MC_D	4.87/4.13	3.12/0.019	3.6/0.421	415/10.47	96.90%
$MCLT_D$	6.08/5.09	3.13/0.026	0.7/0.363	632/11.02	97.40%
MCT	1.98/3.42	3.11/0.011	1.6/0.307	1040/5.395	96.20%
MG_D	5.08/5.42	3.07/0.018	3.7/0.371	285/7.242	96.50%
MGT_D	6.99/5.54	3.10/0.012	0.6/0.181	460/8.558	96.50%
MG_A	11.48/6.13	2.99/0.015	3.7/0.507	223/7.091	96.40%

Table No:5.

The MgO content ranged from 96,2 to 97,4% and the porosity levels were 4% and 1% for non and post impregnated bricks respectively.

3.4. STEEL PLANT No:4, 2 X 102 TONS LD-LBE CONVERTERS

The lining presented here achieved 2056 heats in January 1989, denoting a new plant and Swedish lining life record. The lining life up to 1986 was 1520 heats. The record achieved resulted from redesigning the lining, and its composition is shown in Figure 4.

The 94% of the lining contained sinter "F" and only a small percentage 12% was made out of pitch post impregnated bricks.

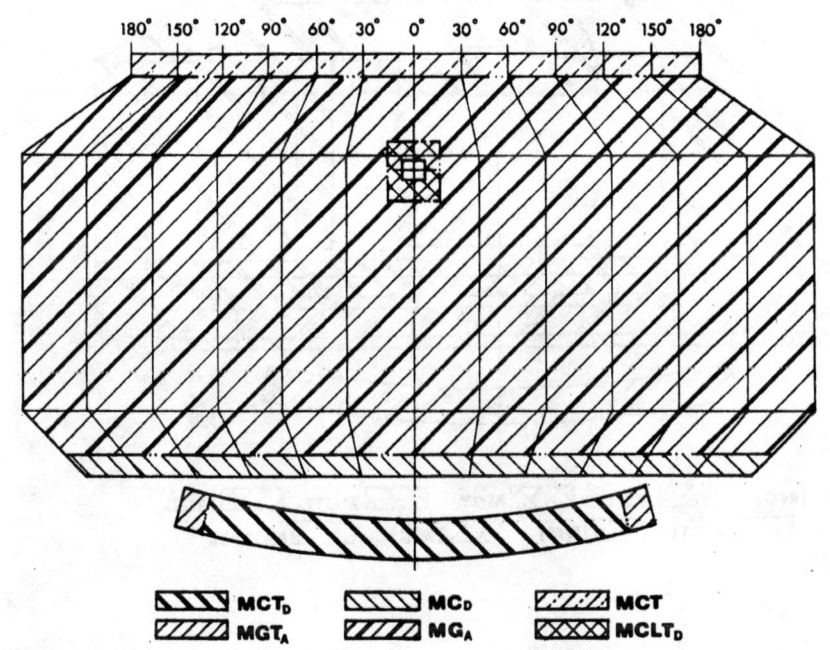

Figure No:4.

The average residual carbon level was 10,9% and the predominant carbon carrier of 89% was graphite. The MgO content of the magnesia part of the bricks ranged from 95,7 to 97,1%. The table 6 shows the quality characteristics.

	R.C./C.V.	B.D./σ	OPEN POR/σ	C.C.S./C.V.	MgO
MC_D	4.62/3.97	3.12/0.021	3.1/0.353	445/4.767	95.70%
MCT_D	6.12/5.23	3.15/0.004	1.0/0.141	435/11.29	96.80%
$MCLT_D$	6.10/5.44	3.13/0.015	1.4/0.231	346/10.48	97.10%
MG_A	11.62/6.37	3.00/0.018	3.6/0.478	247/8.72	96.10%
MGT_A	13.10/7.12	3.02/0.006	0.8/0.263	402/6.21	96.10%
MCT	2.13/0.237	3.14/0.015	1.1/0.152	995/8.44	95.70%

Table No:6.

3.5. STEEL PLANT No:5, 3 X 150 TONS LD-LBE CONVERTERS

A lining record of 1561 heats in March 1989 that is an improvement of 13,8% overshadowed the previous record of the plant. The composition of the lining with 9 qualities and grades, a rather complicated one, is shown in Figure 5.

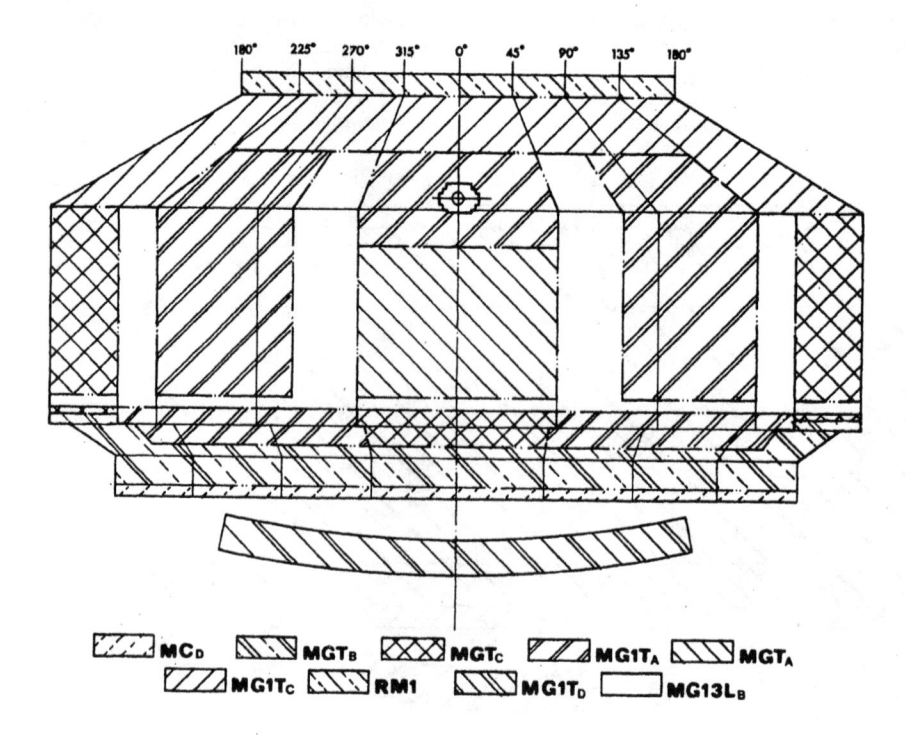

Figure No:5.

Practically all bricks, actually 98% of the whole, were based on "F" sinter. Post impregnated bricks accounted for 80%. The average residual carbon content was 10,8% and 96% of that was graphite bricks. The MgO content was as usually around 95% on an average.

The chemical and physical characteristics are given in the Table 7 and confirm well controlled porosity levels and reproducible production.

	R.C./C.V.	B.D./σ	OPEN POR/σ	C.C.S./C.V.	MgO
MC_D	4.90/7.75	3.12/0.010	2.8/0.465	413/10.073	95.80%
MGT_C	9.33/4.69	3.07/0.021	1.2/0.418	373/11.96	96.20%
MGT_B	12.24/5.16	3.05/0.020	0.9/0.175	351/11.03	96.00%
MGT_A	13.04/3.49	3.04/0.014	1.3/0.405	302/11.68	95.80%
$MG1T_C$	9.46/5.43	3.06/0.024	1.0/0.461	411/11.97	94.10%
$MG1T_D$	6.40/6.12	3.10/0.024	0.7/0.372	429/12.64	94.70%
$MG1T_A$	12.29/4.12	3.03/0.028	1.0/0.460	337/11.95	94.40%
$MG13L_B$	11.14/6.37	3.01/0.011	2.3/0.521	237/4.53	94.30%
RM1	0.93/1.48	3.09/0.018	0.9/0.321	1260/11.47	94.50%

Table No:7.

3.6. STEEL PLANT No:6, 3 X 110 TONS LD-LBE CONVERTERS

The average lining life in the plant was, in 1986, 1222 heats. At the beginning of 1987 an improvement increased the lining life to 1393 heats and in December 1987, the lining life reached 1563 heats.

The composition of the lining is shown in Figure 6.

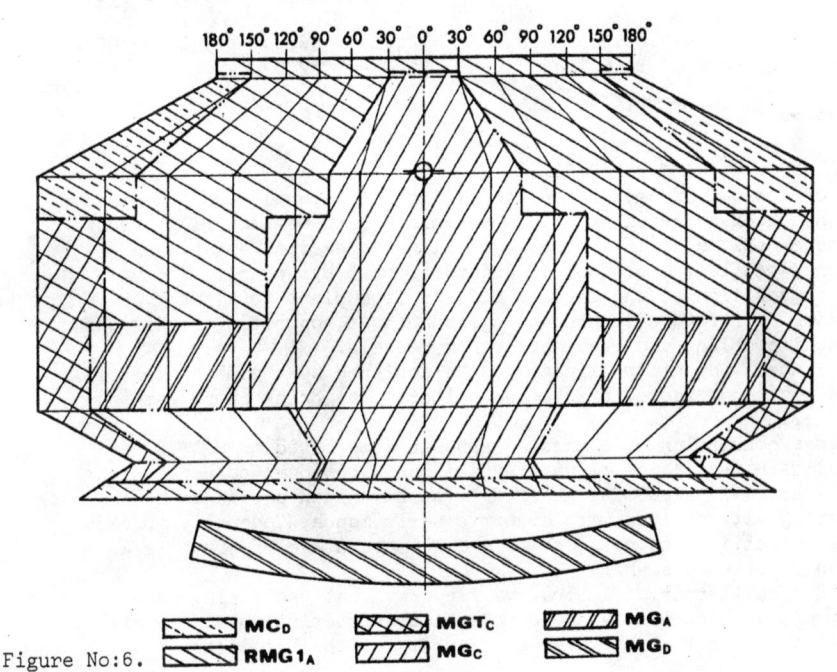

Figure No:6.

The lining is characterized, as the two previous ones 4 and 5, by a higher residual carbon level of approx. 9,5% as compared to the three first linings. In contrast, however, with other linings, the post impregnated part of the lining represented only 13% of the whole. The predominant MgO, close to 85%, was the sinter "F". The MgO content was on an average 96%. The chemical and physical characteristics are found in the table 8.

	R.C./C.V.	B.D./σ	OPEN POR/σ	C.C.S./C.V.	MgO
MC_D	4.68/4.38	3.10/0.024	3.0/0.501	430/11.01	95.90%
MG_D	4.95/5.62	3.08/0.07	4.1/0.282	360/11.29	96.00%
MG_C	7.60/6.92	3.03/0.021	3.5/0.51	290/11.59	95.80%
MGT_C	9.20/6.63	3.06/0.029	0.7/0.327	443/11.91	95.70%
MG_A	11.30/7.20	2.98/0.004	3.9/0.501	260/5.43	95.60%
$RMG1_A$	13.90/7.20	2.92/0.023	2.9/0.42	390/9.46	92.60%

Table No:8.

The main features of the linings discussed, are summarized in the following Table No 9.

LINING FEATURES	No 1	No 2	No 3	No 4	No 5	No 6
MgO %	96,0	96,2	96,8	96,4	95,0	95,8
"F" Sinter %	70,0	25,0	100,0	94,0	98,0	85,0
Avg.Res.Carbon %	7,0	6,8	7,4	10,9	10,8	9,5
Res.Carbon % from graphite bricks	33,3	38,0	73,0	89,0	96,0	85,0
Post-impr.bricks %	91,0	77,0	21,0	12,0	85,0	13,0

Table No:9.

CONCLUSIONS - CLOSING REMARKS:
1) A remarkable increase of the lining life of several European steel plants has been achieved by using magnesia carbon refractories based on sinters deriving from a microcrystalline magnesite variety, in particular MAGFLOT, a product of unique crystal structure.
2) The composition of the linings differs from plant to plant reflecting differences in processing techniques, operating conditions, individual problems and availability of vessels.
3) A balanced wear of the lining can be achieved by a combination of non and post impregnated bricks and selection of brick components and grades.
4) Linings consisting of magnesia-carbon bricks based on MgO sinters with built-in resistance to wear exhibiting a residual carbon content not substantially exceeding the 10% level have been proved to be interesting alternatives with economic performance. Generally speaking, European conditions do not justify a high percentage of graphite as has been reported elsewhere.
5) Field results confirm laboratory findings that the lining performance depends mainly on structural characteristics and related properties rather than the degree of purity of the MgO sinter.

6) All lining developments presented in this report should be
understood as a result of a close cooperation with the end users.

LITERATURE:
1) Z.Foroglou et al., First International Conference on Refractories
   Tokyo, Japan, Nov. 15 - 18, 1983.
2) Z.Foroglou et al., Second International Conference on Refractories
   Tokyo, Japan, Nov. 10 - 13, 1987.
3) J.P.Motte et al., La Revue de Metallurgie-CIT Mai 1988, p.381.
4) Foroglou Z., Malissa H. and Grasserbauer M.,
   Trans.J.Brit. Ceram. Soc. 79, 1980, p.81.

THE USE OF METAL POWDERS IN CARBON-CONTAINING REFRACTORIES

T. RYMON-LIPINSKI
Forschungsinstitut der Feuerfest-Industrie
(Refractory Industry Research Institute)
An der Elisabethkirche 27, D-5300 Bonn 1, Federal Republic of Germany

Summary

The use of antioxidants has recently been introduced as a way of preventing burn-out of carbon from carbon-containing refractory materials. On the basis of previous studies of the reactions of Al, Mg and Si additives in magnesia-carbon materials, the effect of metallic antioxidants can be explained in terms of the production of gaseous antioxidants. The aim of this project was to confirm this assumption on the basis of experiments with powdered molybdenum.

1. INTRODUCTION

Carbon-containing refractory materials are an essential part of modern industry, since carbon has proved itself to be an excellent material for substantially increasing the durability of refractory linings when used in conjunction with various oxides. This is thanks to its desirable properties, which result in an improvement of the properties of the entire composite - e.g. high thermal conductivity, low thermal expansion and resistance to wetting by typical metallurgical slags and liquid metals.
Carbon-containing materials are the best option in systems subjected to great chemical stress. However, carbon also has a weakness, i.e. its low resistance to oxidation, and since the durability of the material depends on the retention of carbon in the pores, efforts are being made to reduce the burn-out rate.
Studies of basic materials (particularly those based on MgO) provided a good starting point for investigating the behaviour of carbon-containing refractory materials in practice. These studies showed, for example, that the magnesium oxide can be reduced by the carbon in such materials. The magnesium vapour resulting from the reaction can diffuse in the direction of the hot side, where it reoxidizes to form a thick layer of secondary MgO. This process leads to the formation of a zoned structure in materials of this kind (Figure 1). According to many authors, the formation of the secondary MgO layer was one of the explanations for the improved resistance of carbon-containing magnesite bricks, in that it prevented the penetration of the corrosive agents into the brick. This would suggest that the addition of metallic magnesium might increase the thickness of the secondary MgO layer, and hence the resistance of such materials to corrosion (Ref.1).
This was confirmed in practical tests (Refs.1 and 2).
However, these tests also showed that the addition of Mg inhibited carbon burn-out. Similar tests were then carried out with other metal additives (Al, Cr, Fe, Fe-Cr). Since these metals have a greater affinity for oxygen than that of carbon, the metal should be oxidized first in a mixture consisting of both components and the carbon protected against

burn-out as a result of the oxygen being thus removed. However, it only proved possible to demonstrate the antioxidizing effect of magnesium in the subsequent laboratory experiments.

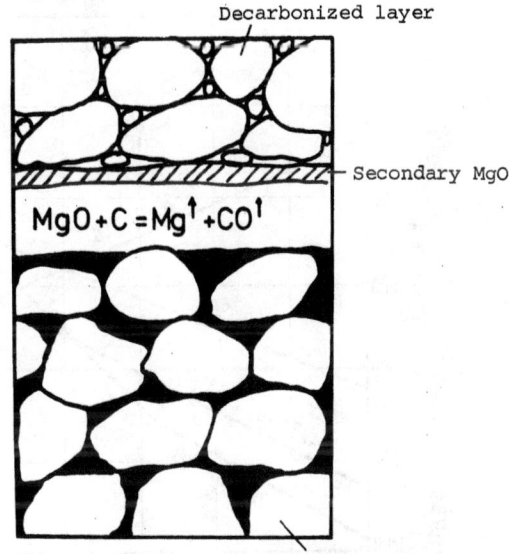

Fig. 1. Zoned structure of an MgO-carbon material after use at high temperatures

2. HOW METALLIC ANTIOXIDANTS WORK

The use of metal-powder additives in carbon-containing refractory bricks is already normal practice nowadays. They are used as antioxidants and often substantially improve the resistance of the carbon to oxidation. Understandably, attempts have been made to find a convincing explanation of how these additives work, since an understanding of the reactions in the bricks with the metal additives could provide a theoretical basis for the search for further and possibly even more efficient antioxidants.

Many authors have attributed the effectiveness of the antioxidants to the fact that they have a higher affinity for oxygen than that of carbon (Refs.1-5). However, the results of previous laboratory tests with various metals, as described above, have cast doubt on this theory and this was one of the reasons why we decided to study the problem. We took the example of carbon-containing magnesite bricks in an oxygen converter and studied the antioxidant properties of three metal powders, i.e. aluminium, magnesium and silicon, which are probably the most common metals used as antioxidants. Thermodynamic analyses were carried out before the experiments proper. The results of these experiments led to a hypothesis involving what we refer to as a **'gaseous antioxidant'**. (See section 6 for details.)

The first results showed that the efficiency of the metals as antioxidants was in fact in inverse correlation to their oxygen affinity, as shown in Figure 2, which contains the Richardson-Ellingham diagram and the results of an oxidation test. It can be clearly seen that silicon, the metal with the lowest affinity for oxygen (i.e. with the lowest free enthalpy of formation of the oxide) was the most efficient antioxidant. By contrast, magnesium, with the highest oxygen affinity, was the least efficient. These results clearly contradicted the widely-held view and showed that the effect of a solid antioxidant has little to do with its oxygen affinity.

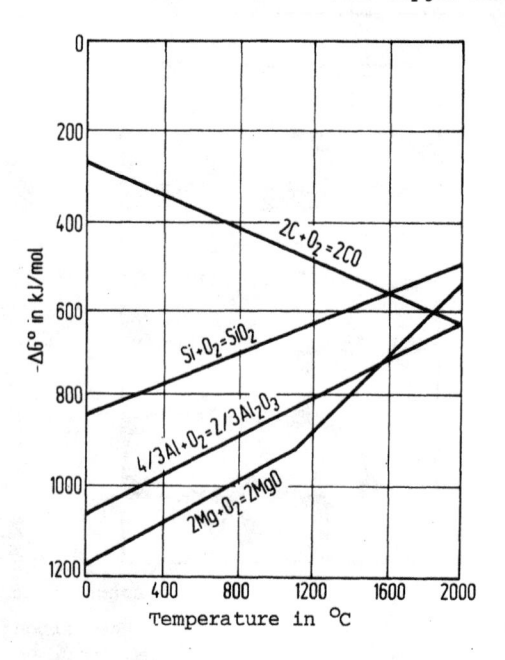

Fig. 2a. Richardson-Ellingham diagram for metals studied

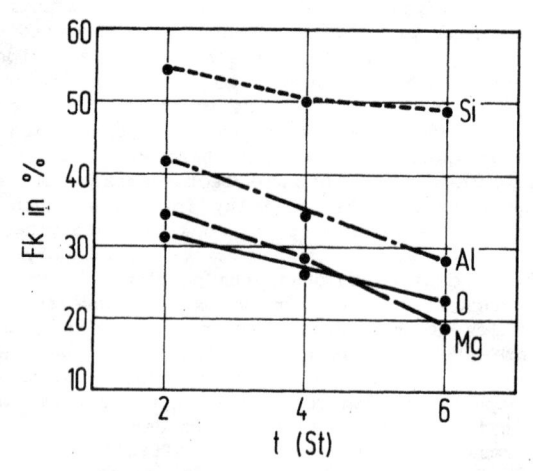

Fig. 2b. Results of oxidation test (Fk = surface of blackcore in %
0 = sample without antioxidant)

The results of both the theoretical considerations and the experiments led to the observation that the carbon in a metalliferous material was protected from oxidation by a reactive gas. Figure 3 shows the situation in the pores of the material. The carbon is surrounded by a cloud of gas, which effectively prevents an oxidant from reaching it. The figure also shows the course of the reaction, which results in reactive gaseous products such as magnesium vapour (in all three systems studied) and SiO (in bricks containing Si).

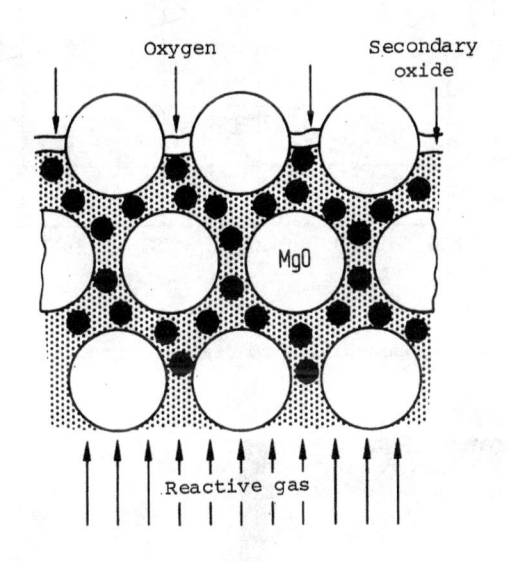

Al : 3MgO + 2Al = 3Mg(g) + Al$_2$O$_3$
Mg : Mg(s) → Mg(l) → Mg(g)
Si : MgO + Si = Mg(g) + SiO(g)

Fig. 3. Possible course of reaction in the pores of a material

Thus it can be assumed that it is not the reaction with a solid metal but with a reactive gas which can prevent oxidation of the carbon. Thus, this gas can be termed a 'gaseous antioxidant'. This best describes the antioxidation mechanism. The gaseous antioxidant is produced mainly inside the brick, i.e. where the atmosphere is virtually neutral. The temperature gradient causes the reactive gas to migrate towards the hot side, where it oxidizes. If the layer of secondary oxide produced in this way is impermeable, the carbon is still further protected from oxidation. Thus, the main function of the metallic additives is to permit the production of a gaseous antioxidant. This hypothesis should now be tested in further experiments.

3. MOLYBDENUM AS AN ANTIOXIDANT

This section describes a series of experiments testing metallic molybdenum as an oxidation inhibitor. As the Richardson-Ellingham diagram shows (Figure 4), molybdenum has a higher affinity for oxygen than that of carbon only up to approximately 800°K. If the theory of higher oxygen affinity of the metal were true, molybdenum would be totally useless for inhibiting the oxidation of carbon. However, the results of a laboratory test (Figure 5) showed clearly that the addition of molybdenum very effectively reduced the carbon burn-out rate.

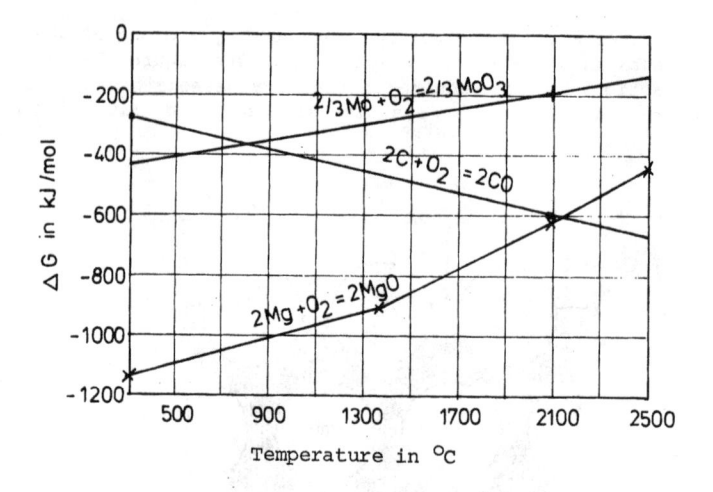

Fig. 4. Richardson-Ellingham diagram for MoO_3 formation

a) MgO carbon sample without additive

b) Similar sample with 5% Mo by weight

Fig. 5. Results of oxidation test
(calcination 2 hours at 1500°C in air)

Cylindrical test samples measuring 50 mm in diameter and height and consisting of 85% MgO (0-2 mm grain size) and 15% graphite (fine crystalline), or 80% MgO, 15% graphite and 5% molybdenum powder (<0.1 mm) were used for the oxidation tests. The test samples were calcined in air for 2 hours at 1 500°C, after which they were cut in two vertically.

The antioxidant properties of the molybdenum can again be explained in terms of the production of a gaseous antioxidant, which is obviously not the molybdenum vapour in a MgO-C-Mo system, since - as shown in Figure 6 - the partial pressure of the metallic molybdenum is too low (Ref.7). However, various oxides are known to be present in the Mo-O system. In addition to the solid oxides, such as MoO_2 and MoO_3, gaseous oxides such as MoO, MoO_2 and MoO_3 also occur.

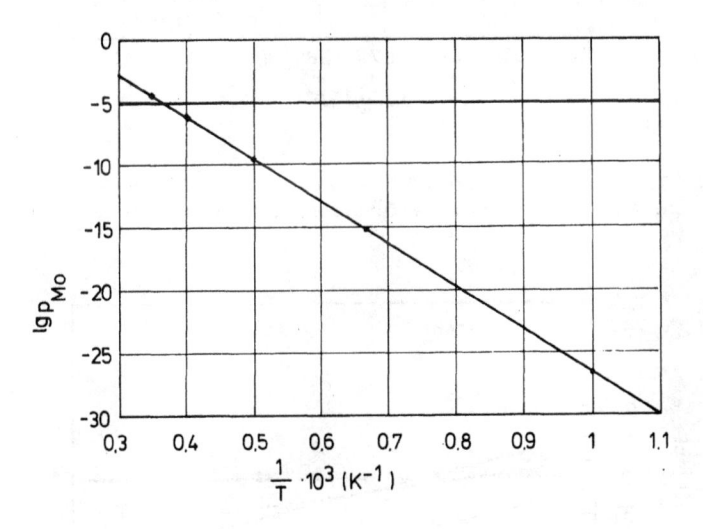

Fig. 6. Mo partial pressure as a function of temperature

In the gaseous phase, and particularly at high temperatures, the latter tends to form polymers of the $(MoO_3)_n$ series (Ref.7). Of the various oxides, gaseous MoO_3 or its polymer forms exhibit the highest partial pressure at temperatures up to approximately 3 000°K (Figure 7).

This pressure reaches a relevant value even at low temperatures. At 1 428°K the solid MoO_3 boils. These facts indicate that the gaseous phase can attain a substantial pressure in the system under consideration. However, since molybdenum has a maximum valency of 6, gaseous MoO_3 cannot act as a gaseous antioxidant, since it can no longer react with the oxygen and hence prevent it penetrating into the brick. However, the oxide is not stable in the presence of carbon, as shown in Figure 8. Even at low temperatures it can be reduced in the gaseous phase, with the formation of gaseous MoO_2 or MoO or even metallic Mo.

In further experiments, the samples were first studied following the oxidation test. The zoned structure is very clear (Figure 5). The outer, light-coloured zone contains no carbon and is fairly porous as a result of the carbon having been burnt out. The presence of magnesium molybdate ($MgMoO_4$) in addition to MgO was detected by means of X-ray diffraction (Cu_k).

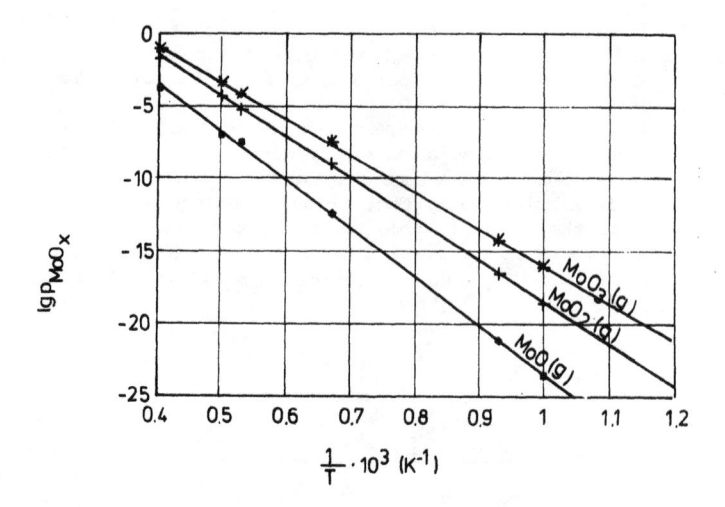

Fig. 7. Partial pressure of the various molybdenum oxides over the Mo-MoO$_2$ system as a function of temperature (Ref.7)

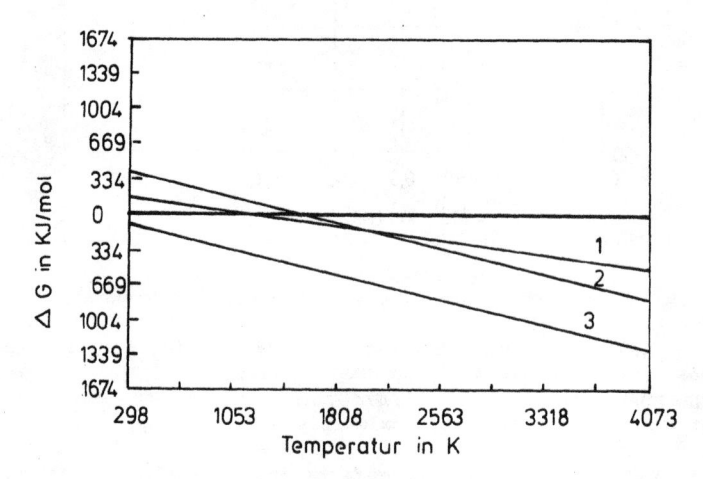

Fig. 8. Free enthalpy of reaction as a function of temperatures for the following reactions

(1) $MoO_3(g)$ + $C(s)$ = $MoO_2(g)$ + $CO(g)$
(2) $MoO_3(g)$ + $2C(s)$ = $MoO(g)$ + $2CO(g)$
(3) $MoO_3(g)$ + $3C(s)$ = $Mo(s)$ + $3CO(g)$.

The inner layer of the sample is black and not as porous as the outer layer. It still contains a great deal of graphite. The X-ray analysis showed that, apart from MgO, this layer contained little magnesium molybdate and very little metallic molybdenum. The two zones are separated by a clearly distinguishable white layer. It was possible to extract this layer for selective X-ray analysis, which showed the presence of MgO and - compared with the other layers - larger quantities of metallic molybdenum. This layer is continuous and very impermeable.

The layer dividing the decarbonized material from the unchanged material was also analysed with an electron microprobe. The results are shown in Figures 9 and 10. It can be seen that the spaces between MgO grains are densely packed with a white substance (Figure 9). A high concentration of molybdenum can also be detected in the boundary layer (Figures 10a and 10b). A chemical analysis of this layer showed that it consisted almost exclusively of metallic molybdenum.

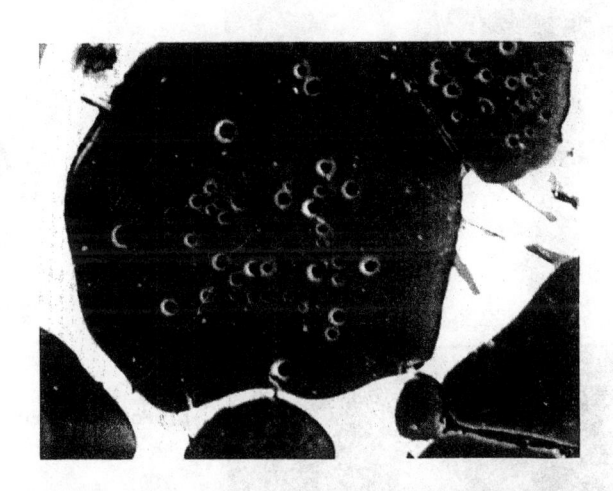

Fig. 9. Results of electron microprobe analysis of sample containing Mo, boundary layer, electron image

These results indicate the following reactions:

(1) $Mo(s) + 3/2 O_2(g) = MoO_3(s)$

(2) $MoO_3(s) \rightarrow MoO_3(g)$

(3) $MoO_3(g) + C(s) = MoO_2(g) + CO(g)$

(4) $MoO_3(g) + 2C(s) = MoO(g) + 2CO(g)$

(5) $MoO_3(g) + 3C(s) = Mo(s) + 3CO(g)$

(6) $MoO_3 + MgO = MgMoO_4$

(7) $MoO(g) + O_2(g) = MoO_3(g)$

(8) $MoO_2(g) + 1/2 O_2(g) = MoO_3(g)$.

They also indicate the following chemical process and explanation of the antioxidizing mechanism of metallic molybdenum. The oxidant (shown here as oxygen) which penetrates into the interior of the brick first reacts with the metallic molybdenum. The resultant MoO_3 reaches a significant partial pressure as soon as the temperature rises above approximately 1 000°K. The result of this is that the oxide in the gaseous phase is transported to those areas where the carbon has not yet been burned out (interior of brick).

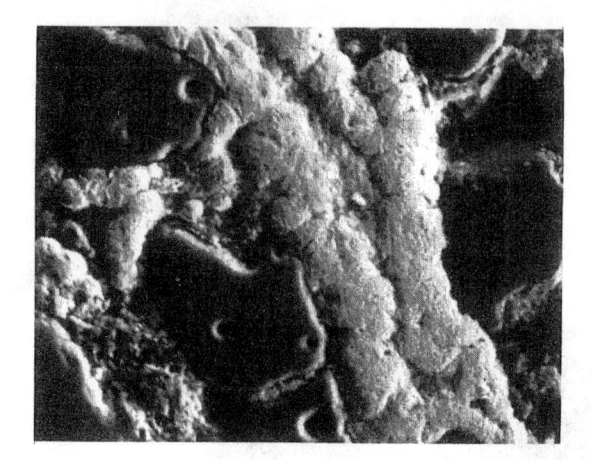

a) electron image

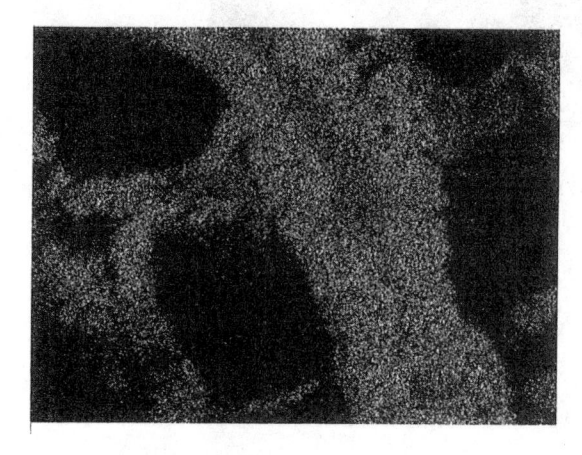

b) Mo-distribution

Fig. 10. Results of electron microprobe analysis of sample containing Mo, another point on boundary layer

Here the carbon reduces the MoO_3 to gaseous MoO_2 or MoO and metallic molybdenum. The reactive gases migrate towards the hot side, where they reoxidize. This binds the penetrating oxidant and hence protects the carbon from burn-out. The precipitated metallic molybdenum forms an impermeable layer, which further increases the antioxidising effect. The MoO_3 gas resulting from the reoxidation can be reduced again at the points where carbon is still present. The cycle continues until the gaseous MoO_3 is exhausted, e.g. in a reaction with MgO (formation of magnesium molybdate). The results reaffirm the importance of a reactive gas for the protection of carbon against oxidation. They show that the formation of such a gas is at least one of the prerequisites for effectively reducing the loss of carbon from refractory material. The formation is an impermeable layer at the boundary between the decarbonized material and the unchanged material can further enhance this process.

REFERENCES

1. RYMON-LIPINSKI, T. New variants in production technology for refractory materials for use in oxygen convertors (in Polish), Zeszyty Naukowe AGH No.472 Ceramika No.41 Krakow 1979.
2. RYMON-LIPINSKI, T. TIZ 107 (1983) No.12, pp.876-79.
3. YAMAGUCHI, A. Taikabutsu Overseas, 4 (1984) No.3, pp.14-18.
4. YAMAGUCHI, A. Taikabutsu Overseas, 7 (1987) No.2, pp.11-16.
5. WATANABE, A., TAKANAGA, S., GOTO, N., ANAN, K. and UCHIDA, M. Taikabutso Overseas, 7 (1987), No.2, pp.17-23.
6. RYMON-LIPINSKI, T. Stahl und Eisen 108 (1988) No.25/26, pp.1263-1267.
7. KULIKOV, L.S. Thermodynamics of Oxides, 'Metallurgia', publishers, Moscow 1986 (in Russian).

We thank the Arbeitsgemeinschaft industrieller Forschungsvereinigungen (AIF) for its financial assistance.

A NEW GENERATION OF PRE-CAST
REFRACTORY PRODUCTS FOR THE STEEL INDUSTRY

JACQUES SCHOENNAHL and CHRISTIAN NATUREL
Savoie Réfractaires, 10 rue de l'industrie
69631 Venissieux Cedex, France

Summary

The spin-off from work done in the field of highly compact concretes has enabled such progress to be made in pre-cast refractory products that their properties are often superior to those of ceramic bricks. Nowadays the use of very high performance materials thus enhances the numerous specific advantages of pre-cast systems over brick or monolithic facings placed on site.
Flexibility of form and shape are two major trump cards for the development of pre-cast products in the steel industry of tomorrow.

1. DEFINITION

Pre-cast refractory products are manufactured ready for use by processing the unshapeds at low temperature in a specialist workshop.
N.B. It will be noted that 'prepacked' sets are pre-cast products.

2. PRINCIPLES OF PRE-CASTING TECHNOLOGY

2.1   Manufacturing Process - Figure 1

Shaping takes place immediately after mixing, using moulds which produce the shape of the final product.
The most widely used technique is vibro-moulding, using vibrating tables or needles. Other techniques are pouring (for insulating materials) and sometimes ramming, vibro-compacting or uniaxial machine pressing.
With vibro-moulding the setting time is a few hours, which gives a very low rate of production per mould, i.e. 1-3 items per day.
Heat treatment is applied so as to give a product which is free from volatile matter and thus in particular the water of constitution.

2.2   Composition - Figure 2

A distinction is made between:

- The basic constituents: aggregate and bond. Conventionally the bond comprises all particles with a diameter of less than $30\,\mu m$. It also contains the hardening agent(s) to develop mechanical resistance in the monolith at ambient temperature.
- Additives: deflocculating agents, surfactants, setting agents (accelerators or inhibitors).
- Optional additives:
  . reinforcing fibres, e.g. steel fibres

```
┌─────────────────────────────────┐
│  PREPARATION OF RAW MATERIALS   │
└─────────────────────────────────┘
            │         ▲
            ▼         │
   ┌──────────────────┐      ┌──────────┐
   │ PREPARATION OF MIXES │◀─│ CHECKING │
   └──────────────────┘      └──────────┘
            │
            ▼
   ┌──────────────────────┐
   │ MIXING (ADDITION OF WATER) │
   └──────────────────────┘
            │
            ▼
      ┌──────────┐
      │ SHAPING  │
      └──────────┘
            │
   ┌─────────────────┐     ▼
   │ PREPARATION     │  ┌─────────┐
   │ OF MOULD        │  │ SETTING │
   └─────────────────┘  └─────────┘
            ▲              │
            │              ▼
      ┌───────────────────┐
      │ RELEASE FROM MOULD│
      └───────────────────┘
            │
            ▼
      ┌────────────────┐
      │ HEAT TREATMENT │
      └────────────────┘
            │
            ▼
      ┌────────────────┐
      │ QUALITY CONTROL│
      └────────────────┘
```

Fig. 1. Principles of the manufacturing process for pre-cast products

- Reinforcing fibres
- Aggregate 30µ to 20mm
- Shrinkage compensator
- **Bond**
  - HARDENER
  - Fines 1µ to 30µ
  - Ultra fines 0.01µ to 1µ
- **Additives**
  - Deflocculating agent
  - setting modifier
- Organic fibrils

Fig. 2. Principles of the composition of pre-cast products

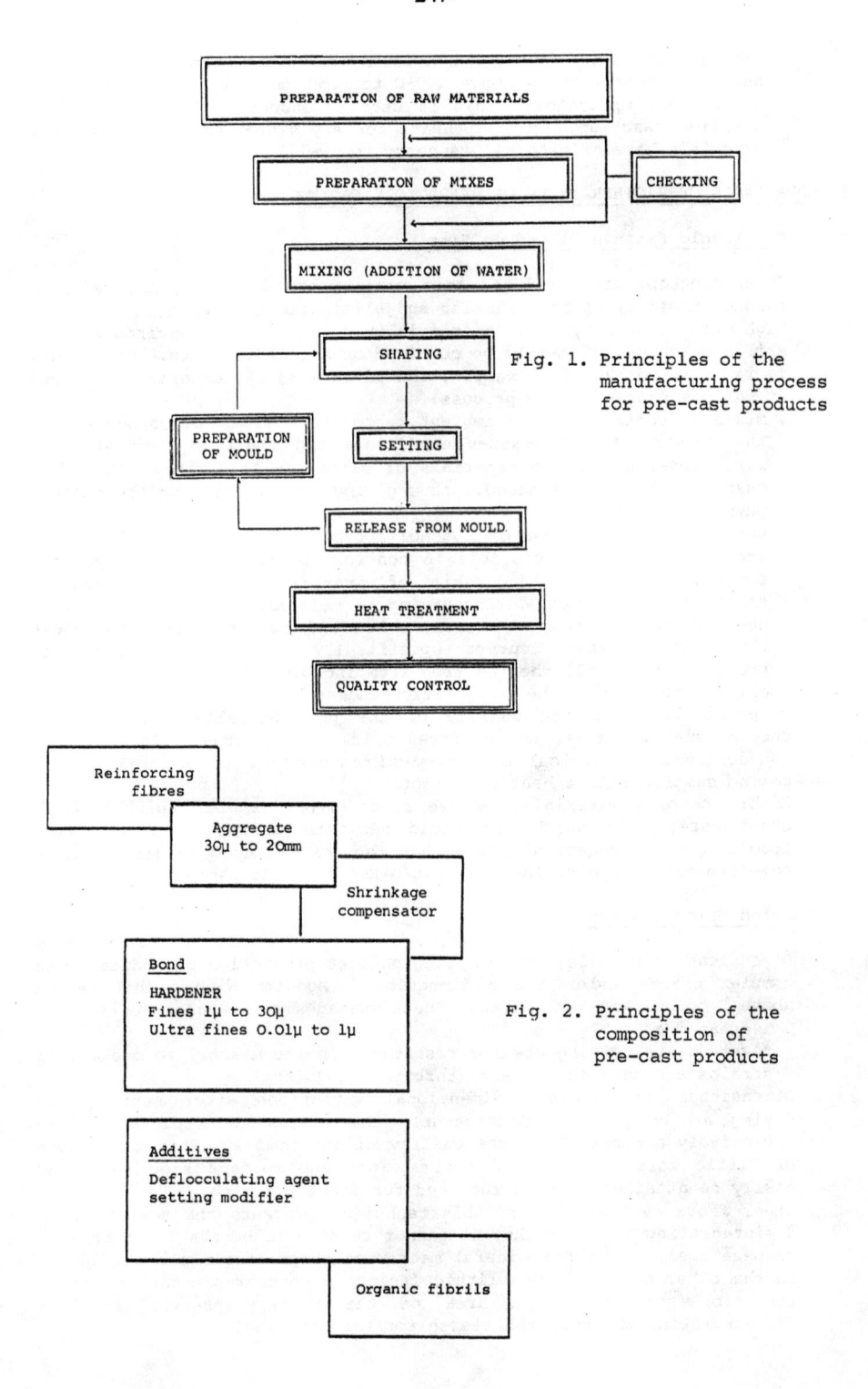

- organic fibrils, to aid dehydration of the monolith (these fibres shrink at temperatures above 100°C to produce capillary ducts.
- shrinkage compensator: this is used to induce a physical/chemical swelling reaction, to compensate for any sintering shrinkage (some materials have a 'natural' tendency to swell).

3. ADVANTAGES OVER OTHER TYPES OF REFRACTORY FACING

3.1 Monolithic Facings Placed on Site

- Interchangeability: reduces to a minimum the down time for relining through rapidity of installation and elimination of drying.
- High product quality: due to specialised methods, an environment and skills which cannot easily be obtained on a building site. This point is illustrated by, for example, the advantages of temperature control at various stages of the process:
  - mixing: control of the ambient temperature frees the process from the constraints of weather conditions and their extreme effects, i.e. freezing of the materials or ultra-rapid setting. The ideal quantity of water is added, thus giving a material with the optimum physical properties.
  - setting: in the case of hydraulic-type setting, control of the atmosphere makes it possible to control the nature of the hydrates formed. In this way the quantity of water fixed can be minimised and the risk of splitting during drying is reduced.
  - heat treatment: special programmable chambers can be used to adjust the heat treatment curves specifically to different types of product, which will then be free from incipient cracks.
- Quality control: the supplier undertakes unequivocally full responsibility for the quality of the product delivered. All the checks usually carried out on fired bricks can be done. In addition, non-destructive physical and chemical checks can be carried out by coring samples before heat treatment.
- Highly complex materials: these often have formulae which require great care, and which one would hesitate to use on some sites. Accordingly, pre-casting means that they can be used in all cases in complete safety and to the greatest benefit of the users.

3.2 Fired Brick Masonry

- Dimensions of modules: pre-casting enables production of modules with complex shapes and/or large dimensions. Modules with a unit weight above 2 tonnes are commonplace. The advantages to the user are:
  - increased assembly speed
  - fewer joints, giving greater resistance in the masonry to mechanical strains and chemical attack (through the joints).
- Dimensional tolerances: dimensional variations after setting and drying are very low. Consequently the dimensional tolerances are exclusively a function of the quality of the moulds. Thus tolerances of little more than 1 mm for dimensions and surface smoothness can easily be attained without the need for final trimming.
- Steel fibre reinforcement: this technique prevents the masonry from disintegrating if it cracks. It cannot be used in bricks fired at high temperatures, with the mineral matrix at temperatures above 1 200°C. On the other hand, in monolithic facings - whether pre-cast or not - the fibres in the cold area of the masonry prevent it from disintegrating whatever the reason for the cracking.

- Moulding costs: pre-cast modules are of interest technically and economically for short or medium runs and large sizes. While very complex or large modules must be pre-cast, an in-depth technical and economic study should be carried out for modules which can be manufactured by conventional machine pressing. For example, for sliding-gate nozzles, the cost of a press mould is about three times that of a vibro-moulding mould.
- Depreciation of the press mould should be evaluated in the light of the time required for executing an order and the difference in labour costs between the two systems.

It will also be noted that with pre-casting, prototype modules or equipment can be made with a minimum investment in moulding.

4. DESIGN OF ITEMS

The general shape, measurements and type of the material obviously depend on its main function.

However, the designer will also attach importance to:

- Optimising the shape of the modules as a function of the areas of preferential wear, so as to minimise the initial cost (Figure 3).
- Facilitating the handling of the modules by integrating the accessories for handling, assembly and even demolition.

5. MATERIALS

5.1 Choice of Type of Material

The progress made in the design of pre-cast formulae has produced properties which are often superior to those of fired bricks.

Thus Table I shows that a pre-cast material, based on a corundum concrete with ultra-low cement content, gives mechanical resistance at high temperature superior to that of one of the best corundum bricks with mullite bond, sintered at very high temperature.

TABLE I. Concrete with ultra-low cement content and fired brick - comparison of physical properties

	Concrete with ultra-low cement content	Corundum brick mullite bond
Al_2O_3 %	92	87
Density	3.30	3.05
Average pore diameter (microns)	0.2	20
Mechanical resistance (MPa)	80	160
Cross bending at 1 500°C	14	6

In addition, the average diameter of the pores, which is one of the physical factors governing resistance to corrosion, is practically 100 times smaller at low temperature. Figure 4 shows that the firing of concrete does not significantly alter the pores under 1 000°C and the diameter is never as large as in brick.

It will also be seen that, in the same way as for concrete, the chemical properties of the bond are a very important factor in controlling

Weight: 2 920 kg

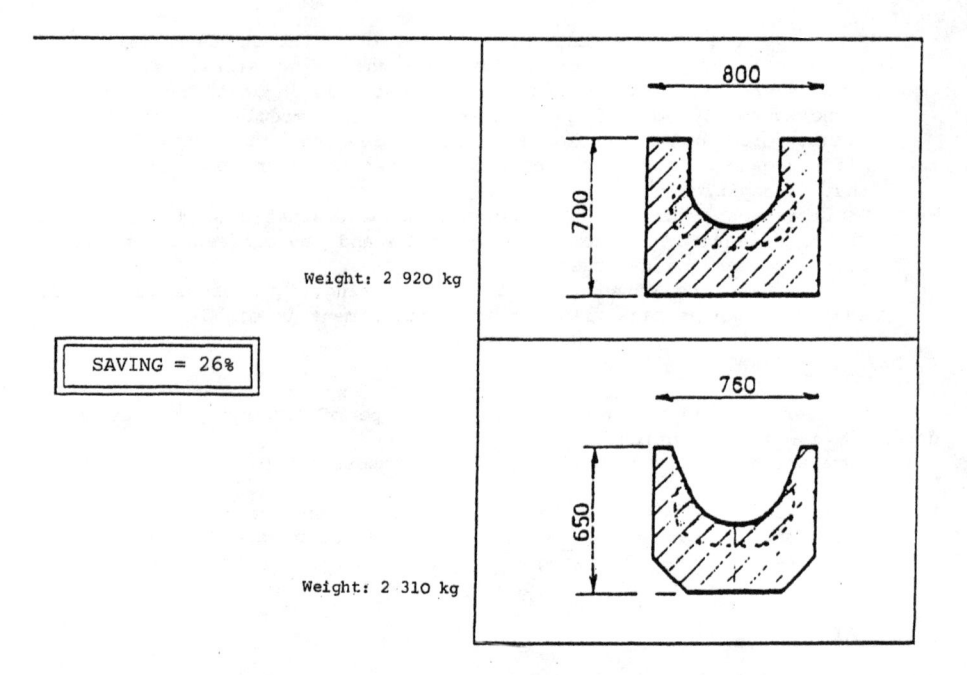

SAVING = 26%

Weight: 2 310 kg

Fig. 3. Design of pouring channels for electric arc furnace

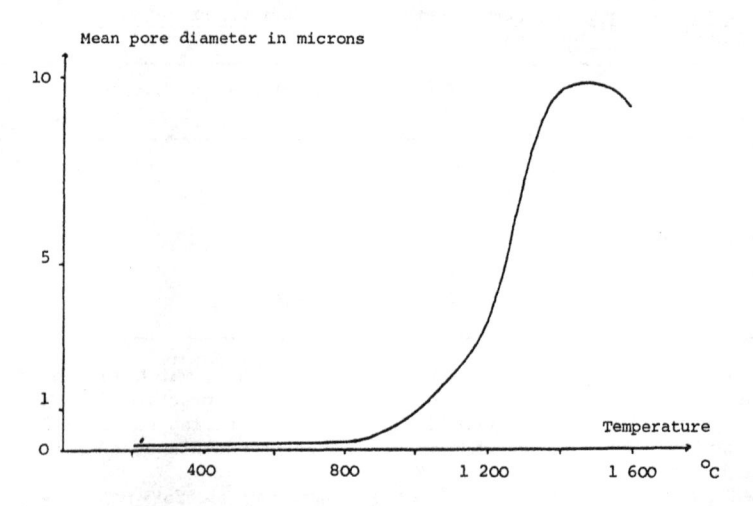

Fig. 4. Corundum-based concrete with ultra-low cement content
Mean pore diameter as a function of temperature

the mechanical properties at high temperature. Table II shows that in spite of a very low CaO content, the SiO_2-Al_2O_3 formula is much less efficient than the Cr_2O_3-Al_2O_3 formula.

TABLE II. Mechanical properties of concretes and pre-cast modules at high temperature: effect of the chemical properties of the bond (corundum aggregate)

Bond	SiO_2-Al_2O_3	Cr_2O_3-Al_2O_3
Al_2O_3 (%)	89	81
SiO_2 (%)	7.5	0.6
Cr_2O_3 (%)	0	10
CaO (%)	0.3	1.2
Cross-bending at 1 600°C (MPa)	3	13
Creep at 1 500°C under 0.2 Mpa (%)	0.6	0

Nevertheless, there is no risk of scaling by inhibiting the reversible dilation or post-swelling, even in the most rigid concretes at high temperatures with the Cr_2O_3-Al_2O_3 formula. When the temperature is first raised, these concretes are capable of absorbing by plastic deformation the stresses which are developed (Figure 5).

In general, the high-performance formulae which have been developed and which have a high initial cost will minimise repair costs, thus proving most advantageous in the long term.

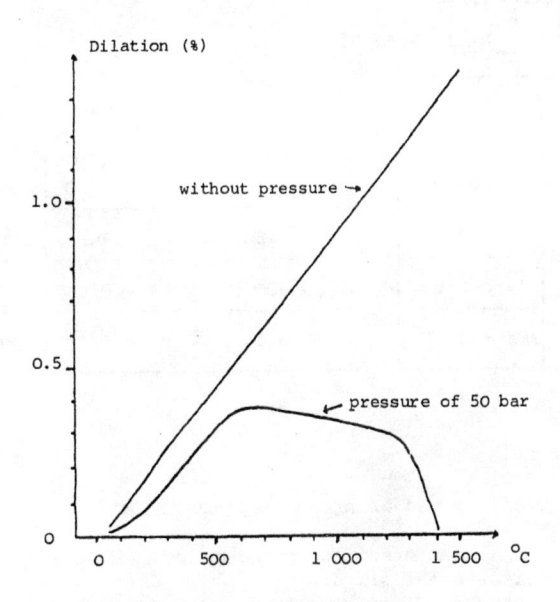

Fig. 5. Corundum-based concrete with Cr_2O_3-Al_2O_3 bond (tempered) Dilation under load at first heating

Figure 6 gives a practical example of a channel in an electric furnace for stainless steel. Although the initial cost is doubled by using a more efficient material, there is a reduction of more than 25% in the cost per tonne of steel because onerous maintenance work is eliminated.

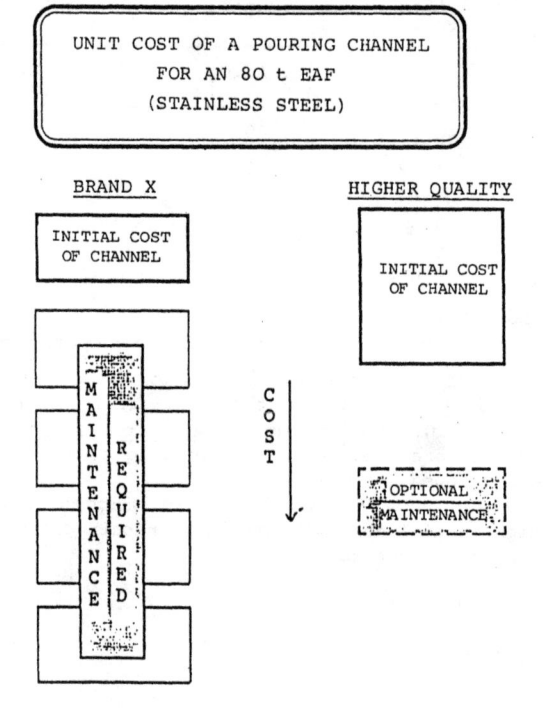

```
            UNIT COST OF A POURING CHANNEL
                  FOR AN 80 t EAF
                  (STAINLESS STEEL)
```

BRAND X   HIGHER QUALITY

INITIAL COST OF CHANNEL   INITIAL COST OF CHANNEL

MAINTENANCE REQUIRED   COST ↓   OPTIONAL MAINTENANCE

Total cost	FF 105 000	FF 30 000	FF 60 000	FF 85 000
No of pourings	300	120	180	240
Production	24 000 t	9 600 t	14 400 t	19 200 t
Unit cost	FF 4.37/t	FF 3.12/t	FF 4.16/t	FF 4.42/t

Fig. 6.

5.2 Typical Properties

There are two main types of pre-cast products, viz:

(1) basic products, the specific property of which is resistance to corrosion by the various steel slags;
(2) high alumina products, which offer the best resistance to thermal shock and particularly to corrosion by cast iron and steel at very high temperatures.

Table III gives the main properties of a typical product of each type.

TABLE III. Main properties of two typical pre-cast products for the steel industry

Type	Magnesia-Chrome	Corundum bond Cr_2O_3-Al_2O_3
MgO (%)	45	0
Al_2O_3 (%)	13.5	81
SiO_2 (%)	4	0.6
Cr_2O_3 (%)	25	10
Density	3.30	3.60
Mechanical resistance (MPa)	80	100
Average pore diameter (microns)	0.2	2.6
PVD at 1 500°C (%)	+0.7	+0.4
Dilation coefficient ($10^{-6} K^{-1}$)	8.6	8.7
Cross bending under heat (MPa)	1 400°C 2	1 600°C 2
Creep at 1 500°C under 0.2 MPa (%)	0.5	0

The corundum product with a Cr_2O_3-Al_2O_3 bond has the best mechanical properties at high temperature at present known for this type of material. Its composition has been closely studied with a view to improving its resistance to thermal shock, the usual weak point of products based on the solid Cr_2O_3-Al_2O_3. Overall it gives exceptional resistance to the impact of jets of steel.

The product based on basic electro-cast magnesia chrome grain has outstanding resistance to corrosion by steel slags.

The particular structure which results from the process of vibro-moulding makes this material comparable with the best bricks which are chemically and mineralogically similar, even when they are shaped using isostatic pressure (Table IV).

TABLE IV. Magnesia-chrome product shaped by isostatic or vibro-moulding pressing - corrosion behaviour

Shaping process	Rate of Corrosion (rotary kiln at 1 600°C)	Mini AOD slag
		CaO = 45.7
Isostatic pressing	100	SiO_2 = 29
Vibro-moulding	103	MgO = 11.6
		Fe_2O_3 = 13
		C = 0.7

6. SOME EXAMPLES OF APPLICATIONS

6.1 Blast Furnace

We shall mention for completeness the surrounds of tuyeres and the ceramic cups, for which the main interest of pre-cast blocks lies in the speed of assembly on site and the stability of the masonry. It should be noted that the use of ceramic cups offers an advantage in steelmaking, in that the temperature of the cast iron is increased or its silicon content reduced.

6.2 Electric Furnace

- Channels: brick facings wear quickly because the bricks are torn out during deslagging operations. Pre-cast linings reinforced with steel fibres are now used.
  They are either:
  . basic, for conventional steels or with high alumina content and a Cr_2O_3-Al_2O_3 bond for stainless steels and steels with very low carbon content.
  Using the best pre-cast materials more than 300 heats are possible before repairs become necessary.
- Small-radius arches: interchangeability is the main property required. Monolithic tops with a unit weight of five tons have been made.

6.3 AOD - Protection of the Lip

Pre-cast blocks replace bricks or graphite tiles, with the advantages of reduced wear and better resistance to mechanical deslagging.

6.4 RH Degasser - Snorkels

Demountable snorkels are an example of pre-cast double layers: internal brickwork with external monolithic facing. The two materials are basic. A complete monolithic facing is now possible, using the basic material described in Table III.

6.5 Continuous Casting: Impact Slabs and Seating Bricks

Materials of the corundum type with a Cr_2O_3-Al_2O_3 bond are used in the most demanding applications which call for mechanical resistance at high temperatures and good resistance to thermal shock.

6.6 Pressure Casting

Only pre-casting technology has made it possible to design a satisfactory seating brick for the blowing component in this new metallurgical process.

7. CONCLUSION

Today it is possible to design pre-cast materials for several applications in the steel industry with intrinsic properties at least equivalent to those of the best ceramic bricks. This is an important factor which in the future will make it possible to make full use of the strong points of pre-casting technology.

SUMMARY OF THE CONTRIBUTIONS FROM REFRACTORY PRODUCERS AND OF THE SESSION ON STANDARDIZATION

G C PADGETT
British Ceramic Research Ltd., Stoke on Trent, UK

There were eight contributions from the European producers on new material developments. The first of these was "Advances in raw materials and manufacturing technology in the production of refractory linings for primary and secondary steelmaking units" by P.Williams, D.Taylor and J.S. Soady. This paper demonstrated the advantages of a producer who has control over the quality of his raw material. The magnesia raw material has larger crystal size, reduced impurities and improved density. This contrasts with other raw materials such as flake graphite which is obtained from politically unstable countries. The second paper by R.D. Schmidt-Whitley was entitled "Clean steel and a clean environment with dolomite refractories". The author showed how dolomite refractories can contribute to clean steel with the additional advantage of low cost. Environmental problems related to the use of coal tar pitch binders were also discussed. Th. Benecke then presented a paper on "The use of non-oxide ceramic materials in metallurgy". This provided a comprehensive review of the non-oxide ceramics which have potential application as refractory materials. Silicon carbide is well established as a bulk refractory. Examples were given of the use of the more exotic boron carbide and boron nitride and prospects for other nitrides and borides were outlined. A paper entitled "Advanced materials for microcrystalline magnesite for the modern steel-making industry", was presented by Z Foroglou. This presentation emphasised the importance of a unique crystal structure and distribution of minor phases rather than absolute purity itself.

Dr P.L. Ghirotti then presented a paper entitled "Evolution of the technology and use of refractories in Italy and in other European countries". This report contained useful statistics on the trends of refractory consumption based partially on information provided by PRE. Italian (and European) consumption figures are getting closer to those achieved in Japan. The main trend being higher quality coupled with a higher price. The next presentation "Refractories to meet future steelmaking requirements has been jointly prepared by C.W. Hardy and P.G. Whiteley. With regard to the blast furnace it is important to have good liaison between the iron maker, the furnace constructor and the refractory producer. An example was given of the development of the new micro-pore carbons which suffer less iron penetration. Quality Assurance is a necessary requirement using the new European Standards in the EN 29000 series. A contribution entitled "The use of metal powders in carbon containing refractories " was presented by M.T. Rymon Lipinski. This proved to be an important fundamental approach to explain the mechanism of the role of anit-oxidants. The most efficient protection is provided by reactive gases (the so-called gas anti-oxidants). The final contribution from the refractory producers was entitled "A new generation of steel-making refractories : pre-cast products". The paper was written by M.J. Schoennahl and was presented by M. Naturel. Pre-cast products are shaped refractories ready for use, obtained by casting and tempering of monolithic materials under factory control. This technology removed some of the uncertainties on the use of monolithics and enables more complex formulations to be used.

The session on the standardization of refractories consisted of three contributions. The first was entitled 'The European Federation of Refractories Manufacturers (PRE) : its technical activity and its role in international standardization" and was presented by Mrs M.Lefebrre. As a result of the technical activity of PRE, fifty documents known as "Recommendations" have been issued. These deal with classification, sampling, dimensions and test methods and are used as a basis for standardization within the framework of the ISO (Technical Committee) TC33 Refractories. The next by B Clavaud was entitled "Monolithic refractories : comments on the methods used for their characterization and quality control". This presentation gave a new approach to the standardization of monolithic refractories. It was proposed to treat them differently to shaped refractories and to include standard procedures for the characterization of installation and pre-firing treatment. The final paper in this session was entitled "Data base on refractory materials for iron and steelmaking" by E. Criado, A.Pastor and R. Sancho. The presentation comprised a description of a bibliographical review of refractory materials on iron and steelmaking for the years 1980-87. The data base contains 2489 information items.

CONCLUDING ADDRESS. THE POINT OF VIEW OF THE
STEEL INDUSTRY AND OF STEEL RESEARCH

Jacques PIRET

Chairman of the ECSC Executive Committee "Refractories"

In a chemical plant, e.g. a oil refinery, it is hardly possible to see any product during processing, except if some incident as a leakage is occurring.
Plants dedicated to iron and steel making and processing are still different from these plants dealing with liquids easier to handle than ours but, nevertheless, the similitudes are much greater than, say 20 years ago :
- the production is achieved by a reduced number of large capacity lines which obviously give their best performances when they are fed with products consistent in composition and quantities ;
- the processes are more reliable and partly automatized ;
- the steel is much less in contact with the atmosphere than before particularly at the final liquid stages ;
- the plants are cleaner.
To make this evolution in the handling of liquid metal possible, more reliable vessel linings, "tubes" and "valves" were essential : the refractory industry provided them participating actively in this way to the modernization of the steel industry.
Nevertheless, the reports presented during this conference show that improvements are still possible. Furthermore, they are required by the evolution of the iron and steel making processes. Hereafter are summarized some leading ideas that may form the point of view of the "steel people" as expressed during this Conference.

1. A better knowledge of the processes and their action on the refractories is necessary in order to improve the lining lives

This subject, one of the most conventional, was treated in detail and for various situations by the ECSC research reports presented during the 1st day. The presentation of M. BEUROTTE (SOLLAC-Dunkerque) is showing that it is still possible to improve, in a quite impressive way, the life of converter linings :
- on one side, by making the users of these linings more sensitive to the actions depending strictly on them which reduce the wear and protect the lining ;
- on the other side, by collaborating with the refractory producers and specially with those proposing improved materials.
Taking a different example, the report of Messrs KLAGES of THYSSEN and SPERL of VDEh is drawing the attention on still many-sided problems to be studied in the field of the blast furnace hearth ; the firm belief that the current situation is not satisfactory cannot be more clearly demonstrated than by the announcement of THYSSEN, one of the most experienced steelmakers in this field, to propose a research programme on this topic.

2. **The demands are still more severe for the purity of the liquid steel, i.e. its cleanliness and its content of residuals**

Some E.C.S.C. research programmes presented in Session I (those of IRSID, C.S.M. and C.R.M.) are partly dedicated to improvements in this direction and a clear comparison of basic and non-basic linings is proposed the producers of refractories.

As emphasized by the report of Dr. BAKER of B.S.C., the evolution in the steel production is still marked by more severe demands for the purity of the metal and also by a greater ratio of production of high grade steels.

The refractories are taking a part in the successful achievement of each step of the metal refining. Here are some especially sensitive examples for which improvements are necessary :
- the lining of the slag line of hot metal ladles (torpedo and normal ladles) utilized in connection with desulphurization and dephosphorization with very aggressive slags ;
- the same slag line problem in steel ladles when a heavy secondary steelmaking is performed with e.g. arc heating, injection,... ;
- the filtration of steel in tundishes with efficiency but without clogging ;
- reliable tundish materials to avoid reintroduction of gases (hydrogen, oxygen, nitrogen) and impurities and to guarantee a longer life (also slag line problems).

3. A great attention is paid to the <u>reliability of the deliveries of refractories and their consistency in the time</u>. In various countries, like France and United Kingdom, policies and control systems are studied together by the users and some producers to guarantee the quality of the material.

4. <u>In the steel industry important developments</u> are foreseen in <u>hot metal production and continuous casting</u>.

In hot metal production, the great objective is to replace as much coke as possible by coal injection coupled wiht oxygen enrichment of the blast. Various processes are developed, using the current blast furnace or alternative vessels. New designs and working conditions but not necessarily new qualities will be the request for the refractories.

As pointed out by Dr. BAKER, more important changes can result from the evolution of continuous casting : direct linking with the rolling mill, thin product casting, etc... We think that new feeding systems withstanding more severe conditions and new refractory or ceramic devices carrying out new functions will be exciting challenges.

CONCLUSIONS. WORK FOR THE FUTURE

The number of participants in the Conference and the content of the papers presented are showing the need for additional research efforts in the field of refractories. The topics treated are providing guidelines for the research programmes to be undertaken in the future :

- a better comprehension of the wear mechanism of the refractories in connection with the process parameters, searching for a better behaviour ; the blast furnace hearth and the heavily stressed slag lines are specially concerned ;
- the contribution of the refractories to an improved steel cleanliness, not only avoiding negative effects but also using the refractories as an active agent of cleanliness ;
- the development of new refractory tools adapted to the evolution of the production routes, a special attention being paid to continuous casting (direct linking with rolling mill and thin product casting).

ACKNOWLEDGEMENTS

This Conference was organized by the Commission of the European Communities within the general frame of its "Steel Research - ECSC" programme. We express again our thanks to Mr. A. GARCIA ARROYO, Director of Research Technology and to Dr. P.R.V. EVANS, Head of Division of Steel Research, for making this meeting possible and for giving their personal support.

Such a conference can be a success only with the assistance of an efficient technical organization. We want to thank sincerely Mr. NICOLAY and Mrs EISEN of Division XIII and Mr. POOS of Division IX, all from Luxembourg, for the optimal conditions encountered during the two days of the Conference itself and during the reception offered by the Commission on the evening of the first day. These thanks are also addressed to Mr. AMAVIS and his team Messrs ARTELT, DIAZ and Mrs VAN LAETEM and to the interpretors who successfully struggled with our jargon.

In addition to their know-how, the organizers gave the members of the technical committee the impression of taking pleasure in preparing this Conference. Meetings conducted throughout in such a friendly atmosphere are the best catalyst of the European Fact.

* * * * *

CLOSING ADDRESS

R AMAVIS
Directorate Technical Research
Directorate-General Science, Research and Development
Commission of the European Communities

Ladies and Gentlemen,

I am pleased to say that after these two days of hard work, the Commission has become more aware of the problems encountered and the opportunities for progress in iron and steelmaking in relation to the use and manufacture of refractory products.

Some first thoughts have been advanced as well as proposals and suggestions concerning joint actions that have to be made in order (a) to improve product quality and (b) to reduce production costs.

We feel that the two way dialogue between the manufacturers of steelworks refractories and iron and steel producers is an essential basis for joint research and development projects.

We now have the responsibility to diffuse widely and rapidly the results of this conference to make aware the needs and the ways of stimulating scientific and technical progress in this area as well as to assist in planning future R&D effort.

In the modern world, both the iron and steel and the refractory industries must continue to be efficient and competitive. This is dictated by market forces!

The Commission of the European Communities is also particularly pleased with this conference, the quality of the presentations and also with the discussions taking place among those present.

Before finally bringing this conference to a close we would therefore like to thank all of the speakers, both from the platform and the floor, and all of those representatives of iron and steel producers and refractory suppliers present. Also, we are grateful to those who have prepared this conference and contributed to its success and, in particular, the technicians and our worthy interpreters who have had to work very hard and have nevertheless done so particularly well.

I therefore hope that you return safely to your various places of work, that you will recall with pleasure and satisfaction the two days that you spent in Luxembourg and that you will have derived some benefit from them for your future activities in your various areas of competence.

To conclude, I must inform you that the documents relating to this conference will be sent to you all within the next four or five months at the latest.

Having said that, I declare that the Luxembourg conference on the 'Use of Refractory Products in Iron and Steelmaking' has come to an end.

Zusammenfassungen

SPANNUNGSBERECHNUNGEN FÜR FEUERFESTAUSKLEIDUNGEN MIT HILFE DER FINITE ELEMENTANALYSE

J. Butter, J. de Boer
Hoogovens Groep BV, Ijmuiden

Das thermische Spalling stellt eine wesentliche Abnutzungserscheinung bei keramischen Feuerfestmaterialien dar. In einigen Fällen ergeben sich aus dem Versagen aufgrund interner Spannungen weitreichende Konsequenzen :sogar die Abschaltung einer ganzen Anlage ist nicht auszuschließen.
Ein Forschungsvorhaben sollte die quantitative Bestimmung dieses Phänomens ermöglichen. Ziel war die Entwicklung eines flexiblen Modells für die temperaturabhängigen Werkstoffeigenschaften. Die Auswertung der Ergebnisse sollte eine einfache Interpretation erlauben.
Das Modell basiert auf Standardsoftware nach der Methode der finiten Elemente, die Sofware mußte jedoch für die spezifischen Eigenschaften poröser Werkstoffe modifiziert werden.
Das endgültige Modell wurde mit Hilfe von Laborsimulationen und Messungen vor Ort erprobt.
Das Modell wurde auf verschiedene kritische Punkte der Feuerfestauskleidung von Öfen in integrierten Hüttenwerken angewandt. Beispielhaft werden die Ergebnisse eines EGKS-Forschungsvorhabens über den Einsatz eines mathematischen Modells für den keramischen Brenner eines Koksofens genannt. Das zweite Beispiel zeigt jüngere Ergebnisse eines Vorhabens über die Abnutzung von Rührelementen in Sauerstoff-Aufblas-Konvertern. Hierbei handelt es sich um ein gemeinschaftliches EGKS-Vorhaben von Britisch Steel und Hoogovens.

VERSCHLEISSMECHANISMEN DER FEUERFESTEN AUSKLEIDUNG VON KONVERTERN UND PFANNEN UND REPARATURVERFAHREN

C. Guenard, P. Tassot
IRSID, Maizières-les-Metz

1. Bodenabnutzung Bodenblasender Konverter

Diese in Zusammenarbeit mit dem Stahlwerk Rehon der USINOR Durchgeführte Arbeit hat zu einer Steigerung der durchschnittlichen Lebensdauer des ersten Bodens von 250 auf 400 Schmelzen geführt, und zwar durch:

- Steuerung der Propantemperatur,
- technologische Verbesserung der Blasdüse,
- Verringerung der Anzahl thermischer Lartspiele der Feuerfestauskleidung durch Vorheizen der Blasdüsen-Schutzgase,

- systematische Beschichtung nach jeder Schmelze,
- Verringerung des Risikos einer Blasdüsenverstopfung durch Verminderung des Durchsatzes von Verwirbelungsstickstoff;
- Steuerung der gleichmäßigen Verteilung von Propan an allen Düsen,
- Qualitätssteigerung des eingesetzten Propans.

Der erzielte Fortschritt führte zu einer Einsparung von 4,6 FF/t Stahl-schmelze.

2. Thermomechanische Spannungen in Pfannenauskleidungen

Stahlpfannen unterliegen einem thermischen Lastspiel, das Rißbildung und Abplatzungen der Feuerfestauskleidung verursacht.
Labor- und Industrieerprobung boten Einblicke in die Rißbildungsmechanismen von Feuerfestauskleidungen aus Dolomit, Dolomit-Kohlenstoff, Magnesit-Chrom sowie Stoffen mit hohem Aluminiumoxidgehalt.

3. Messung der Restdicke von Auskleidungen

Zur Verfolgung der Dickenentwicklung von Auskleidungen und Spritzplatten wurde die Laser-Interferometrie eingeführt.

4. Reparatur von auskleidungen durch spritzverfahren

Die halbnasse Spritzreparatur wurde auf Feuerfest-auskleidungen angewandt. Eine Untersuchung der verschiedenen Parameter (Wasser- und Pulverdurch- sätze, Partikelgeschwindigkeiten, Abstand Lanze-Wandung usw.) erlaubten eine Verbesserung dieser Methode.
Daneben wurde für die Instandsetzung der Auskleidungen einiger Pilot- und Industrieanlagen die Flammspritztechnik eingesetzt.

ENERGIEEINSPARUNG DURCH EINSATZ UNGEBRANNTER FEUERFESTER BAUSTOFFE

M. Koltermann
Hoesch Stahl AG, Dortmund

Ziel des Forschungsvorhabens war, die Möglichkeiten des Einsates von ungebrannten Tonerdesilikatsteinen in Torpedopfannen und Stahlgießpfannen zu untersuchen. Die Eigenschaften dieser Steine sollten gebrannten Steinen gegenübergestellt werden. Es wurden Andalusitsteine mit Phosphat- und Pechbindung und Zusatz von Kohlenstoff im Labor bis zu Temperaturen von 1800 °c untersucht und Verschlackungsversuche in einem Induktionsofen durchgeführt. Die Betriebsergebnisse mit diesen Steinen ergaben in 200 t - Torpedopfannen die gleichen Haltbarkeiten wie mit gebrannten Steinen. In Böden basischer Pfannen wurden mit phosphatgebundenen Andalusitsteinen die gleichen Haltbarkeiten erreicht wie mit gebrannten Andalusitsteinen. Zusätzlich durchgeführte Versuche mit ungebrannten Andalusit-Zirkonsilikatsteinen erbrachten keinen Erfolg.
Der Kostenvorteil beim Einsatz ungebrannter Steine hängt wesentlich von der Entwicklung der Energie- und Bindemittelpreise ab.

EINTAUCHAUSGÜSSE MIT BORNITRID-ZUSATZ FÜR STRANGGUSSANLAGEN

J. Piret
Centre de Recherches Métallurgiques, Lüttich

Zunächst wurden im Labor der Abnutzungsmechanismus von Ausgüssen aus Aluminiumoxid-Graphit im Bereich des Gießpulvers untersucht. Mittels eines beschleunigten Korrosionstests ließ sich zeigen, daß die Zugabe von Bornitrid in die Masse die Abnutzung der Ausgüsse stark verminderte. Eine Reihe von Industrieerprobungen bestätigten die guten Resultate aus dem Labor. Derzeit finden diese Ausgüsse breite Verwendung. Neben dem interessanten Aspekt der Beständigkeit gegen Korrosion durch die Schlacke zeigte sich in der industriellen Praxis auch eine Überlegenheit hinsichtlich der Widerstandsfähigkeit gegenüber Wärmeschock, werd Verstopfung durch AL_2O_3-Einschlüsse sowie hinsichtlich der Qualität des Gießprodukts und der Kosten.

ERFAHRUNGEN MIT KALZIUMOXID ALS FEUERFESTEM BAUSTOFF IN DER STAHLINDUSTRIE

E. Marino
Centro Sviluppo Materiali, Rom

Dank seiner Stabilität und Feuerbeständigkeit könnte Kalziumoxid sich hervorragend als Feuerfestmaterial für die Stahlindustrie eignen.
Herstellung und Einsatz von Feuerfestmaterialien auf der Basis von Kalziumoxid werden jedoch durch die Hydrierneigung des CaO behindert.
Im Rahmen zweier Forschungsvorhaben mit finanzieller Unterstützung durch die EGKS wurden Feuerfestkomponenten unter Verwendung stabilisierten Kalziumoxids mit entsprechender Eignung für die Herstellung, Lagerung und Verwendung ohne nennenswerte Hydrierungsprobleme entwickelt.
Das erste Vorhaben behandelte die Herstellung und Werkserprobung von Feuerfestziegeln auf der Basis stabilisierten Kalziumoxids. Hierbei bestätigte sich auch der positive Einfluß dieses neuen Materials auf die Pfannenbehandlung von Stahl.
Im zweiten Vorhaben ging es um die Entwicklung und Erprobung von Kalkdosierdüsen für den Strangguß. Es zeigte sich, daß diese Düsen den Strangguß Al-beruhigter Stähle ohne Verstopfungsprobleme durch Aluminiumoxidablagerungen ermöglichen. Zahlreiche Versuche an Brammenstranggußanlagen haben die Wirksamkeit dieser Düsen demonstriert, die das vollständige Vergießen von Stahlschmelzen auf eine Art und Weise erlauben, die bei Verwendung normaler Düsen undenkbar ist.

FESTLEGUNG DER SCHNELLSTMÖGLICHEN AUFHEIZUNG VON FEUERFEST ZUGESTELLTEN AGGREGATEN IN DER EISEN-UND STAHLINDUSTRIE

D.Wolters
Thyssen Stahl AG, Duisburg

Das Vorhaben wurde vom 1. Oktober 1974 bis zum 30. Juni 1982 von der Abteilung Forschung/Feuerfesttechnologie der Thyssen Stahl AG durchgefürht. Ein naturgasbefeuerter zylindrischer Ofen von 5 m länge und 2,5 m Durchmesser wurde mit einer dreischichtigen Auskleidung aus verschiedenen Ziegeltypen entsprechend der üblichen Praxis zugestellt: zunächst, an der Ofenwand, 125 mm Dicke Feuerleichtsteine; als dauerhaftes Sicherheitsfutter 125 mm dicke Steine aus hochfeuerfestem Ton A III (A 30) und/oder Hochleistungs-Feuerleichtsteine; als Verschleißfutter sodann 250 mm dicke Steine aus feuerfestem Ton AO (A 40), Magnesit, Silika, Mullit oder Korund. Ziel dabei war es, während der Aufheizphase die radiale Ausdehnung des Ofengehäuses wie auch die axialen Ausdehnungen und Drücke der einzelnen Schichten zu ermitteln. Die Aufheizraten wurden in Abhängigkeit von den spezifischen Materialien variiert. In einigen Fällen wurden begleitend auch einzelne Steine vor und nach den Tests untersucht, daneben wurden akustische Emississionmessungen durchgeführt. Im Verlaufe der 27 Prüfläufe waren verschiedene konstruktive Änderungen an der Ofeneinheit erforderlich.

Bei diesen Tests zeigte sich, daß sich bei einer derartigen Ofenstruktur alle Ziegelarten rascher aufheizen lassen, als normal üblich ist. Hierfür sind Leistungsfähige Brenner erforderlich. Die Mörtelfugen absorbieren weniger Expansion, als bislang angenommen, aus diesem Grund sind Expansionsfugen in axialer und radialer Richtung erforderlich. Di e raschere Aufheizrate wurde bislang jedoch in der Praxis noch nicht realisiert. Konstruktionsingenieure weisen darauf hin, daß für den Aufheizvorgang die gesamte Konstruktion zu berücksichtigen ist. Insbesondere darf kein überhöhter Temperaturgradient auftreten. Die Relativbewegungen zwischen den einzelnen Ausmauerungsschichten und insbesondere in der Nähe des Stahlgefäßes sind so klein wie möglich zu halten.

FLAMMSPRITZEN VON ISOLIERENDEN SCHUTZSCHICHTEN IN VERTEILERRINNEN

J. Piret
Centre de Recherches Métallurgiques, Lüttich

Die Technik des Sauerstoff-/Naturgas-Flammspritzens wurde für die Aufbringung von isolierenden Schutzschichten in Verteilerrinnen angewandt.

Gegenüber den aktuellen Verfahren mit Kaltverteiler mit isolierender Schutzauskleidung (Platten, Schlickerguß) bringt diese Technik folgende Vorteile :

- Bildung einer rein keramisch abbindenden Isolierauskleidung, d.h. völlig ohne Wasserstoff im Bindemittel;

- Verringerung der thermischen Verluste beim Angießen, da die Auskleidung nicht nur isolierend wirkt, sondern auch durch das Einbringverfahren einen hohen Enthalpiegehalt mit sich bringt;

- die Möglichkeit, Dichte und Dicke der Auskleidungsschicht in Abhängigkeit von den in den einzelnen Zonen der Verteilerrinne auftretenden Belastungen zu wählen.

DIE VEREINIGUNG EUROPÄISCHER FEUERFESTHERSTELLER (PRE): AKTIVITÄTEN UND STAND DER ARBEITEN IM BEREICH DER INTERNATIONALEN NORMUNG VON FEUERFESTERZEUGNISSEN

M. Lefèbvre
Fédération Européenne des Fabriquants
de Produits Réfractaires (PRE), Paris

Seit ihrer Gründung im Jahre 1953 bemüht sich die Vereinigung Europäischer Feuerfesthersteller (PRE), die Unternehmen aus 14 Ländern Westeuropas umfaßt, um die Erleichterung von Kontakten zwischen Herstellern und Anwendern durch Vereinheitlichung und Harmonisierung der Verfahren zur Charakterisierung von Feuerfesterzeugnissen. Die Arbeit im technischen Bereich mündete in der Erstellung von 50 Dokumenten, "EMPFEHLUNGEN" genannt, die sich mit Klassifizierung, Probenahme, Abmessungen und Testverfahren für Feuerfesterzeugnisse befassen und als Grundlage für die Normungsarbeit des Technischen Ausschusses "ISO TC 33 - FEUERFESTERZEUGNISSE" dienen. Die Veröffentlichungen finden daneben auch Berücksichtigung durch die Normungsinstanzen der Mitgliedsländer bei der Überarbeitung alter Dokumente oder der Erstellung neuer Normen.

Derzeit konzentrieren sich die technischen Bemühungen des PRE auf die Verbesserung überkommener Verfahren für traditionelle Werkstoffe und ihrer Anpassung an neue Produkte unter Berücksichtigung aktueller Qualitätsanforderungen seitens der Anwender.

UNGEFORMTE FEUERFESTE BAUSTOFFE: ÜBERLEGUNGEN ZUR KLASSIFIKATION UND PRÜFUNG

B. Clavaud
Lafarge Réfractaire, St. Priest

Als europäischer Hersteller technologisch hochstehender ungeformter Feuerfesterzeugnisse mit weltweiten Lizenznehmern befassen wir uns mit Möglichkeiten, die im Laborversuch erzielten Werte mit der Einsatzerprobung in Übereinstimmung zu bringen, um optimale Verfahren zur Definition und Beibehaltung eines vorgegebenen Qualitätniveaus für unsere Produkte zu erzielen. Ganz offensichtlich können die bisherigen Verfahren (ASTM, PRE, ISO) diese drei Aspekte nicht zufriedenstellend oder wirtschaftlich zusammenfassen.

In den dargestellten Beispielen liegt der Schwerpunkt auf dem Problem der Klassifizierung der maximalen empfohlenen Betriebstemperatur.

DATENBANK FÜR FEUERFESTWERKSTOFFE FÜR DIE EISEN-UND STAHL INDUSTRIE

E. Criado, A. Pastor,
Instituto de Ceramica y Vidrio, C.S.I.C., Madrid;
R. Sancho,
Instituto de Información y Documentación en Ciencia y Tecnologia (CSIC), Madrid

Die wichtigsten amerikanischen und europäischen Datenbanksysteme bieten derzeit keinen Zugang zu Datenbasen in der UdSSR, die weltweit führender Produzent von Stahl und Feuerfestmaterialien ist. Das gleiche gilt für Datenbanken in Japan, dem Land, aus dem die meisten Erfindungen auf dem Gebiet der Feuerfeststoffe stammen, was aus dem hohen Anteil (53 %) an Patenten aus Japan abzulesen ist. Von den sechs leicht zugänglichen Datenbanken, die konsultiert wurden, waren Chemical Abstracts (USA) und am zweite Stelle Pascal (Frankreich) am ertragreichsten bei Abfragen über Feuerfesterzeugnisse für die Stahlerzeugung. Bei den meisten (62%) Veröffentlichungen handelte es sich um Zeitschriftenartikel, gefolgt von Patenten (30%). Konferenzbeiträge machte lediglich 6% aus, technische Berichte nur 2 %. Die Anteile von Dissertationen und Büchern sind praktisch vernachlässigbar.

Taikabutsu (Japan) mit 215 Beiträgen und Ogneupory (UdSSR) mit 205 waren im untersuchten Zeitraum die produktivisten Zeitschriften.
Die UdSSR stand an erster Stelle bei der Anzahl veröffentlicher Zeitschriften und Referate. Die Bundesrepublik Deutschland, die USA und Japan veröfftenlichten in etwa gleichen Anzahlen Zeitschriften über

Feuerfeststoffe, Japan stand an zweiter Stelle bei der Veröffentlichung von Referaten. Für den Berichtszeitraum wurden Hinweise auf 87 verschiedene Konferenzen gefunden; insgesamt wurden 160 Beiträge vorgelegt. Die meisten Beiträge entfielen auf :

- das Internationale Kolloquium über Feuerfestwerkstoffe (BRD)
- die Steelmaking Conference (USA)
- die National Open Hearth and Basic Oxygen Steel Conference (USA)

Die ständige Zunahme an Informationen und das steigende technologische Interesse an Feuerfestwerkstoffen für die Stahlindustrie sowie die weit verstreuten Informationsquellen sind Gründe für die Einrichtung einer einzigen interdisziplinären Datenbank für die Erfassung aller weltweit veröffentlichten Informationen in diesem wichtigen Bereich.

ANFORDERUNGEN AN DIE FEUERFESTEN WERKSTOFFE IN EINEM MODERNEN HÜTTENWERK

G. Klages
Thyssen Stahl AG, Duisburg
H. Sperl
Verein Deutscher Eisenhüttenleute, Düsseldorf

In der Vergangenheit wurden häufig Erkenntnisse zur Verbesserung der feuerfesten Zustellung von Schmelz- und Transportgefäßen in der Stahlindustrie empirisch ermittelt, d.h. durch den Einbau von Versuchsfeldern wurden die Eignung von feuerfesten Werkstoffen in speziellen Anwendungsfällen geprüft. Diese Arbeitsweise war und ist natürlich zeit- und kostenaufwendig.

Mit dem erarbeiteten Kenntnisschatz und den zur Verfügung stehenden Meß- und Prüftechniken ist es heute möglich, aus dem auf dem Markt vorhandenen Angebot an feuerfesten Stoffen mit den unterschiedlichsten Gebrauchseigenschaften den für den jeweiligen Verwendungszweck optimalen Feuerfestwerkstoff auszuwählen.

Während es verhältnismäßig einfach ist, die chemischen und physikalischen Eigenschaften der Feuerfestprodukte zu ermitteln, ist die Definition aller möglichen Beanspruchungsarten in den verschiedenen Zonen metallurgischer Gefäße sehr schwierig. Ohne präzise Kenntnis dieses Beanspruchungsprofils, das letztlich die Haltbarkeit der Feuerfeststoffe bestimmt, läßt sich allerdings auch keine optimale Zustellung erzielen.

Anhand von Beispielen (Hochofen, Roheisentransportpfanne mit Entschwefelung, Sauerstoffaufblaslonverter mit Boden-spülung, Stahlgießpfanne mit Sekundärmetallurgie) wird versucht, eine derartige Ereignisanalyse durchzuführen und die hieraus entwickelte feuerfeste Zustellung zu erläutern.

ENTWICKLUNG BEI DER STAHLERZEUGUNG UND ENTSPRECHENDE ANFORDERUNGEN AN FEUERFESTE MATERIALIEN

R. Baker
British Steel plc, Rotherham

Nach wie vor gehen die Anforderungen an die Stahlerzeugung in Richtung auf reinere Produkte mit engeren Toleranzen für die Zusammensetzung, sowie gleichmäßigeren physikalischen Eigenschaften, und all dies bei Kosten, die sowohl gegenüber anderen Stahlherstellern als auch gegenüber Konkurenzwerkstoffen wettbewerbsfähig sind.

Die Antwort auf diese Herausforderungen sind eine Reihe von Entwicklungen der Stahlerzeuger, die auf die Verbesserung der Steuerung von Zusammensetzung und Qualität, Steigerung des Stahlausbringens und Verminderung der Energiekosten abzielen. Die bedeutung der Laufenden wie auch der ins Auge gefaßten Entwicklüngen für die Feuerfesthersteller zeigt sich besonders in den folgenden Bereichen:

a) Mehrstufige Arbeitsgänge zur präziseren Steuerung der Anteile von Schwefel, Phosphor, Sauerstoff, Wasserstoff, Kohlenstoff und Stickstoff;
b) Entwicklungen in den wichtigsten Primärprozessen im Hinblick auf die genauere Einstellung der Zusammensetzung, daneben aber auch auf eine weitere Verminderung der Kosten;
c) Entwicklungen in der Stranggießanlage zur Verminderung der Reoxidierung, Verhinderung der Wiedereinbringung gelöster Gase, Verbesserung der Durchsatzsteuerung sowie für Einsparungsgewinne durch Steigerung des Anteils von Sequenzgüssen,
d) Entwicklungen an Gießanlagen sowie zwischen Gießerei und Walzwerk zur Verringerung von Kühlintensität und Temperaturverlusten im Guß-Halberzeugnis sowie Steigerung des Anteils der direkt an das Walzwerk weitergeleiteten Gießschmelze.

VERÄNDERUNGEN BEIM EINSATZ FEUERFESTER BAUSTOFFE IN DER FRANZÖSISCHEN STAHLINDUSTRIE

M. Beurotte,
Sollac, Dunkerque

Hauptziele der französischen Eisen- und Stahlindustie sind die Verbesserung der Produktqualität und die Verringerung der Produktionskosten.

Insgesamt haben Feuerfesterzeugnisse zu einer Kostenverringerung beigetragen und sich gleichzeitig an die Entwicklung in der traditionellen Stahlerzeugung sowie an neue Prozesse angepaßt.

Im laufenden Jahrzehnt hat der Verbrauch an Feuerfesterzeugnissen in der französischen Eisen- und Stahlindustrie deutlich abgenommen. Dies ist zum Teil auf die Schließung der ältesten Anlagen zurückzuführen, weithaus mehr aber auf Fortschritte in den verschiedenen Entwicklungs-richtungen, die hier dargestellt werden:

- Aufbau von Forschungsstrukturen in der Stahlindustrie.
- Aufbau von Partnerschaften mit Zulieferern.
- Kooperation zwischen Anwender und Hersteller von Feuerfesterzeugnissen sowie Forschungsanstrengungen für langlebigere Feuerfesterzeugnisse.
- Verträge zwecks Qualitätssteigerung mit Unternehmen der Feuerfestzustellung.

ZUSAMMENFASSUNG

Steigende kundenanforderungen versucht die Eisen- und Stahlindustrie mit der Entwicklung leistungsfähigerer, gleichmäßigerer und höherwertigerer Produkte zu erfüllen.

Dies bringt eine kontinuierliche Weiterentwicklung des metallurgischen Prozesses mit sich.

Feuerfeste Baustoffe müssen sich dieser Entwicklung anpassen und die steigenden Anforderungen erfüllen. Dies läßt sich in kurzer Zeit nur erreichen bei enger Zussamenarbeit zwischen den beteiligten Partnern.

In den letzten Jahren haben bedeutende konzeptionelle Änderungen stattgefunden. Zur Bewahrung von Wettbewerbsfähigkeit und hoher Leistungsfägigkeit muß die Forschung in dieser Richtung weitergeführt werden.

FORTSCHRITTE IM HINBLICK AUF ROHSTOFFE UND HERSTELLUNGSTECHNOLOGIEN BEI DER ERZEUGUNG VON FEUERFESTEN BAUSTOFFEN FÜR DIE PRIMÄR- UND SEKUNDÄRMETALLURGIE

P. Williams, D. Taylor und J.S. Soady
Steetley Refractories Ltd, Worksop

Bei der Entwicklung und Herstellung basischer feuerfester Erzeugnisse für die Primär- und Sekundärmetallurgie ist eine kritische Bewertung der "Umgebung" des Stahlerzeugungsprozesses anzustelllen. Diese Umgebung umfaßt thermische, mechanische und chemische Belastungen, die auf der gemeinsamen Basis der Zeit einwirken. Bei der Stahlverarbeitung sind diese "Stress"-Faktoren inhärent und liegen außerhalb des Einflußbereichs des Herstellers von Feuerfesterzeugnissen. Dieser ist jedoch nach wie vor gezwungen, die Qualität, Eignung und Beständigkeit seiner Erzeugnisse unter den harten Bedingungen der Stahlerzeugung zu verbessern und mit den Entwicklungen in wirtschaftlich vertretbarer Weise Schritt zu halten.

Das Papier beschreibt die Philosophie der Steetley Refractories Ltd., die sich mit basischen Feuerfesterzeugnissen aus Magnesiumoxid,

Dolomit und Magnesiumoxid-Chrom unter den Gesichtspunkten der Rohstoffauswahl und der Fertigungssteuerung befaßt. Bei der Auswahl der Rohstoffe geht es unter anderem um die relative Verfügbarkeit und die Entwicklung und Nutzung verbesserter Magnesiumoxide mit erhöhter Schtüttdichte, größeren Periklas-Kristallabessungen und verringerte Verunreinigungen, in Verbindung mit Verfügbarkeit, Größenklassierung und Kohlenstoffgehalt natürlicher Schuppengraphite. Die Entwicklung von Bindemittelsystemen haben den Magnesiumoxid- und Dolomit-Feuerfesterzeugnissen mit Kohlenstoffanteil eine weitere Verbreitung insbesondere in der Sekundärmetallurgie erschlossen.

Daneben wurden formalisierte Gütesicherungssysteme eingerichtet, um dem Kunden in der Stahlindustrie einsatzgerechte Produkte garantieren zu können. Statistische Prozeßsteuerungstechniken für Ein- und Mehrfachpressen haben zu höherer Gleichförmigkeit der Produkte geführt. Angesichts des in jüngster Zeit gestiegenen Bedarfs für feuerfeste Erzeugnisse und der Einführung modernster Pressentechnologie für gleichbleibende Produktqualität bei hohem Ausstoß waren Kapitalinvestitionen erforderlich. Durch Zusammenarbeit mit Rohstofflieferern und Endabnehmern konnten die Hersteller von Feuerfesterzeugnissen die Anforderungen infolge der technologischen Entwicklungen bei der Stahlerzeugung überwiegend erfüllen.

EINSATZ VON DOLOMIT FÜR DIE HERSTELLUNG HOCHREINER STÄHLE

R.D. Schmidt-Whitley
Didier S.I.P.C., Paris

Feuerfeste Erzeugnisse aus Dolomit finden in der Stahlindustrie zunehmende Verbreitung im Hinblick auf Forderungen nach höheren Stahlqualitäten.

Die Vorteile eines inerten, praktisch SiO_2-freien feuerfesten Stoffes wie Dolomit werden beschrieben für Pfannenfrischprozesse wie Desoxidation, Entschwefelung, Legierung und Einschlußverminderung. Umweltprobleme im Zusammenhang mit der Verwendung von Kohle teer-Pechlo indern werden erörtert. Es wird dargestellt, wie hochreine Stähle und Umweltfreundlichkeit mit neu entwickelten, geformten oder ungeformten Feuerfeststoffen aus Dolomit bei annehmbaren Kosten möglich sind.

NICHTOXIDISCHE SONDERKERAMISCHE STOFFE IN DER METALLURGIE

Th. Benecke
Elektroschmelzwerk, Kempten

Allgemeine und spezifische Hochtemperatureigenschaften und Kombinationen solcher Eigenschaften von nichtoxidischen sonderkeramischen Stoffen für die Metallurgie, beispielsweise:

- Siliziumkarbid (SiC), löslich in Eisen- und Stahlschmelzen, dennoch aber feuerfest: Legierung und Desoxidierung von Eisen- und Stahlschmelzen, Antioxidans für Al_2O_3/SiC/C-Monolithe und Torpedopfannensteine im Hochofenbereich.

- Borkarbid (B_4C), neues Antioxidans für kohlenstoffgebundene feuerfeste Erzeugnisse in Strangguß und Konverter.

- Bornitride (BN), nicht benetzbar durch Stahlschmelze, leicht zu bearbeiten wie Graphit, im Gegensatz zu Graphit aber in der Stahlschmelze nicht löslich:

BN-Pulver für Feuerfestauskleidungen im Strangguß;
BN-Sinterteile als Bruchringe für den Horizontalstrangguß;
BN für Neuentwicklungen beim endabmessungsnahen Gießen.

Perspektiven anderer Nitride und Boride.

ENTWICKLUNGEN BEI DER HERSTELLUNG UND VERWENDUNG VON FEUERFESTEN WERKSTOFFEN IN ITALIEN UND ANDEREN EUROPÄISCHEN LÄNDERN

P.L. Ghirotti
Nuova Sanac Spa,. Genua

Der Bericht soll einen Überblick über die derzeitige Entwicklung auf dem Sektor feuerfeste Erzeugnisse und deren Anwendung in Europa und Italien geben. Der mengenmäßige Rückgang der weltweiten Nachfrage nach feuerfesten Erzeugnisse und die Tendenz zu technisch anspruchsvolleren Produkten dürften sich in den nächsten Jahren fortsetzen.
In den letzten zehn Jahren (1977-1987) hat sich in Italien der spezifische Verbrauch von Feuerfeststoffen in der Stazhlindustrie von 20,2 Kg/t auf 13,3 Kg/t vermindert.
In anderen Abnehmerindustrien (Zement, Glas, Nichteisenmetalle, Keramik und sonstige) sind die Verbrauchswerte beständiger.
Besondere Bedeutung gewinnen hierbei - angesichts der jüngsten Entwicklungen in der Stahlerzeugung mit den Zielen Kostenreduzierung,

Qualitätssteigerung und Produktions-flexibilität - Feuerfestwerkstoffe für Roheisen-Vorbehandlungsgefäße (Torpedopfannenwagen und Roheisen-pfannen), Konverter Auskleidungen mit kombinierten Blastechniken (Auf- und Bodenblasen) und insbesondere feuerfeste Materialien für Nachbehandlungs-Gießpfannen sowie feuerfeste Werkstoffe für den Strangguß.

Für Pfannen Auskleidungen gilt es, die Rolle völlig basischer Pfannen bezüglich der optimalen technisch-wirtschaftlichen Bilanz zu klären und zu verifizieren.

Ein weiterer Schwerpunkt ist die Entwicklung künftiger metallurgischer Prozesse, die nicht nur dem ständigen Ziel reinerer Stähle bei niedrigeren Kosten, sondern auch den Problemen von Rohstoffquellen und Energieeinsparung (Schrotteinsatz im kontinuierlichen Prozeß, "Erzschmelztechnologie", direkte Kombination von Strangguß und Warmwalzen- CC-DR) gewindmet sind.

Für die Feuerfestindustrie kann die rechnerunterstützte Fertigung (CAM) und die Automatisierung zu einer wesentlichen Steigerung von Produktqualität und Produktivität führen, und damit umfangreichere und flexiblere Möglichkeiten zur Bereitstellung von Werkstoffen zu annehmbaren Kosten schaffen.

Wo immer Steine und monolitische Werkstoffe eingesetzt werden, finden technische Keramik und Spitzentechnologie mehr und mehr Verwendung.

Die künftige Enwicklung wird sich zunehmend auf spezifische Anwendungen, Produkte mit Zeit- und Energieeinsparung bei Herstellung und Zustellung, Berücksichtigung ökologischer Aspekte bei Produktion und Zustellung, zunehmend mechanisierte und automatisierte Zustellung sowie Ofenreparaturtechniken konzentrieren.

All dies wird noch stärker als in der Vergangenheit eine unablässige Teamarbeit zwischen Ingenieuren und Technikern der Stahlindustrie und der Ofen- und Feuerfesthersteller erfordern.

FEUERFESTE BAUSTOFFE FÜR DIE ANFORDERUNGEN DER STAHLINDUSTRIE IN DER ZUKUNFT

C.W. Hardy
Britisch Steel plc, Teeside;
P.G. Whiteley
GR-Stein Refractories Ltd, Worksop

In absehbarer Zukunft wird der Hochofen-/Sauerstoffstahl-konverter-Prozeß weltweit die wichtigste Quelle für Stahlschmelze sein. Zunehmend werden an die Produktions-einheiten Forderungen hinsichtlich hoher Produktivität und niedrigem Kosten-/Heizenergieaufwand gestellt werden, die ihrerseits strengere Anforderungen an die Feuerfestwerkstoffe mit sich bringen.

Wichtige Entwicklungen bei Feuerfesterzeugnissen betreffen die Verlängerung der Reisedauer von Hochöfen durch Werkstoff- und Konstruktionsverbesserungen sowie Anwendung neuer Reparaturtechniken. Die Schmelzbadvorbereitung wird weitere Fortschritte bei Gießanlagen-

Werkstoffen und Feuerfest-auskleidungen für Transport- und Zwischengefäße erfordern. Die Dauer der Ofenreise von Sauerstoffstahl-konvertern ist zu verlängern, Betriebsbedingungen sind weniger hart zu gestalten, und Feuerfestoffe könnten durch Wasserkühlung unterstützt werden. Die Sekundärmetallurgie gewinnt an Bedeutung und führt zu höheren Anforderungen an die Feuerfestwerkstoffe. Verbesserte Auskleidungsstoffe werden eingeführt, Schlüsselaspeckt ist jedoch die Schlackenzusammensetzung.

Die Forderung nach gleichbleibender Qualität und Lieferkontinuität stärken die Bedeutung unabhängiger Gütesicherung und statistischer Prozeßsteuererungstechniken während der Fertigung.

FORTSCHRITTLICHE MATERIALIEN AUS MIKROKRISTALLINEM MAGNESIT FÜR DIE MODERNE STAHERZEUGUNG

Dr. Z. Foroglou
Financial Mining Industrial and Shipping Corporation, Athen

Eine Reihe von Steinen aus Magnesiumoxid und Kohlenstoff wurde speziell für die Auskleidung von Gefäßen für die moderne Stahlerzeugung entwickelt, auf der Grundlage von Magnesiumoxidklinkern aus natürlichem mikrokristallinem Magnesit.

Auskleidungen aus derartigen Sinterwerkstoffen zeigten bemerkenswerte Leistungsfähigkeit in verschiedenen LD-Stahlwerken wie SOLLAC DUNKERQUE und SOLMER in Frankreich, LULEA und OXELOESUND in Schweden, PIOMBINO und BAGNLOLI in Italien usw

Diese industriellen Anwendungen vervollständigen die auf den Feuerfest-Konferenzen 1983 und 1987 in Tokio vorgestellten Arbeiten und bekräftigen die damals ausgedrückte Ansicht, daß die Überlegenheit von Magflot Sinters gegenüber synthetischen MgO-Produkten auf die optimale chemische Zusammensetzung und Verteilung zugehöriger Kleinphasen sowie ihre einzigartige Kristallstruktur und nicht so sehr auf die absolute Reinheit des Materials zurückzuführen ist.

Weiter zeigt sich, daß Magnesiumoxidklinker aus mikrokristallinem Magnesit Leistungsfähiger als MgO-Kohlenstoff-Feuerfeststoffe sind und einen flexiblen Einsatz verschiedener Klinkersorten für wirtschaftliche Lösungen von Feuerfestproblemen in modernen Stahlgefäßen ermöglichen.

EINSATZ VON METALLPULVERN UND KOHLENSTOFFHALTIGEN FEUERFESTSTOFFEN

T. Rymon Lipinski
Forschungsinstitut der Feuerfestindustrie, Bonn

Metallzuschläge zu kohlenstoffhaltigen Feuerfeststoffen werden als Antioxidantien eingesetzt, die den Kohlenstoff vor Oxidation schützen. Bislang aber war der Mechanismus dieser Erscheinung nicht ausreichend erklärt.
Der Bericht gibt die jüngsten Ergebnisse unserer experimentellen Arbeit an Magnesiumoxid-/Kohlenstoff-Werkstoffen wieder, wie sie in der Sauerstoffstahlerzeugung eingesetzt werden. Aluminium, Magnesium und Silizium in Pulverform wurden als Zuschläge zu Feuerfeststoffen geprüft. Aufgrund thermochemischer Analysen und experimenteller Erprobung ließ sich eine Erklärung des Schutzmechanismus der Antioxidantien finden. Nach diesem Modell ist ein wirksamer Schutz des Kohlenstoffs gegen Oxidation auf reaktive Gase (sog. Gas-Antioxidantien) zurückzuführen, die (z.B. Mg-Dampf und SiO-Gas) im Inneren der Feuerfeststeine als Reaktionsprodukte zwischen den Hauptkomponenten des Materials (MgO, C) und den Metallzuschlägen entstehen. Zur Verifizierung des vorgeschlagenen Modells für den Oxidationsschutz des Kohlenstoffs wurde daneben auch Untersuchungen mit anderen Metallen durchgeführt.

EINE NEUE GENERATION VON VORGEFERTIGTEN FEUERFESTEN WERKSTOFFEN FUR DIE STAHLINDUSTRIE

M.J. Schoennahl
Savoie Réfractaires, Vénissieux

Hierbei handelt es sich um vorgeformte, einsatzfertige Feuerfestauskleidungen, die durch Gießen und Härten monolitischer Werkstoffe in einer speziellen Werkstatt entstehen.
Vorteile dieser Technologie sind:
- gegenüber Ziegelauskleidungen:
* höhere Wärmeschockfestigkeit
* kurze Lieferzeit
* schnelle Zustellung
- gegenüber monolitischen Werkstoffen mt Flüssigeinguß:
* höhere Qualität
* schnelle Zustellung
* keine Trochnungszeiten

Zahlreiche Fortschritte machen die vorgefertigten Elemente in vielen Fällen sogar den gebrannten Ziegeln überlegen.

Daneben ist die Flexibilität in Form und Abmessungen ein wichtiger Pluspunkt bei den neuen Prozeßentwicklungen in der Eisen- und Stahlindustrie.

Résumés

CALCUL DES CONTRAINTES DANS LES REVETEMENTS REFRACTAIRES PAR LA METHODE DES ELEMENTS FINIS

J.A.M. Butter et J. de Boer
Hoogovens Groep BV, Ijmuiden

L'écaillage thermique constitue un important phénomène d'usure des céramiques réfractaires. Dans certains cas, les conséquences d'une dégradation due à des contraintes internes sont d'une très grande portée; cela peut même aller jusquà l'arrêt de toute installation.

Des recherches ont été réalisées pour quantifier ce phénomène. L'objectif était de mettre au point un modèle flexible dans lequel il est possible de faire intervenir les propriétés des matériaux qui varient avec la température. Les résultats devaient être évalués de telle sorte que leur interprétation puisse se faire facilement.

Le modèle repose sur un logiciel de calcul standard par la Méthode des Eléments Finis. Il a toutefois fallu optimiser ce logiciel pour prendre en compte les propriétés spécifiques des matériaux poreux.

Le modèle final a été expérimenté au moyen de simulations de mesures en laboratoire et de mesures in situ.

Il a été appliqué à plusieurs points critiques du revêtement de fours dans l'aciérie intégrée. Sont fournis comme exemple les résultats d'un projet CECA sur l'utilisation d'un modèle mathématique pour le brûleur en céramique d'un four à coke. Le deuxième exemple concerne les résultats récents de recherches sur l'usure des éléments de brassage utilisés dans des convertisseurs à oxygène. Il s'agissait d'un projet CECA réalisé en coopération entre BS et Hoogovens.

MECANISMES D'USURE DES REVETEMENTS DE CONVERTISSEURS ET DE POCHES A ACIER ET METHODES DE REPARATION

C. Guenard, P. Tassot
Institut de Recherches de la Sidérurgie française,
Maizières-lès-Metz

1. Usure de fonds de convertisseurs à soufflage par le fond

Ces travaux, réalisés en collaboration avec l'aciérie d'Usinor à Rehon, ont permis de porter la durée de vie moyenne du premier fond de 250 à 400 charges par les moyens suivants :

- commande de la température du propane,
- amélioration de la technologie des tuyères,
- réduction des cycles thermiques imposés aux réfractaires par chauffage des gaz de protection des tuyères,
- tartinage systématique après chaque charge,

- réduction des risques de bouchage des tuyères par réduction du débit de l'azote de brassage,
- contrôle de l'équirépartition du propane entre les tuyères,
- contrôle de la qualité du propane utilisé.

Les différents progrès réalisés ont permis une économie de 4,6 francs par tonne d'acier liquide.

2. Contraintes thermomécaniques dans les revêtements des poches à acier

Les poches à acier sont soumises à un cycle thermique qui entraîne la fracturation et l'écaillage des revêtements.

Des essais en laboratoire et sur une installation industrielle ont permis de comprendre les mécanismes de fracturation de la dolomie, de la dolomie-carbone, de la magnésie-chrome et des réfractaires à haute teneur en alumine.

3. Mesure de l'épaisseur résiduelle des revêtements

On a recouru à l'interférométrie laser pour suivre l'évolution des épaisseurs des revêtements et des panneaux gunités.

4. Réparation des revêtements par projection

Le gunitage par voie humide a été appliqué à la réparation des revêtements réfractaires. Une étude des différents paramètres (débits d'eau et de poudre, vitesse des particules, distance entre la lance et la paroi, etc.) a permis d'améliorer cette méthode.

La projection à travers une flamme a également été utilisée pour réparer les revêtements de certaines installations pilotes et à l'échelle industrielle.

ECONOMIE D'ENERGIE GRACE A L'UTILISATION
DE BRIQUES REFRACTAIRES NON-CUITES

Manfred Koltermann
Hoesch Stahl A.G., Dortmund

L'objectif de recherche était d'étudier les possibilités d'utilisation de briques silico-alumineuses non-cuites dans les poches torpilles et les poches de coulée. Les propriétés de ces briques devaient être comparées avec celles des briques cuites. Des briques d'andalousite liées avec du phosphate et du brai, et additionnées de carbone, ont été exposées, lors d'essais en laboratoire, à des températures pouvant aller jusqu'à 1800°C. On a également effectué des essais de scorification dans un four à industion.

En poche torpilles de 200 t, ces briques non cuites ont présenté une durabilité équivalente à celle des briques cuites. Dans le fond des poches de coulée basiques, on a obtenu avec les briques d'andalousite liée avec du phosphate les même durabilités qu'avec les briques d'andalousite cuites. Des essais supplémentaires éffectués avec des briques d'andalousite-silicate de zirconium non cuites se sont soldés par un échec.

Le gain de coûts lié à l'utilisation de briques non cuites est essentiellement fonction de l'évolution des prix de l'énergie et des liants.

BUSETTES PLONGEANTES DE COULEE CONTINUE ENRICHIES AU NITRURE DE BORE

J. Piret
Centre de Recherche Métallurgiques, Liège

Nous avons tout d'abord étudié, par des essais de laboratoire, le mécanisme d'usure des busettes du type alumine-graphite dans la zone du laitier de lubrification. Grâce à notre test de corrosion accéléré, nous avons pu montrer que l'apport de nitrure de bore en mélange dans la masse réduisait fortement l'usure des busettes. Plusieurs essais industriels ont confirmé les bons résultats obtenus à l'échelle du laboratoire. Actuellement, ces busettes sont utilisées couramment. A côté de l'intérêt au niveau de la résistance à la corrosion par le laitier, la pratique industrielle a montré qu'elles constituent un bon choix au point de vue de la résistance aux chocs thermiques, du bouchage par les inclusions d'Al_2O_3, de la qualité du produit coulé et de leur coût.

UTILISATION D'OXYDE DE CALCIUM COMME MATERIAU REFRACTAIRE DANS LA SIDERURGIE

E. Marino
Centro Sviluppo Materiali, Rome

En raison de sa stabilité et de sa réfractarité, l'oxyde de calcium pourrait constituer un excellent matériau réfractaire pour les aciéristes.

La tendance du CaO à s'hydrater a toutefois empêché la préparation et l'utilisation des réfractaires à base d'oxyde de calcium.

Au cours de deux projets de recherche réalisés avec l'aide financière de la CECA, on a mis au point des composants réfractaires en utilisant de l'oxyde de calcium stabilisé possédant les caractéristiques appropriées des points de vue de la fabrication, du stockage et de l'utilisation et ne présentant aucune difficulté importante du point de vue de l'hydratation.

Le premier projet a porté sur la fabrication et l'expérimentation en usine de briques de poches fabriquées avec de l'oxyde de calcium stabilisé. Ces essais ont également confirmé l'influence positive de ce nouveau type de revêtement dans le traitement de l'acier en poche.

Le deuxième projet a porté sur des travaux de mise au point et d'essai de busettes d'alimentation en chaux pour la coulée continue. Les travaux réalisés ont montré qu'il est possible grâce à ces busettes de réaliser la coulée continue d'acier calmé à l'aluminium sans se heurter aux difficultés de bouchage provoquées par les dépôts d'alumine. De nombreux essais réalisés sur les lignes de coulée continue pour billettes ont démontré l'efficacité de ces busettes qui permettent la coulée complète de charges d'acier d'une façon que les busettes normales ne permettent absolument pas.

DETERMINATION DE LA VITESSE OPTIMALE D'ECHAUFFEMENT DES MAÇONNERIES REFRACTAIRES UTILISEES EN SIDERURGIE

D. Wolters
Thyssen Stahl AG, Duisburg

Cette étude a été réalisée entre le 1er octobre 1974 et le 30 juin 1982 par l'Unité de recherche sur la technologie des réfractaires de la société Thyssen Stahl. Un four cylindrique de 5 mètres de longueur et 2,5 mètres de diamètre, chauffé au gaz naturel, a été pourvu d'un revêtement composé de trois couches de briques différentes disposées comme dans la pratique, c'est-à-dire parois de la cuve revêtues d'une maçonnerie réfractaire de 125 mm d'épaisseur en briques isolantes légères, revêtement de sécurité permanent de 125 mm en chamottes AIII (A 30) et/ou en briques isolantes légères de haute qualité; revêtement interne d'usure de 250 mm en chamottes AO (A 40), magnésie, silice, mullite ou briques de corindon. Il s'agissait de déterminer la dilatation radiale de la cuve ainsi que la dilatation radiale et les pressions subies par chacune des trois couches de revêtement pendant la montée de la température. On a étudié la relation entre la variation de la vitesse d'échauffement et le comportement spécifique de chacun des matériaux. Dans certains des cas, on a procédé à l'examen des briques isolées avant et après les essais, ainsi qu'à des mesures d'émissions acoustiques. De nombreuses modifications de structures ont dû être apportées au four au cours de la réalisation de ces 27 essais.

Ces essais ont démontré que tous les types de briques constituant la maçonnerie de ce four pouvaient être chauffés plus rapidement que dans la pratique, à condition de disposer de brûleurs efficaces.
Les joints de mortier ayant démontré une capacité d'absorption de la dilation moins importante que prévue, la pose de joints de dilatation en direction axiale et radiale s'avère nécessaire.

Cependant, la possibilité d'augmenter la vitesse d'échauffement mise en évidence par ces essais n'a pas été exploitée. Les ingénieurs d'études ont en effet fait remarquer qu'au cours de la montée de température, on n'avait pas pris en considération le comportement de l'intégralité de la structure. En tout état de cause, un gradient thermique trop élevé n'est pas souhaitable. Les mouvements relatifs entre les différentes assises, et en particulier celles recouvrant la cuve d'acier doivent être maintenus au niveau le plus faible possible.

REVETEMENTS DE PROTECTION ISOLANTS POUR PANIERS REPARTITEURS DE COULEE CONTINUE PLACES PAR GUNITAGE A LA FLAMME

J. Piret
Centre de Recherches Métallurgiques, Liège

La technique du gunitage dans une flamme oxygène/gaz naturel a été utilisée pour constituer le revêtement de protection de paniers répartiteurs.

Par rapport aux procédés utilisés actuellement de paniers froids à protection interne isolante (plaques ou slurry casting), cette méthode apporte les avantages suivants :

- constitution de revêtement isolant à prise purement céramique, c'est-à-dire rigoureusement exempt d'hydrogène dans le liant ;
- une réduction des pertes thermiques en début de coulée car non seulement le revêtement est isolant mais son mode de placement implique un contenu enthalpique élevé ;
- la possibilité de choisir la densité et l'épaisseur de la couche du revêtement de protection en fonction de l'intensité des sollicitations rencontrées d'une zone du panier à l'autre.

LA FEDERATION EUROPEENNE DES FABRICANTS DE PRODUITS REFRACTAIRES (PRE) : SON ACTIVITE TECHNIQUE ET SA PLACE DANS LES TRAVAUX DE NORMALISATION INTERNATIONAUX

Monique Lefèbvre,
Fédération européenne des fabriquants
de produits réfractaires

Dès sa création en 1953, la Fédération Européenne des Fabricants de Produits Réfractaires (PRE), qui regroupe les producteurs de quatorze pays d'Europe de l'Ouest, s'est attachée à faciliter les échanges entre fabricants et utilisateurs en unifiant et harmonisant les méthodes de caractérisation des réfractaires. Son activité technique a abouti à la rédaction de cinquante documents appelés "RECOMMANDATIONS", qui concernent la classification, l'échantillonnage, les dimensions et les méthodes d'essai des produits réfractaires et qui servent de base aux travaux de normalisation réalisés au sein du Comité Technique "ISO TC 33 - "MATERIAUX REFRACTAIRES". Ces documents sont également pris en considération par les Instances de Normalisation des pays membres lors de la révision d'anciens documents ou la rédaction de nouvelles normes.

Actuellement, les travaux techniques des PRE visent à améliorer les méthodes anciennes définies pour les matériaux traditionnels en les adaptant aux nouveaux produits et en tenant compte des nouvelles exigences de qualité requises par les utilisateurs.

LES REFRACTAIRES NON-FACONNES
REFLEXIONS SUR LES METHODES DE CARACTERISATION ET DE CONTROLE

B. Clavaud
Lafarge Réfractaires, St-Priest

Producteur Européen de réfractaires non-façonnés à haute performance, bailleur de licence à l'échelle mondiale, nous nous interrogeons sur les méthodes à utiliser pour, à la fois, concilier la caractérisation nécessaire en laboratoire, la réelle estimation de ces caractéristiques après mise en oeuvre sur le terrain et la voie à suivre pour contrôler et définir le niveau de qualité des lots de production. On constate, en effet, une inadaptation fréquente de nos méthodes (ASTM, PRE, ISO) à intégrer fidèlement et économiquement ces trois exigences.

Quelques exemples seront pris, on insistera sur le problème de la classification en terme de valeur limite de classe de température.

BASE DE DONNEES RELATIVE AUX MATERIAUX REFRACTAIRES
DESTINES A LA SIDERURGIE

E. Criado, A. Pastor, Instituto de Cerámica
y Vidrio (CSIC), Madrid ; R. Sancho, Instituto
de Información y Documentación en Ciencia y
Tecnologia (CSIC), Madrid.

Les principales bases de données américaines et européennes n'offrent actuellement pas accès aux bases de données d'URSS, premier producteur mondial d'acier et de réfractaires. Cet accès n'est pas non plus possible pour les bases de données au Japon, pays d'origine de la plupart des inventions dans le domaine des matériaux réfractaires, ce que reflète le chiffre élevé (53%) de brevets déposés au Japon.

Parmi les six bases de données facilement accessibles qui ont été consultées, Chemical Abstrcts (U.S.A.), suivie de Pascal (France) ont été les plus productives en réponse aux recherches d'information en matière de matériaux réfractaires destinés à la sidérurgie.

La majeure partie (62%) des publications étaient des articles de revue,

réunions ne représentaient que 6%, les rapports techniques à peine 2%. La part des thèses de doctorat et des livres était pratiquement négligeable.

Taikabutsu (Japon) avec 215 études, et Ogneupory (URSS) avec 205 études se sont révélées les revues les plus riches en information au cours de la période étudiée.

L'URSS occupait le premier rang en nombre de revues et études publiées. La république fédérale d'Allemagne, les Etats-Unis et le Japon ont publiés à eux trois un nombre similaire de revues traitant des réfractaires, le Japon occupant le second rang parmi les producteurs d'études. 87 réunions différentes tenues au cours de la période ont été inventoriées. Au total 160 communications ont été présentées à ces réunions. La plupart ont été présentées :

- à l'International Colloquium of Refractories (RFA)
- à la Steelmaking Conference (Etats-Unis)
- à la National Open Hearth and Basic Oxygen Steel Conference (Etats-Unis)

L'augmentation progressive du volume d'information et l'intérêt technologique croissant pour les matériaux réfractaires employés en sidérurgie d'une part, des sources d'information disponible largement disséminées d'autre part, ces deux ensembles de facteurs militent en faveur de la création d'une base de données interdisciplinaire unique pour rassembler toute l'information publiée à l'échelle mondiale dans ce domaine stratégique.

EXIGENCES RELATIVES A LA QUALITE DES REFRACTAIRES UTILISES DANS LES PROCEDES SIDERURGIQUES MODERNES

G. Klages
Thyssen Stahl AG, Duisburg
H. Sperl
Verein Deutscher Eisenhüttenleute, Düsseldorf

Dans le passé, les connaissances permettant d'améliorer les revêtements réfractaires de poches de fusion et de transport utilisées en sidérurgie ont souvent été acquises par des méthodes empiriques, c'est-à-dire par la mise en place de champs d'expérimentation permettant de tester si tel ou tel réfractaire était adapté à un type particulier d'application. Evidemment, une telle méthode de travail était et reste particulièrement coûteuse et laborieuse.

Aujourd'hui, la somme des connaissances acquises ainsi que les techniques de mesure et d'essai disponibles permettent de sélectionner le matériau le mieux adapté à l'utilisation prévue parmi le choix de réfractaires de qualité diverses proposé sur le marché.

Si la détermination des propriétés physico-chimiques des réfractaires apparaît relativement aisée, il est en revanche très difficile de définir tous les types de contraintes qui s'exercent aux différents endroits des cuves métallurgiques. Sans connaissance précise de ce diagramme de contraintes qui détermine en fin de compte la durabilité du matériau, il n'est pas possible de parvenir à une qualité optimale du garnissage.

On a donc tenté d'effectuer une analyse de ce type à partir d'exemples concrets (haut fourneau, poche de transport de fonte avec désulfuration, convertisseur à soufflage d'oxygène par le haut et brassage du fond, poche de coulée de l'acier avec traitement en poche), et de décrire le garnissage réfractaire développé à partir de cette analyse.

EXIGENCES RELATIVES AUX REFRACTAIRES DUES AUX DEVELOPPEMENTS DE LA SIDERURGIE

Dr R. BAKER
British Steel plc,
Rotherham

L'aciériste est sans cesse contraint de fabriquer un produit pur, d'une composition plus précise, avec des propriétés physiques plus homogènes et à un coût qui le rende compétitif à la fois avec les produits des autres fabricants et avec les matériaux concurrents.

Pour relever ces défis, l'aciériste a procédé à un certain nombre de changements visant à améliorer le contrôle de la composition et de la qualité, à accroître le rendement final et à réduire les frais énergétiques. Les changements auxquels on est en train de procéder et ceux que l'on envisage concernant le fournisseur de réfractaires portent plus précisément sur les domaines suivants :

a) échelonnement des opérations pour mieux contrôler les niveaux de soufre, de phosphore, d'oxygène, d'hydrogène, de carbone et d'azote;
b) changements dans les principaux processus primaires dans le sens d'un contrôle plus étroit de la composition, mais aussi d'une nouvelle réduction des coûts;
c) changements à hauteur de l'atelier de coulée continue pour réduire la réoxydation, éviter la réintroduction de gaz dissous, améliorer la régulation et réaliser des gains de rendement en resserrant la séquence des opérations de coulée.
d) changements à hauteur de la machine de coulée et entre celle-ci et le laminoir pour réduire l'intensité du refroidissement, freiner la déperdition de température dans le semi-produit et accroître la proportion de la production acheminée à chaud jusqu'au laminoir.

LA MUTATION "REFRACTAIRES" DANS LA SIDERURGIE FRANCAISE

M. Beurotte
Sollac - Dunkerque

La Sidérurgie Française s'est fixée comme objectifs principaux d'améliorer la qualité des produits qu'elle propose, tout en réduisant leurs coûts de fabrication.

Les réfractaires dans leur ensemble ont apporté leur contribution à la réduction des coûts tout en s'adaptant pour répondre à l'évolution des filières traditionnelles de fabrication de l'acier et aux développements de nouveaux procédés.

Au cours de la présente décennie, la consommation de produits réfractaires dans la sidérurgie française a diminué de façon significative. Cette amélioration résulte, en partie, de l'arrêt des installations les plus vétustes mais surtout, des progrès obtenus à partir de plusieurs axes de développement que le présent exposé se propose de développer.

1. Développement de structures de recherche en sidérurgie

2. Développement d'un partenariat avec les fournisseurs

3. Bonne coopération entre exploitant et réfractoriste et recherche de longue performance

4. Mise en oeuvre des matériaux réfractaires

Il nous semble primordial de poursuivre dans cette voie si l'on veut rester performants et compétitifs.

AMELIORATIONS DES MATIERES PREMIERES ET DE LA TECHNOLOGIE DE FABRICATION DES REFRACTAIRES POUR LA METALLURGIE PRIMAIRE ET SECONDAIRE

P. Williams, D. Taylor, J.S. Soady
Steetley Refractories Ltd, Worksop

Dans la mise au point et la fabrication de réfractaires basiques pour les opérations de métallurgie primaire et secondaire en masse, il faut s'attacher à évaluer l'"environnement" sidérurgique. Cet environnement réunit des contraintes thermiques, mécaniques et chimiques qui se superposent dans le temps. Ces facteurs de "contrainte" sont inhérents au traitement de l'acier et échappent au fabricant de réfractaires qui doit cependant continuer à améliorer la qualité, les aptitudes et l'homogénéité des produits réfractaires pour faire face à cet environnement sidérurgique difficile et suivre l'évolution dans des conditions qui restent acceptables du point de vue commercial.

Le présent document décrit la méthode utilisée par Steeley Refractories Ltd., qui consiste à aborder la question des réfractaires à base de magnésie, dolomie et magnésie-chrome du point de vue de la sélection des matières premières et du contrôle de la fabrication. S'agissant de la sélection des matières premières, il faut noter les possibilités relatives d'obtention, la mise au point et l'utilisation de magnésies améliorées d'une densité apparente accrue, avec des cristaux de périclase de plus grandes dimensions et moins d'impuretés, ainsi que les possibilités d'obtention, le calibrage et les teneurs en carbone des graphites lamellaires naturels. Les développements intervenus en matière de liants ont permis de trouver de plus larges applications pour les réfractaires carbonés à base de dolomie et de magnésie, en particulier dans la métallurgie secondaire.

Des systèmes codifiés d'assurance de qualité ont également été mis en oeuvre pour garantir au client sidérurgiste la livraison de produits correspondant à leur destination. Des techniques de contrôle statistique des processus ont été mises au point pour des presses à une ou plusieurs cavités et ont permis d'obtenir une plus grande homogénéité des produits. Il a fallu procéder à des investissements en capital pour faire face à la forte demande récente de produits réfractaires et aux tout derniers progrès des techniques de pressage mises en oeuvre pour maintenir l'homogénéité des produits avec des rendements élevés. En travaillant avec les fournisseurs de matières premières et les consommateurs, le fabricant de réfractaires est parvenu à satisfaire la majorité des exigences créées par l'évolution des techniques sidérurgiques.

AMELIORATION DE LA PROPRETE DE L'ACIER ET DE L'ENVIRONNEMENT PAR L'UTILISATION DES REFRACTAIRES EN DOLOMIE

R.D. Schmidt - Whitley,
Didier SIPC, Paris

Les réfractaires en dolomie sont de plus en plus utilisés en sidérurgie pour répondre à la demande d'amélioration de la qualité des aciers. Ce document décrit les avantages que présente l'utilisation d'un réfractaire inerte à très faible teneur en SiO_2, tel que la dolomie, dans les opérations d'affinage en poche comme la désoxydation, la désulfuration, l'alliage et la réduction des inclusions. Les problèmes écologiques liés à l'utilisation de liants composés de brai de goudron y sont examinés, et on montre comment la propreté de l'acier et de l'environnement peuvent être améliorées grâce à de nouveaux réfractaires en dolomie, moulés et non moulés, d'un prix raisonnable.

UTILISATION DE CERAMIQUES SPECIALES NON OXYDEES EN METALLURGIE

Th. Benecke,
Elektroschmelzwerk, Kempten

Description des propriétés générales et spécifiques à haute température et des combinaisons de propriétés de céramiques non oxydées en métallurgie.

Exemples :

- Carbure de silicium (SiC) : soluble dans la fonte et l'acier liquide, mais néanmoins réfractaire. Utilisé pour la réalisation d'alliages et la désoxydation de la fonte et de l'acier liquide. Prévient l'oxydation des monolithes $Al_2O_3/SiC/C$ et des briques garnissant les poches torpilles utilisées en aval des hauts fourneaux.
- Carbure de Bore (B_4C) : Prévient l'oxydation des réfractaires sidérurgiques à liaison carbonée employés dans les installations de coulée continue et les convertisseurs.
- Nitrure de bore (BN) : non mouillable par l'acier liquide et aussi facile à mettre en oeuvre que le graphite, mais, contrairement à ce dernier, non soluble dans l'acier liquide:

   Utilisation de BN en poudre pour la fabrication de réfractaires des installations de coulée continue;
   Utilisation de BN fritté comme anneau de rupture pour les cokes à haute teneur en carbone;
   Utilisation de BN pour la réalisation de pièces de fonderie proches des dimensions finales.

   Perspective d'utilisation d'autres nitrures et brorures.

EVOLUTION DE LA TECHNOLOGIE ET DE L'EMPLOI DES REFRACTAIRES EN ITALIE ET DANS D'AUTRES PAYS EUROPEENS

P.L. Ghirotti
Nuova Sanac SpA., Gênes

Ce rapport a pour objet de présenter un bilan de l'état actuel de la recherche dans le secteur des produits réfractaires et de leur applications en Italie et dans les autres pays européens. Dans les prochaines années, il est vraisemblable que le volume global de la demande de matériaux réfractaires diminuera et que les besoins continueront de s'orienter vers des matériaux plus sophistiqués.

Au cours des dix dernières années (de 1977 à 1987), la consommation spécifique de réfractaires de la sidérurgie italienne est passée de 20,2 kg/t à 13,3 kg/t.

En revanche, la consommation des autres industries utilisatrices (ciment, verre, métallurgie des non-ferreux, industries céramiques et autres industries) est restée plus stable.

Les efforts déployés récemment par l'industrie sidérurgique en vue de diminuer les coûts, améliorer la qualité des produits et la flexibilité de la production, auront des répercussions sur l'importance accordée aux revêtements réfractaires des cuves de prétraitement de la fonte (poches torpilles et poches à fonte), des convertisseurs faisant appel à un procédé de soufflage mixte (soufflage par le haut et par le fond), et surtout aux réfractaires utilisés dans les opérations de postaffinage, de coulée en poche et de coulée continue.

En ce qui concerne les revêtements de poches, le rôle des revêtements entièrement basiques devra être analysé afin de vérifier si l'équilibre technico-économique optimal est réalisé.

On devra suivre de près l'évolution des nouveaux procédés métallurgiques, dont beaucoup ne visent pas uniquement à obtenir des aciers plus purs et plus propres à faible coût, mais ont aussi pour objectif de résoudre le problème des sources d'approvisionnement en matières premières et de réaliser des économies d'énergie (utilisation de ferrailles dans les procédés en continu, techniques de fusion des minerais, coulée continue associée au laminage à chaud).

En ce qui concerne l'industrie des réfractaires, les systèmes de fabrication assistée par ordinateur (FAO) et l'automatisation pourraient entraîner une amélioration considérable de la qualité des matériaux et de la productivité, et ouvrir ainsi de plus larges possibilités de fournir des produits à des prix raisonnables.

Là où des revêtements monolithiques ou briquetés sont employés, on devra recourir plus fréquemment aux techniques de la céramique industrielle et à la haute technologie.

L'évolution future dans le secteur des réfractaires sera de plus en plus orienté vers des applications spécifiques, la mise au point de produits permettant de réaliser des économies de temps et d'énergie au niveau de la production et des installations, la protection de l'environnement, la technologie des installations mécaniques et automatisées et les techniques de réparation des fours.

Plus encore que par le passé, ces nouveaux défis nécessiteront une collaboration permanente entre les ingénieurs et les fabricants de fours et de matériaux réfractaires.

MATERIAUX REFRACTAIRES REPONDANT AUX BESOINS FUTURS DE L'INDUSTRIE SIDERURGIQUE

C.W. Hardy
British Steel plc, Teessite
P.G. Whiteley
GR-Stein Refractories Ltd, Worksop

Dans les prochaines années, la filière basée sur le haut fourneau et le convertisseur à oxygène continuera de fournir la majeure partie de l'acier liquide produit dans le monde. Les unités de production se verront placées devant la nécessité d'accroître toujours davantage leur productivité et d'améliorer leur rapport coût/rendement thermique, ce qui entraînera une sollicitation accrue des réfractaires.

Ce rapport présente les innovations importantes en matière de réfractaires qui ont permis d'allonger la durée de campagne des hauts fourneaux grâce à l'amélioration des matériaux et des équipements et à l'application de nouvelles techniques de réparation courante. Le prétraitement de la fonte nécessitera de nouvelles améliorations, à la fois des matériaux de la halle de coulée et du garnissage réfractaire des poches de transport et/ou de transfert de la fonte. La durée normale de campagne des convertisseurs à oxygène devra être allongée. Les conditions opératoires devront être moins sévères et les réfractaires pourraient être renforcés par refroidissement à l'eau. Les procédés de métallurgie secondaire prennent une importance croissante et font peser sur les réfractaires de plus grandes exigences qualitatives. De nouveaux matériaux réfractaires plus performants sont mis sur le marché, mais la composition du laitier est considérée comme un élément fondamental.

Le recours à des organismes indépendants pour réaliser les contrôles de qualité, et l'application de méthodes statistiques de contrôle en cours de fabrication (Statistical Process Control) se développent de plus en plus pour répondre à la demande de produits de qualité uniforme et de continuité de l'approvisionnement.

NOUVEAUX MATERIAUX DE POINTE A BASE DE MAGNESITE MICROCRISTALLINE POUR LA SIDERURGIE MODERNE

Z. Foroglou
Financial Mining Industrial
and Shipping Corporation, Athènes

Une série de briques carbonnées à base de magnésie spécialement conçues pour le garnissage des cuves sidérurgiques modernes a été élaborées à partir de magnésite microcristalline naturelle.

Des résultats remarquables ont été obtenus avec ces nouveaux revêtements dans plusieurs aciéries modernes exploitant le procédé LD, comme la SOLLAC DUNKERQUE et SOLMER en France, LULEA et OXELOESUND en Suède, PIOMBINO et BAGNOLI en Italie, et d'autres encore.

Ces applications in situ complètent les travaux présentés lors de la Conférence sur les réfractaires qui s'est déroulée entre 1983 et 1987 à Tokyo, et viennent confirmer la théorie présentée à cette occasion, selon laquelle la supériorité des briques Magflot par rapport aux produits synthétiques de haute qualité à base de MgO s'explique davantage par l'optimisation de leur composition chimique et de la répartition de leurs phases mineures associées et par leur structure cristalline particulière, que par la pureté absolue du matériau.

Ces résultats démontrent également que les briques de magnésie élaborées à partir de magnésite microcristalline sont plus performante que les réfractaires carbonés à base de MgO, et permettent de disposer d'une palette de matériaux de qualités diverses apportant des solutions économiques aux différents problèmes de réfractaires des creusets sidérurgiques modernes.

ADDITION DE POUDRES METALLIQUES AUX REFRACTAIRES CARBONES

M.T. Rymon Lipinsky
Forschungsinstitut der Feuerfestindustrie, Bonn

Les substances métalliques ajoutées aux réfractaires carbonés sont utilisées comme antioxydants pour protéger le carbone contre l'oxydation. Le mécanisme de ce phénomène n'a cependant pas encore été expliqué de façon satisfaisante.

Le présent document présente les récents résultats de nos expériences concernant les matériaux à teneur en magnésie-carbone couramment utilisés dans les convertisseurs à l'oxygène. Des poudres d'aluminium, de magnésium et de silicium ajoutées aux composés réfractaires ont fait l'objet d'essais. A partir d'analyses thermochimiques et de résultats d'expériences, une explication du mécanisme de la protection assurée par les antioxydants sera présentée. D'après le modèle avancé, l'efficacité de la protection du carbone contre l'oxydation est due à des gaz réactifs (les antioxydants gazeux). Les antioxydants gazeux (par exemple la vapeur de magnésium et l'oxyde de silicium gazeux) se forment à l'intérieur de la brique comme produits de réaction avec les principaux composants du matériau (MgO, C) et les additifs métalliques. Des vérifications ont été faites avec d'autres métaux afin de confirmer le modèle proposé pour la protection du carbone contre l'oxydation.

UNE NOUVELLE GENERATION DE MATERIAUX REFRACTAIRES
POUR L'ACIERIE : LES MATERIAUX PREFABRIQUES

J. Schoennahl
Savoie Réfractaires, Vénissieux

Ce sont des produits réfractaires façonnés, prêts à l'emploi, obtenus par mise en forme et traitement thermique d'un matériau non-façonné, dans un atelier spécialisé;

Les avantages de cette technologie sont :
- Face aux revêtements en briques :

. une meilleure résistance aux chocs thermiques et mécaniques,
. un délai de livraison plus court
. la rapidité du montage
- Face aux revêtements monolithiques mis en oeuvre sur site
. un niveau de qualité plus élevé
. la rapidité du montage
. la suppression du séchage

Ces matériaux ont fait l'objet de tels progrès que leur niveau de qualité dépasse bien souvent celui des briques céramiques.

De plus, la flexibilité de forme et de format qu'ils permettent est un des éléments déterminant pour le développement de nouveaux procédés sidérurgiques.

LIST OF PARTICIPANTS

ACQUARONE G.
POLITECNICO DI TORINO
Corso Duca degli Abruzzi 24
I - 10129 TORINO

ALIPRANDI G.
ISTITUTO INGEGNERIA CHIMICA
GENOVA UNIVERSITY
Via Opera Pia, 15
I - 16145 GENOVA

ALLENSTEIN J.
WAMBESCO ROHSTOFFHANDELSGES. GmbH
Steinsche Gasse 4
D - 4100 DUISBURG 1

ALTPETER S.
RADEX AUSTRIA AKTIENGESELLSCHAFT
Millstätterstrasse 10
A - 9545 RADENTHEIN

AMAVIS R.
DG XII
CCE
200, rue de la Loi
B - 1049 BRUXELLES

AMOSS D.L.
BRITISH STEEL P.L.C.
Scunthorpe Works
P.O. Box 1
UK - SCUNTHORPE,HUMBERSIDE DN16 1BP

ANDRESEN K.
RADEX DEUTSCHLAND AG
Postfach
D - 5401 URMITZ bei KOBLENZ

ANTONOPOULOS V.
FINANCIAL MINING INDUSTRIAL
AND SHIPPING CORPORATION
18-20 Sikelias Street
GR - 11741 ATHENS

ARTELT P.
DG XII
CCE
200, rue la Loi
B - 1049 BRUXELLES

ATTWOOD R.N.
MONOCON REFRACTORIES LTD
Old Denaby
UK - DONCASTER,S.YORKSHIRE DN12 4LQ

AYO ATELA J.L.
ARISTEGUI MATERIAL REFRACTARIO S.A.
Barrio Florida 60
E - 20120 HERNANI - GUIPUZCOA

BACIC N.
Montanistika Askerceva 20
YU - 61000 LJUBLJANA

BADIA E.
ALTOS HORNOS DE VIZCAYA
C/ Carmen 2
BARACALDO
E - VIZCAYA

BAKER R.
BRITISH STEEL TECHNICAL
Swinden House
Moorgate
UK - ROTHERHAM, S.YORKSHIRE S60 3AR

BARNFIELD M.
THOR CERAMICS LTD
Stanford St.
P.O. Box 3
UK - CLYDEBANK, SCOTLAND G81 1RW

BAUER J.M.
SAVOIE REFRACTAIRES
4, rue de l'Industrie
B.P. 76
F - 69633 VENISSIEUX CEDEX

BAUMGARTEN H.
VEITSCHER MAGNESITWERKE AG
Schubertring 10-12
A - 1010 WIEN

BAUTISTA M.
BARRO MEX, S.A. DE C.V.
Via Morelos N°224
MEXICO - TULPETLAC,EDO.DE MEX.55400

BECKER J.
Centrale d'Achats
ARBED S.A.
L - 2930 LUXEMBOURG

BELL P.F.
BRITISH STEEL P.L.C.
Scunthorpe Works
P.O. Box 1
UK - SCUNTHORPE,HUMBERSIDE DN16 1BP

BENECKE T.
ELEKTROSCHMELZWERK KEMPTEN GmbH
Herzog Wilhelm Str.16
D - 8000 MÜNCHEN 2

BERGMAN A.
INSTITUT FÜR ALLGEMEINE METALLURGIE
Technische Univ. Clausthal
Robert Koch Strasse 42
D - 3392 CLAUSTHAL-ZELLERFELD

BERTRAND B.
METALLURGIQUE ET MINIERE
DE RODANGE-ATHUS S.A.
B.P. 24
L - 4801 RODANGE

BEUROTTE M.
SOLLAC DUNKERQUE
Rue du Comte J. Grande Synthe
B.P. 2508
F - 59381 DUNKERQUE Cedex 1

BIANCHI S.
Settore Refrattari
ASSOPIASTRELLE
Via Porta degli Archi, 3/18
I - 16121 GENOVA

BONGERS A.
N.V. GOUDA VUURVAST
Goudkade 16
NL - 2802 AA GOUDA

BOUDARD A.
ARISTEGUI MATERIAL REFRACTARIO S.A.
Barrio Florida 60
E - 20120 HERNANI - GUIPUZCOA

BRONCKART E.
Division Chertal
S.A. COCKERILL
B - 4470 VIVEGNIS

BRONKHORST R.N.
Export Department
CULLINAN REFRACTORIES LTD
1 Premier Street
S.AFRICA - OLIFANTSFONTEIN 1665

BUNSE H.
SANATECH FEUERFESTE PRODUKTE GmbH
Saarwerdenstrasse 5
D - 4000 DÜSSELDORF

BURGER F.
FEDERATION EUROPEENNE DES
FABRICANTS DE PRODUITS REFRACTAIRES
Forchstr. 95
Case Postale 151
CH - 8032 ZURICH

BUTSTRAEN G.
SIDERMIN FRANCE
104, rue de l'Ecole Maternelle
F - 59140 DUNKERQUE

BUTTER I.
HOOGOVENS IJMUIDEN
P.O. Box 10000, building 3J.22
NL - 1970 CA IJMUIDEN

CAMARDO N.
SIDERMIN ITALIA
Via Vallecamonica 19A
I - BRESCIA

CARLET
FOSECO
B.P. 1
F - 08350 DONCHERY

CARSWELL G.P.
STEETLEY REFRACTORIES LTD
Steetley Works
Worksop
UK - NOTTINGHAMSHIRE S80 3EA

CESSELIN P.
A.T.S.
Immeuble Elysées
Cedex 35
F - 92072 PARIS LA DEFENSE

CHRYSSIKOPOULOS A.
FINANCIAL MINING INDUSTRIAL
AND SHIPPING CORPORATION
18-20 Sikelias Street
GR - 11741 ATHENS

CIRASSE
PICARD & BEER
306-310 Avenue Louise
B - 1050 BRUXELLES

CLAVAUD B.
LAFARGE REFRACTAIRES MONOLITHIQUES
Chemin de Chassieu
B.P. 630
F - 69804 SAINT PRIEST CEDEX

COLLE H.
FA. LAEIS BUCHER GmbH
Postfach 2560
D - 5500 TRIER

COLSON M.
S.A. STEPHAN PASEK & CIE
13, chaussée de Dinant
B - 5198 ANHEE SUR MEUSE

CORBIER A.
REFRACOL
79, rue de la Gare
F - 59770 MARLY-LEZ-VALENCIENNES

CRIADO E.
INSTITUTO DE CERAMICA Y VIDRIO
C S I C
Carretera de Valencia km 24300
E - 28500 ARGANDA DEL REY

CRONERT W.
MARTIN & PAGENSTECHER GmbH
Schanzenstrasse 31
D - 5000 KÖLN 80

CROONA U.
HÖGANÄS AB
S - 26383 HÖGANÄS

DALY N.
QUIGLEY COMPANY OF EUROPE LTD
Tivoli Industrial Estate
IRL - TIVOLI, CORK

DAUSSAN G.
LE LABORATOIRE METALLURGIQUE
29-33 route de Rombas
B.P. 66
F - 57146 WOIPPY CEDEX

DAVIES B.
UNIVERSAL ABRASIVES LTD
Doxey Road
Stafford
UK - STAFFS ST16 1EA

DE BOER J.
Dept. Head Refractories & Ceramics
HOOGOVENS IJMUIDEN
P.O. Box 10000, building 3J.22
NL - 1970 CA IJMUIDEN

DE LAFARGE P.
LAFARGE REFRACTAIRES MONOLITHIQUES
Chemin de Chassieu
B.P. 630
F - 69804 SAINT PRIEST CEDEX

DE LATTRE E.
SAMBRE ET DYLE
Bd Pierre Mayence 19
B - 6000 CHARLEROI

DE LORGERIL J.
SOLLAC / FOS
F - 13776 FOS SUR MER CEDEX

DEABRIGES J.
Service Marketing
ALUMINIUM PECHINEY
B.P. 43
F - 13541 GARDANNE CEDEX

DECKER A.
C.R.M.
11, rue Solvay
B - 4000 LIEGE

DEL PINO MUEL L.E.
DIDIER S.A.
Avda Conde Santa Barbar,a s/n
E - 33420 LUGONES (ASTURIAS)

DELUNARDO A.
Division Chertal
S.A. COCKERILL
B - 4470 VIVEGNIS

DESBARRES J.
DIDIER WERKE DUISBURG
25 rue Joffre
B.P. 107
F - 57103 THIONVILLE

DESCAMPS
FOSECO
B.P. 1
F - 08350 DONCHERY

DHONDT R.
SIDMAR
John Kennedylaan 51
B - 9020 GENT

DIAZ P.
DG XII
CCE
200, rue la Loi
B - 1049 BRUXELLES

DIEDERICH G.
FEUERFESTWERK BAD HÖNNINGEN GmbH
Am Hohen Rhein 1
D - 5462 BAD HÖNNINGEN

DRAGAN P.
MAGNOHROM
YU - 36000 KRALJEVO

DRAMAIS R.
BELREF
N°100, rue de la Rivierette
B - 7330 SAINT GHISLAIN

DUBOTS D.
Abrasifs & Réfractaires
PECHINEY ELECTROMETALLURGIE
Usine SERS-CHEDDE
F - 74190 LE FAYET

DUCHATEAU J.L.
FOSECO
B.P. 1
F - 08350 DONCHERY

DUFOUR A.
SOLLAC
F - 13776 FOS SUR MER

DULLISCH R.
RADEX WEST GmbH
Postfach
D - 5401 URMITZ bei KOBLENZ

DUMONT C.
DOLEXI SARL
B.P. 28
F - 78590 NOISY-LE-ROI

DURAND P.
SAVOIE REFRACTAIRES
2, rue de l'Industrie
B.P. 76
F - 69633 VENISSIEUX CEDEX

DUSSAULX M.
SAVOIE REFRACTAIRES
10, rue de l'Industrie
B.P. 1
F - 69631 VENISSIEUX CEDEX

EBERS S.
FOSBEL GmbH
Joseph-Ruhr-Strasse 8
D - 05350 ENSKIRCHEN

EDEUS G.
SKAMOL A/S
Ostergade 58-60
DK - 7900 NYKOBING MORS

ELLIS D.
DSF REFRACTORIES LTD
Friden, Martington
Near Buxton
UK - DERBYSHIRE SK17 0DX

ELWELL M.
CULLINAN REFRACTORIES
Private Bag X1
S.AFRICA - OLIFANTSFONTEIN 1665

EMFIETZOGLOU V.
FINANCIAL MINING INDUSTRIAL
AND SHIPPING CORPORATION
Fourni
GR - 34004 MANTOUDI

EVANS P.
DG XII
CCE
200, rue de la Loi
B - 1049 BRUXELLES

FIERENS L.
N.V. SIDMAR
John Kennedylaan 51
B - 9020 GENT

FISCH G.
EUREFCO SARL
4-6, rue de la Libération
L - 5631 MONDORF-LES-BAINS

FLEISCHER J.
DOLOMITWERKE WÜLFRATH
Mörikestr. 14
D - 4044 KAARST 1

FOROGLOU Z.
FINANCIAL MINING INDUSTRIAL
AND SHIPPING CORPORATION
18-20 Sikelias Street
GR - 11741 ATHENS

FRANCHI A.
DOLOMITE FRANCHI S.P.A.
Via S. Croce, 13
I - 25122 BRESCIA

GARCIA ARROYO A.
DG XII
CCE
200, rue de la Loi
B - 1049 BRUXELLES

GAUCHE A.
SIDERMIN S.A.
2, rue de l'Avenir
CH - 2800 DELEMONT

GAVAGE J.E.
PAUL BEQUET S.P.R.L.
138, Chaussée d'Andenne
B - 5202 BEN AHIN

GAVHURE M.M.
THE ZIMBABWE IRON AND STEEL COMPANY
Private Bag 2
ZIMBABWE - REDCLIFF

GAYAT G.
SOLLAC
F - 13776 FOS SUR MER

GEBHARDT F.
FORSCHUNGSINSTITUT DER
FEUERFEST-INDUSTRIE
An der Elisabethkirche 27
D - 5300 BONN 1

GEBHARDT H.
MICROTHERM EUROPA NV
Hoge Heerweg 69
B - 2700 ST-NIKLAAS

GERLACH R.J.
WAMBESCO ROHSTOFFHANDELSGES. GmbH
Steinsche Gasse 4
D - 4100 DUISBURG 1

GHIROTTI P.L.
NUOVA SANAC S.p.A.
Via Martin Piaggio, 13
I - 16122 GENOVA

GIMENEZ S.
REFRACTA
Apartado de Correos 19
C/. Marquès del Turia 1
E - 46930 QUART DE POBLET (VALENCIA)

GIROLDI P.
COMAP S.A.
8, Place de la République
B.P. 55
F - 57102 THIONVILLE CEDEX

GOERES G.
SIDEX S.A.
109, av. du X Septembre
L - 2551 LUXEMBOURG

GOERIGK M.
EKW EISENBERGER KLEBSAND-WERKE GmbH
Postfach 1220
D - 6719 EISENBERG/PFALZ

GOURLAOUEN J.C.
LABORATOIRES DE REFRACTAIRES
ET MATERIAUX CERAMIQUES S.A.
71 avenue du Général Leclerc
B.P. 3013
F - 54012 NANCY CEDEX

GRAAS P.J.
LABORLUX
1, Avenue des Terres Rouges
B.P. 349
L - 4004 ESCH/ALZETTE

GRANITZKI K.E.
BASALT FEUERFEST GmbH
Kölnstrasse 112 - 114
D - 5205 ST. AUGUSTIN 2 - HANGELAR

GRIFFIN R.
QUIGLEY CO.
640 N. 13 Street
USA - EASTON PA 18042

GRIFFIN-APPADOO G.
FLOGATES LTD
Sandiron House
UK - BEAUCHIEF, SHEFFIELD S7 2RA

GRILLET
FOSECO
B.P. 1
F - 08350 DONCHERY

GUENARD C.
IRSID
B.P. 320
F - 57214 MAIZIERES LES METZ CEDEX

GUILLIER R.
Division Abrasifs & Réfractaires
PECHINEY ELECTROMETALLURGIE
Tour Manhattan
Cedex 21
F - 92087 PARIS LA DEFENSE

HAASSER A.
ETS. HAASSER
F - 67620 SOUFFLENHEIM

HAGENBURGER K.
HAGENBURGER Feuerfeste Produkte
Obersülzer Strasse 16
D - 6718 GRÜNSTADT 1

HAJEK K.
RADEX AUSTRIA AG
A - 9545 RADENTHEIN

HAMANA T.
SUMITOMO METAL INDUSTRIES LTD
Königsallee 48
D - 4000 DÜSSELDORF 1

HANSE E.
VESUVIUS CORP.
68, rue de la Gare
B.P. 19
F - 59750 FEIGNIES

HARDY C.W.
BRITISH STEEL TECHNICAL
Teesside Laboratories
P.O. Box 11
GRANGETOWN
UK - MIDDLESBROUGH, CLEVELAND TS66UB

HARDY L.K.
GR-STEIN REFRACTORIES LTD
Lowground Brickworks
Sandy Lane
UK - WORSOP, NOTTS S81 3EX

HAY P.B.
NAYLOR, BENZON & CO.LTD
Brettenham House
Lancaster House
UK - LONDON WC2E 7EN

HECQ J.C.
FORGES DE CLABECQ
B - 1361 CLABECQ

HEIN K.
SILCA GmbH
Auf dem Hüls 6
D - 4020 METTMANN

HEINTGES G.
FEUERFESTWERK BAD HÖNNINGEN GmbH
Am Hohen Rhein 1
D - 5462 BAD HÖNNINGEN

HERMES J.J.
Service Travaux Neufs/Constructions
ARBED ESCH-BELVAL
B.P. 142
L - 4002 ESCH/ALZETTE

HERNANDEZ OTER J.
MAGNESITAS NAVARRAS S.A.
B.P. 182
E - 31080 PAMPLONA (NAVARRA)

HETTLER
FOSECO
B.P. 1
F - 08350 DONCHERY

HEY A.
KSR INTERNATIONAL LTD
Sandiron House
UK - BEAUCHIEF, SHEFFIELD S7 2RA

HOELPES G.
PAUL WURTH S.A.
32, rue d'Alsace
B.P. 2233
L - 1022 LUXEMBOURG

HOFFMANN J.
ARBED ESCH-SCHIFFLANGE
B.P. 141
L - 4002 ESCH/ALZETTE

HORWITZ G.
SYGMAFER
37, rue des Déportés
F - 57070 METZ

HOYLE D.
KSR INTERNATIONAL LTD
Sandiron House
UK - BEAUCHIEF, SHEFFIELD S7 2RA

HUBERT J.
GEOTRONICS
P.A. Les Portes de la Forêt
F - 77090 COLLEGIEN

HUBERT P.
TERRES REFRACTAIRES DU BOULONNAIS
B.P. 9 NESLES
F - 62152 NEUFCHATEL HARDELOT

HUGHES J.
BRITISH STEEL, SEAMLESS TUBES
Clydesdale Works
Clydesdale Road
UK - BELLSHILL, LANARKSH.SCOT.ML42RR

HÄRKKI J.
INSTITUT FÜR METALLURGIE,TU - HELSINKI
Vuorimiehentie 2
SF - 02150 ESPOO

JACKSON B.
DYSON REFRACTORIES LTD
381 Fulwood Road
UK - SHEFFIELD S10 3GB

JOHANN A.
FEUMAS FEUERFESTE MASSEN GmbH
Industriegelände
Postfach 58
D - 6632 SAARWELLINGEN

JOHNSON H.
CONSOLIDATED CERAMIC PRODUCTS/COMAT
838 Cherry Street
USA - BLANCHESTER, OHIO 45107

JUDONG H.
S.A. DOLOMIES DE MARCHE-LES-DAMES
21, Avenue Rogier
B - 4000 LIEGE

JUHL L.
SKAMOL A/S
Ostergade 58-60
DK - 7900 NYKOBING MORS

JUMA K.
FOSECO INTERNATIONAL LTD
285 Long Acre
UK - NECHELLS, BIRMINGHAM B7 5JR

JUNG M.
FOSECO GmbH
Gelsenkirchener Strasse 10
D - 4280 BORKEN

KENT J.
STEETLEY REFRACTORIES LTD
Steetley Works
Worksop
UK - NOTTINGHAMSHIRE S80 3EA

KESSEL I.S.
BRITISH STEEL
Welsh Laboratories
UK - PORT TALBOT,WEST GLAM.SA13 2NG

KHARIUK N.
STEETLEY REFRACTORIES LTD
Steetley Works
Worksop
UK - NOTTINGHAMSHIRE S80 3EA

KLAGES G.
THYSSEN STAHL AG
Kaiser-Wilhelm-Str. 100
Postfach 110561
D - 4100 DUISBURG 11

KLEEBLATT S.
CERAFER SàRL
18, rue Mont Royal
L - 8255 MAMER

KLEIN W.
MARTIN & PAGENSTECHER GmbH
Zum Eisenhammer 23
D - 4200 OBERHAUSEN 1

KNAUDER J.
RADEX AUSTRIA AKTIENGESELLSCHAFT
Millstätterstrasse 10
A - 9545 RADENTHEIN

KNEIP E.
ECOLE PROFESSIONNELLE / DIFFERDANGE
B.P. 60
L - 4503 DIFFERDANGE

KNIERIM B.H.
SABLIERES DE SAMBRE ET DYLE
Boulevard Pierre Mayence 19
B - 6000 CHARLEROI

KNIZEK I.O.
BARRO MEX, S.A. DE C.V.
Via Morelos N°224
MEXICO - TULPETLAC,EDO.DE MEX.55400

KOCH A.
UNIMETAL
B.P. 3
F - 57360 AMNEVILLE

KOEGEL O.
EUROPE COMMERCE Sàrl
10 A, rue des Roses
L - 2445 LUXEMBOURG

KOLTERMANN M.
HOESCH STAHL AG
Rheinische Strasse 173
Postfach 9 02
D - 4600 DORTMUND

KRAUSE O.
A. FLEISCHMANN GmbH
Bahnhofstrasse 6-8
D - 6521 OFFSTEIN

KREBS R.
RATH DEUTSCHLAND GmbH
Krefelder Strasse 682
D - 4050 MÖNCHENGLADBACH 1

KUCAB B.
PROMOREF
121, rue du Président Wilson
F - 92300 LEVALLOIS PERRET

KÖHLAN H.D.
STAHLWERKE PEINE-SALZGITTER AG
Postfach 411180
D - 3320 SALZGITTER

KÖHLER-UHL I.
KRUPP STAHL AG
Postfach 101220
D - 5900 SIEGEN

KÖNIG G.
MARTIN & PAGENSTECHER GmbH
Bruchfeld 33
D - 415 KREFELD - LINN

KÖSTER V.
BADISCHE STAHLWERKE
AKTIENGESELLSCHAFT
Weststr. 31
D - 7640 KEHL/RH.

KÜHN M.
FORSCHUNGSGEMEINSCHAFT
EISENHÜTTENSCHLACKEN
Bliersheimer Strasse 62
D - 4100 DUISBURG 14

LAIRE C.
SOLLAC
17, Avenue des Tilleuls
F - 57191 FLORANGE CEDEX

LAMUT J.
Montanistika Askerceva 20
YU - 61000 LJUBLJANA

LASQUIBAR I.
ARISTEGUI MATERIAL REFRACTARIO
Barrio Florida 60
E - HERNANI - GUIPUZCOA

LAURENT F.
BELREF
Rue de la Hamaide
B - 7340 TERTRE

LE DOUSSAL H.
SOCIETE FRANCAISE DE CERAMIQUE
23, rue de Cronstadt
F - 75015 PARIS

LEE B.J.
MORGANITE THERMAL CERAMICS LTD
Liverpool Road
UK - NESTON, SOUTH WIRRAL L64 3RE

LEFEBVRE M.
F.E.F.P.R.
44, rue Copernic
F - 75116 PARIS

LETE E.
COCKERILL SAMBRE S.A.
1, rue de l'Usine
B - 6090 CHARLEROI (COUILLET)

LOBEMEIER D.
Oxygenstahlwerk II
THYSSEN STAHL AG
Postfach 110561
D - 4100 DUISBURG 11

LOBGEOIS
FOSECO
B.P. 1
F - 08350 DONCHERY

LOMBOIS T.
REFRATEC
176, rue Charles de Gaulle
F - 57290 SEMERANGE

LOPEZ DE NOVALES
PRODUCTOS DOLOMITICOS S.A.
Revilla de Camargo
E - 39600 CANTABRIA

LORANG L.
REFRALUX Sàrl
B.P. 142
L - 4002 ESCH/ALZETTE

MAKSYM
STOPING AG
Zugerstrasse 76a
CH - 6340 BAAR

MARECHAL P.
PECHINEY ELECTROMETALLURGIE
Usine SERS-CHEDDE
F - 74190 LE FAYET

MARINO E.
CENTRO SVILUPPO MATERIALI
Via Di Castel Romano 100
I - 00129 ROMA

MARSH R.
SHEERNESS STEEL CO. PLC.
UK - KENT ME12 1TH

MARTENS M.
S.C.R.-SIBELCO S.A.
Quellinstraat 49
B - 2018 ANTWERPEN

MARTIN J.
LE LABORATOIRE METALLURGIQUE
29-33 route de Rombas
B.P. 66
F - 57146 WOIPPY CEDEX

MARTIN R.
Recherches et Développement
ARGILES ET MINERAUX AGS
Clerac
F - 17270 MONTGUYON

MARTINEZ W.A.
QUIGLEY CO.
640 N. 13 Street
USA - EASTON PA 18042

MASSE F.
SOLLAC DUNKERQUE
Rue du Comte Jean
Grande-Synthe
B.P. 2508
F - 59381 DUNKERQUE CEDEX 1

MATHER T.
DYSON REFRACTORIES LTD
381 Fulwood Road
UK - SHEFFIELD S10 3GB

MATHIEU P.
PAUL BEQUET S.P.R.L.
138, Chaussée d'Andenne
B - 5202 BEN AHIN

MATSUKIZONO M.
KUROSAKI REFRACTORIES EUROPE
Immermannstrasse 45
D - 4000 DÜSSELDORF 1

MENDIOLA APODACA F.
MAGNESITAS NAVARRAS S.A.
B.P. 182
E - 31080 PAMPLONA (NAVARRA)

MONTANARI E.
BASITAL REFRATTARI BASICI SPA
Via Chiaravagna, 145
I - 16153 GENOVA-SESTRI PONENTE

MOREL J.
UGINE SAVOIE
Centre de Recherches
F - 73400 UGINE

MOSSER B.
Abrasifs & Réfractaires
PECHINEY ELECTROMETALLURGIE
Tour Manhattan
Cedex 21
F - 92087 PARIS LA DEFENSE

MULLER J.R.
Centrale d'Achats
ARBED S.A.
L - 2930 LUXEMBOURG

MULLER R.
METALLURGIQUE ET MINIERE
DE RODANGE-ATHUS S.A.
B.P. 24
L - 4801 RODANGE

MUTSAARTS P.
Division Tecnozzle
SA BELREF
Rue de la Riviérette, 100
B - 7330 SAINT GHISLAIN

MÜLLER H.A.
WIENEN & THIEL
Cranger Strasse 68
Postfach 201044
D - 4650 GELSENKIRCHEN-BUER

NAUJOKAT E.
MARTIN & PAGENSTECHER GmbH
Bruchfeld 33
D - 4150 KREFELD 12

NEWBOUND D.
STEETLEY REFRACTORIES LTD
Steetley Works
Worksop
UK - NOTTINGHAMSHIRE S80 3EA

NICOLAY D.
DG XIII/C3
Bâtiment Jean Monnet
Rue Alcide Gasperi
L - 2920 LUXEMBOURG

NILSSON K.
GEOTRONICS AB
P.O. Box 64
S - 18211 DANDERYD

NOKERMAN A.
BELREF
Rue de la Hamaide
B - 7340 TERTRE

OBERBACH M.
DIDIER-WERKE AG FEUERFEST
Rathausallee 4
D - 4100 DUISBURG 46

OFFENSTADT R.
FEUMAS FEUERFESTE MASSEN GmbH
Industriegelände
Postfach 58
D - 6632 SAARWELLINGEN

OGER J.
PAUL BEQUET SPRL
15, rue Colombière
B - 4240 ST. GEORGES

OSSWALD J.P.
SOLLAC
17, avenue des Tilleuls
F - 57191 FLORANGE

OTS J.M.
SIDERMIN S.A.
2, rue de l'Avenir
CH - 2800 DELEMONT

OUAYOUN J.P.
COMPTOIR DES PRODUITS MAGNESIENS
25, rue de Clichy
F - 75009 PARIS

OVYN E.
S.A. DOLOMIES DE MARCHE-LES-DAMES
21, Avenue Rogier
B - 4000 LIEGE

OWEN A.J.
SCHOOL OF MATERIALS
UNIVERSITY OF LEEDS
UK - LEEDS LS2 9JT

PADGETT G.
BRITISH CERAMIC RESEARCH Ltd
Queens Road, Penkhull
UK - STOKE-ON-TRENT ST4 7LQ

PARIDAENS S.
THOR CERAMIC Ltd, FRANCE - BENELUX
Av. Charles-Quint, 213 Bte 5
B - 1080 BRUXELLES 8

PARKIN N.A.
DSF REFRACTORIES LTD
Friden, Martington, near Buxton
UK - DERBYSHIRE SK17 0DX

PAVSE M.
Montanistika Askerceva 20
YU - 61000 LJUBLJANA

PIATKOWSKI-GRZYMALA W.
Meninghoferweg 26
D - 5820 GEVELSBERG

PICKENHAHN F.
RADEX DEUTSCHLAND AG
D - 5401 URMITZ

PIRET J.
CENTRE DE RECHERCHES METALLURGIQUES
Abbaye du Val-Benoit
B - 4000 LIEGE

POHL S.
SIPO FF. TECHNIK GmbH
Richard Wagner Strasse 12
D - 6635 SCHWALBACH

POIRSON G.
TERRES REFRACTAIRES DU BOULONNAIS
B.P. 9 NESLES
F - 62152 NEUFCHATEL HARDELOT

POOS A.
DG IX/E
Bâtiment Jean Monnet
Rue Alcide Gasperi
L - 2920 LUXEMBOURG

POPOVIC D.
METALL-CHEMIE HAMBURG
Elstrasse 4
D - HAMBURG 76

POUSSEUR E.
PRODUITS REFRACTAIRES POUSSEUR
B.P. 10
F - 08320 VIREUX-MOLHAIN

PROVOST G.
SOLLAC CRDM
B.P. 2508
F - 59381 DUNKERQUE Cedex 1

RADOMIR J.
MAGNOHROM
YU - 36000 KRALJEVO

RAMON V.
VECTUR S.A.
Ercilla 17-5°
E - 48009 BILBAO

RANSBOTIJN P.
MICROTHERM EUROPA NV
Hoge Heerweg 69
B - 2700 ST-NIKLAAS

RASSEL P.
PAUL WURTH S.A.
32, rue d'Alsace
B.P. 2233
L - 1022 LUXEMBOURG

RETRAYT L.
RADEX DEUTSCHLAND AG
D - 5401 URMITZ bei KOBLENZ

RIGAUD M.
ECOLE POLYTECHNIQUE
Université de Montréal
Case Postale 6079 - Succ. A
CANADA - MONTREAL H3C 3A7

ROHRBACHER M.
COMAP S.A.
8, Place de la République
B.P. 55
F - 57102 THIONVILLE CEDEX

ROONEY P.
FIBRE TECHNOLOGY LTD
Brookhill Road
Pinxton
UK - NOTTINGHAM NG16 6NT

ROOTHOOFT-VAN LAETHEM F.
DG XII
CCE
200, rue de la Loi
B - 1049 BRUXELLES

ROSSET P.
CREUSOT - MARREL
Usine de Chateauneuf
F - 42800 RIVE DE GIER

RUDOLPH G.
CARBOREF GmbH, DÜSSELDORF
Sperberstrasse 60
D - 4044 KAARST

RYDER A.
PREMIER PERICLASE LTD
Boyne Roaddylaan 51
Drogheda
IRL - CO.LOUTH

RYMON-LIPINSKI T.
FORSCHUNGSINSTITUT DER
FEUERFEST-INDUSTRIE
An der Elisabethkirche 27
D - 5300 BONN 1

RÄDE D.
BAYER AG WERK UERDINGEN
Rheinuferstrasse 7-9
D - 4150 KREFELD-UERDINGEN

SANDBERG B.
ELKEM MATERIALS A/S
P.O. Box 126
N - VAAGSBYGD, 4602 KRISTIANSAND

SANTAMBRIA J.M.
DEGUISA
C/ Barreta Aldamar n°4
E - 48001 BILBAO

SCHEIDT F.
IRSID
B.P. 320
F - 57214 MAIZIERES-LES-METZ

SCHERRMANN G.
SAARSTAHL AG
Postfach 101980
D - 6620 VÖLKLINGEN

SCHMIDT-WHITLEY R.D.
DIDIER - SOCIETE INDUSTRIELLE DE
PRODUCTION & DE CONSTRUCTIONS
102, rue des Poissonniers
F - 75018 PARIS

SCHNEIDER H.
FORSCHUNGSINSTITUT DER
FEUERFEST-INDUSTRIE
An der Elisabethkirche 27
D - 5300 BONN 1

SCHNEIDER M.
IRSID
F - 57210 MAIZIERES-LES-METZ

SCHOENNAHL J.
SAVOIE REFRACTAIRES
B.P. 1
F - 69631 VENISSIEUX CEDEX

SCHOLLAERT A.
DETRICK BELGE S.A.
Rue du Magistrat 2
B.P. 11
B - 1050 BRUXELLES

SCHOUMACHER C.
ARBED S.A.
Division de Differdange
L - 4503 DIFFERDANGE

SCHRODER M.
REFRACOL
79, rue de la Gare
F - 59770 MARLY-LEZ-VALENCIENNES

SCHRÖER H.
KRUPP STAHL AG
Postfach 141980
D - 4100 DUISBURG 14

SCHULTZ LOUP M.B.
CERAMICA DO LIZ S.A.
Rua Manuel Simoes da Maia, Apartado 15
P - 2400 LEIRIA

SCHWARZ C.
DIDIER-WERKE AG FEUERFEST
Rathausallee 4
D - 4100 DUISBURG 46

SCHÄFER U.L.
POSSEHL ERZKONTOR GmbH
Graf-Adolf-Platz 1,
Postfach 200605
D - 4000 DÜSSELDORF 1

SCHÄUFFELE F.
A. FLEISCHMANN GmbH
Schlosserstrasse 23-25
D - 6000 FRANKFURT/MAIN

SEUTIN M.
CARBOREF AG
Chamerstrasse 79
CH - 6303 ZUG

SIMEONI P.
CAM PRODOTTI SPA
Via Camperio 9
I - 20123 MILANO

SISKENS C.A.M.
Refractory Department
HOOGOVENS IJMUIDEN B.V.
Postbus 10000
NL - 1970 CA IJMUIDEN

SKINNER P.
PETER SKINNER LIMITED
Psalter House
121 PSALTER LANE
UK - SHEFFIELD S11 8YR

SMITH P.L.
G.R. - STEIN REFRACTORIES Ltd
Central Research Laboratories
Sandy Lane
UK - WORKSOP S80 3EU

SMITH S.
DAVY McKEE
Ashmore Housemile Zola
Richardson Road
UK - STOCKTON, CLEVELAND

SOADY J.
STEETLEY REFRACTORIES LTD
Steetley Works
Worksop
UK - NOTTINGHAMSHIRE S80 3EA

SOLMECKE R.O.
MANNESMANNRÖHREN-WERKE AG
Hüttenwerke Huckingen
Postfach 251167
D - 4100 DUISBURG 25

SPERL H.
VEREIN DEUTSCHER EISENHÜTTENLEUTE
Sohnstrasse 65
D - 4000 DÜSSELDORF 1

STEVENS J.
S.A. STEPHAN PASEK & CIE
13, chaussée de Dinant
B - 5198 ANHEE SUR MEUSE

STOURNARAS C.J.
CERECO S.A.
P.O. Box 146
GR - 34100 CHALKIS

STRADTMANN J.
DOLOMITWERKE GmbH
Wilhelmstrasse 77
D - 5603 WÜLFRATH

TAMMERMANN E.
SOCIETE FRANCAISE DE CERAMIQUE
23, rue de Cronstadt
F - 75015 PARIS

TASSOT P.
IRSID
Voie Romaine
B.P. 320
F - 57214 MAIZIERES-LES-METZ

TAVERNIER H.
IRSID
Voie Romaine
B.P. 320
F - 57214 MAIZIERES-LES-METZ

TEMPINI L.
L. TEMPINI & C. SRL
Via Govine N°14
I - 25055 PISOGNE (BS)

THEUER W.
STEULER INDUSTRIEWERKE GmbH
Georg-Steuler Strasse 175
P.O. Box 1448
D - 5410 HÖHR-GRENZHAUSEN

THOMAS A.
INTERNATIONAL METALS S.A.
3-5, Place Winston Churchill
B.P. 484
L - 2014 LUXEMBOURG

THOMPSON I.
ALCAN CHEMICALS LTD
Chalfont Park, Gerrards Cross
UK - BUCKINGHAMSHIRE SL9 0QB

TIRLOCQ J.
I.N.I.S.Ma.
Avenue Gouverneur Cornez, 4
B - 7000 MONS

TONDEUR M.
PAUL BEQUET S.P.R.L.
138, Chaussée d'Andenne
B - 5202 BEN AHIN

TSIKAS B.
FINANCIAL MINING INDUSTRIAL
AND SHIPPING CORPORATION
Fourni
GR - 34004 MANTOUDI

TSUYUGUCHI K.
KUROSAKI REFRACTORIES EUROPE
Immermannstrasse 45
D - 4000 DÜSSELDORF 1

VAN DER WIEL K.
Refractory Maintenance Department
HOOGOVENS IJMUIDEN B.V.
Postbus 10000
NL - 1970 CA IJMUIDEN

VAN DINTER M.
KRÄMER & CO. GmbH
Postfach 110309
D - 4100 DUISBURG 11

VAN ERCK P.
Avenue Montesquieu, 28
B - 4220 JEMEPPE/SERAING

VAN HAUWERMEIREN E.
PLIBRICO S.A.
Rue Arthur Maes 65
B - 1130 BRUXELLES

VAN HUMBEECK J.
BELREF S.A.
Rue de la Rivièrette, 100
B - 7330 SAINT-GHISLAIN

VAN ROY F.
PICARD & BEER
306-310 Avenue Louise
B - 1050 BRUXELLES

VATERLAUS A.
ARVA AG, TECHNOLOGICAL CONSULTING
Bahnhofstrasse 38
CH - 8803 RÜSCHLIKON

VEYRET J.B.
IRSID
Voie Romaine
B.P. 64
F - 57210 MAIZIERES-LES-METZ

VIADIEU T.
Service 0954
REGIE NATIONALE DES USINES RENAULT
8-10 Avenue Emile Zola
F - 92109 BOULOGNE BILLANCOURT CEDEX

VIKMAN R.
AB SANDVIK STEEL
S - 81181 SANDVIKEN

WALDHANS T.
REFRATECHNIK GmbH
Rudolf-Winkel-Strasse 1
D - 3400 GÖTTINGEN

WEEDON I.
MYTAG AG
Wiesentalstrasse 198
CH - 7000 CHUR

WEYAND S.
RADEX WEST GmbH
Postfach
D - 5401 URMITZ bei KOBLENZ

WHITELEY P.G.
GR - STEIN REFRACTORIES Ltd
Central Research Laboratories
Sandy Lane
UK - WORKSOP S80 3EU

WIDONG R.
Service Travaux Neufs/Constructions
ARBED ESCH-BELVAL
B.P. 142
L - 4002 ESCH/ALZETTE

WILDGOOSE J.N.
KSR INTERNATIONAL LTD
Sandiron House
Beauchief
UK - SHEFFIELD S7 2RA

WILLIAMS P.
STEETLEY REFRACTORIES LTD
Steetley Works
Worksop
UK - NOTTINGHAMSHIRE S80 3EA

WILLIAMS R.E.
MORGANITE THERMAL CERAMICS LTD
Liverpool Road
UK - NESTON, SOUTH WIRRAL L64 3RE

WINCHURCH K.
ROBERT LICKLEY REFRACTORIES LTD
P.O. Box 24
UK - DUDLEY, WEST MIDLANDS DY1 2UF

WINTER I.
DYSON REFRACTORIES LTD
381 Fulwood Road
UK - SHEFFIELD S10 3GB

WOLTERS D.
Forschung Zentrale Feuerfesttechnik
THYSSEN STAHL AG
Postfach 110561
D - 4100 DUISBURG 11

YAGUCHI P.M.
MORGAN TOCERA LIMITED
Norton
UK - WORCESTER WR5 2PU

ZABALETA I.
REFRACTA
Apartado de Correos 19
C/. Marquès del Turia 1
E - 46930 QUART DE POBLET (VALENCIA)

ZOGLMEYR G.
DOLOMITE FRANCHI S.P.A.
Via S. Croce, 13
I - 25122 BRESCIA

ZOTTL G.
SIMMERING-GRAZ-PAUKER AG
Brehmstrasse 16
A - 1110 WIEN

ZUYDERDUYN D.C.J.
HOOGOVEN IJMUIDEN
Postbus 10000 / 4C-01
NL - 1970 CA IJMUIDEN

INDEX OF CONTRIBUTORS

AMAVIS, R., 260

BAKER, R., 142
BENECKE, T., 187
BEUROTTE, M., 152
BUTTER, J.A.M., 11

CLAVAUD, B., 109
CRIADO, E., 118

DE BOER, J., 11

FOROGLOU, Z.E., 225

GARCIA-ARROYO, A., 3
GHIROTTI, P.L., 195
GUENARD, C., 22

HARDY, C.W., 211

JACQUEMIER, M., 109

KLAGES, G., 133
KOLTERMANN, M., 31

LE DOUSSAL, H., 109
LEFEBVRE, M., 97

MARINO, E., 59

NATUREL, C., 246

PADGETT, G.C., 255
PASTOR, A., 118
PIRET, J., 7, 47, 81, 257

RYMON-LIPINSKI, T., 236

SANCHO, R., 118
SCHMIDT-WHITLEY, R.D., 175
SCHOENNAHL, J., 246
SOADY, J.S., 163
SPERL, H., 133

TASSOT, P., 22
TAYLOR, D., 163

WHITELEY, P.G., 211
WILLIAMS, P., 163
WOLTERS, D., 69

Printed in the USA
CPSIA information can be obtained
at www.ICGtesting.com
LVHW011544240923
759164LV00007B/842

9 781851 664917